Dynamics of Detonations and Explosions: Detonations

Edited by
A. L. Kuhl
Lawrence Livermore National Laboratory
Los Angeles, California

J.-C. Leyer
Université de Poitiers
Poitiers, France

A. A. Borisov
USSR Academy of Sciences
Moscow, USSR

W. A. Sirignano
University of California
Irvine, California

Volume 133
PROGRESS IN
ASTRONAUTICS AND AERONAUTICS

A. Richard Seebass, Editor-in-Chief
University of Colorado at Boulder
Boulder, Colorado

Technical papers presented from the Twelfth International Colloquium on Dynamics of Explosions and Reactive Systems, Ann Arbor, Michigan, July 1989, and subsequently revised for this volume.

Published by the American Institute of Aeronautics and Astronautics, Inc., 370 L'Enfant Promenade, SW, Washington, DC 20024-2518.

American Institute of Aeronautics and Astronautics
Washington, D.C.

Library of Congress Cataloging-in-Publication data

International Colloquium on Dynamics of Explosions and Reactive
Systems (12th : 1989 : Ann Arbor, Mich.)
Dynamics of detonations and explosions—detonations / edited by A.L. Kuhl ...
[et al.].
p. cm.—(Progress in astronautics and aeronautics; v. 133)
"Technical papers presented from the Twelfth International Colloquium on
Dynamics of Explosions and Reactive Systems, Ann Arbor, Michigan, July 1989,
and subsequently revised for this volume."
Includes index.
ISBN 0-930403-97-5
1. Explosions—Congresses. 2. Detonation waves—Congresses.
I. Kuhl, A. L. II. American Institute of Aeronautics and Astronautics.
III. Title. IV. Series.
TL507.P75 vol. 133 [QD516] 629.2 s—dc20 [541.3'61] 91-7933

Table of Contents

Preface xix

Chapter I. Gaseous Detonations **1**

Chronology of Early Research on Detonation Wave **3**
P. A. Bauer, *University of Paris, Paris, France*, and E. K. Dabora, *University of Connecticut, Storrs, Connecticut*, and N. Manson, *Laboratoire d'Energétique et de Détonique, URA 193 au CNRS, ENSMA, Poitiers, France*

Detonability of Hydrocarbon Fuels in Air **19**
H. D. Beeson, R. D. McClenagan, C. V. Bishop, and F. J. Benz, *NASA White Sands Test Facility, Las Cruces, New Mexico*, and W. J. Pitz and C. K. Westbrook, *Lawrence Livermore National Laboratory, Livermore, California*, and J. H. S. Lee, *McGill University, Montreal, Quebec, Canada*

Detonation of Cryogenic Gaseous Hydrogen-Oxygen Mixtures.............**37**
M. Plaster, R. D. McClenagan, and F. J. Benz, *NASA White Sands Test Facility, Las Cruces, New Mexico*, and J. E. Shepherd, *Rensselaer Polytechnic Institute, Troy, New York*, and J. H. S. Lee, *McGill University, Montreal, Quebec, Canada*

Detonation Characteristics of Gaseous Methane-Oxygen-Nitrogen Mixtures at Extremely Elevated Initial Pressures...........................**56**
P. Bauer, H. N. Presles, and M. Dunand, *Laboratoire d'Energétique et de Détonique, URA 193 au CNRS, ENSMA, Poitiers, France*

Correlation Between Chemical Kinetics and Detonation Structure for Gaseous Explosive Systems...**63**
C. Paillard, G. Dupré, A. Aiteh, and S. Youssefi, *National Centre of Scientific Research and University, Orléans, France*

Influence of Tetrafluoromethane on Hydrogen-Oxygen-Argon Detonations **77**
E. Nzeyimana and P. J. Van Tiggelen, *Université Catholique de Louvain, Louvain-la-Neuve, Belgium*

Chapter II. Detonation: Initiation and Transmission **89**

Initiation of Hydrogen-Air Detonations by Turbulent Fluorine-Air Jets **91**
S. B. Murray and I. O. Moen, *Defense Research Establishment Suffield, Defense Technologies Division, Ralston, Canada*, and P. A. Thibault, *Combustion Dynamics Ltd., Medicine Hat, Canada*, and R. Knystautas and J. H. S. Lee, *McGill University, Montreal, Quebec, Canada*, and A. Sulmistras, *Concordia University, Montreal, Quebec, Canada*

Detonation of Unconfined and Semiconfined Charges of Gaseous Mixtures **118**
A. A. Borisov and S. V. Khomik, *USSR Academy of Sciences, Moscow, USSR*, and V. N. Mikhalkin, *Moscow Institute of Physical Engineering, Moscow, USSR*

Initiation of a Detonation Wave Due to Multistage Self-Ignition **133**
S. M. Frolov, B. E. Gelfand, and S. A. Tsygranov, *USSR Academy of Sciences, Moscow, USSR*

Critical Energy of Direct Detonation Initiation in Gaseous Mixtures **142**
A. A. Borisov and S. V. Khomik, *USSR Academy of Sciences, Moscow, USSR*, and V. R. Mikhalkin and E. V. Saneev, *Moscow Institute of Physical Engineering, Moscow, USSR*

Limit Criterion of Detonation in Circular Tubes **156**
G. Dupré, *National Centre of Scientific Research and University, Orléans, France*, and O. Péraldi, J. Joannon, J. H. S. Lee, and R. Knystautas, *McGill University, Montreal, Quebec, Canada*

Effects of a Negative Step of Fuel Concentration on Critical Diameter of Diffraction of a Detonation **170**
D. Desbordes, *Laboratoire d'Energétique et de Détonique, URA 193 au CNRS, ENSMA, Poitiers, France* and A. Lannoy, *Electricité de France, Direction des Etudes et Recherches, Saint-Denis, France*

Oblique Detonation at Hypersonic Velocities **187**
E. K. Dabora, *University of Connecticut, Storrs, Connecticut*, and D. Desbordes and C. Guerraud, *Laboratoire d'Energétique et de Détonique, URA 193 au CNRS, ENSMA, Poitiers, France* and H. G. Wagner, *Max Planck Institute on Flow Research, Göttingen, Germany*

Numerical Simulation of Layered Detonations **202**
D. A. Jones, *Materials Research Laboratory, Melbourne, Australia*, and M. Sichel, *University of Michigan, Ann Arbor, Michigan*, and R. Guirguis, *Naval Surface Warfare Center, White Oak, Maryland*, and E. S. Oran, *Naval Research Laboratory, Washington, DC*

Chapter III. Nonideal Detonations and Boundary Effects 221

Photographic Study of the Structure and Propagation Mechanisms of Quasidetonations in Rough Tubes 223
A. Teodorczyk, J. H. S. Lee, and R. Knystautas, *McGill University, Montreal, Quebec, Canada*

Calculation of the Velocity of Gaseous Detonation in a Rough Tube Based on Measurements of Shock Wave Attenuation 241
S. M. Frolov, B. E. Gelfand, and S. P. Medvedev, *USSR Academy of Sciences, Moscow, USSR*

Studies of the Mechanisms of Detonation Propagation in Porous Structures 256
G. O. Thomas, D. H. Edwards, and S. H. M. Jones, *University College of Wales, Aberystwyth, England, United Kingdom*

Propagation of Gaseous Detonations Over Liquid Layers 268
A. A. Borisov and A. H. Mailkov, *USSR Academy of Sciences, Moscow, USSR*, and V. V. Kosenkov and V. S. Aksenov, *Moscow Institute of Physical Engineering, Moscow, USSR*

Surface Detonations and Indirect Ignition Processes 279
B. Plewinsky, W. Wegener, and K.-P. Herrmann, *Bundesanstalt für Materialforschung und -Prüfung, Berlin, Germany*

Chapter IV. Multiphase Detonations 295

Detonation of Unconfined Large-Scale Fuel Spray-Air Clouds 297
W. B. Benedick and S. R. Tieszen, *Sandia National Laboratories, Albuquerque, New Mexico*, and R. Knystautas and J. H. S. Lee, *McGill University, Montreal, Quebec, Canada*

Lean Detonation Limit of Sensitized Kerosene Sprays in Air 311
E. K. Dabora, *University of Connecticut, Storrs, Connecticut*

Explosive Decomposition of Hydrazine by Rapid Compression of a Gas Volume 325
R. L. Bunker and D. L. Baker, *NASA White Sands Test Facility, Las Cruces, New Mexico*, and J. H. S. Lee, *McGill University, Montreal, Quebec, Canada*

Detonation Structure of Corn Starch Particles-Oxygen Mixtures 342
F. Zhang and H. Groenig, *Shock Wave Laboratory, RWTH Aachen, Germany*

Influence of Dust Parameters on Detonation Velocity, Structure, and Limits 356
O. Dahab, *Minia University, Minia, Egypt*, and M. Kapuściński, *Marine University, Szczecin, Poland*, and P. Wolański, *Warsaw University of Technology, Warsaw, Poland*

Multiple Gauges for In Situ Measurements of Pressure and Particle Velocity in Condensed Materials **371**
P. A. Urtiew and L. M. Erickson, *Lawrence Livermore National Laboratory, Livermore, California*

Author Index for Volume 133 .... **383**

List of Series Volumes **385**

Table of Contents for Companion Volume 131

Preface xvii

Chapter I. Ignition Dynamics 1

Detailed Numerical Modeling of H_2-O_2 Ignition by Hot Spots 3
U. Maas and J. Warnatz, *Universität Stuttgart, Stuttgart, Germany*

Ignition Phenomena in H_2-O_2 Mixtures 19
C. Treviño, *Universidad Nacional Autónoma de México, Mexico*

Ignition of Fresh Mixtures Injected into Burned Gases 44
S. Kadowaki, Y. Ohta, M. Furutani, and K. Terada, *Nagoya Institute of Technology, Nagoya, Japan*, and H. Takahashi, *Meijo University, Nagoya, Japan*

Analysis of Ignition Mechanism of Combustible Mixtures by Short-Duration Sparks 55
M. Kono, K. Ishii, K. Niu, T. Tsukamoto, and Y. Ujiie, *University of Tokyo, Tokyo, Japan*

Flat Plate Boundary Layer Ignition with Fuel Thermal Diffusion 71
P. L. García-Ybarra and J. L. Castillo, *Universidad Nacional de Educación a Distancia, Madrid, Spain*

Chapter II. Diffusion Flames and Shear Effects 87

Analysis of the Structure of Counterflow Hydrogen-Air Diffusion Flames 89
V. Tangirala and K. Seshadri, *University of California, San Diego, La Jolla, California*, and C. Treviño, *Universidad Nacional Autónoma de México, Mexico*, and M. D. Smooke, *Yale University, New Haven, Connecticut*

Flame Structure of Axisymmetric Hydrogen-Air Diffusion Flames 111
S. Fukutani, N. Kunioshi, and H. Jinno, *Kyoto University, Yoshida-honmachi, Sakyo-ku, Kyoto, Japan*

Numerical Modeling of the Structure and Properties of Tubular Strained Laminar Premixed Flames 125
G. Dixon-Lewis, *University of Leeds, Leeds, England, United Kingdom*, and V. Giovangigli, *Ecole Polytechnique et CNRS, Palaiseau, France*, and R. J. Kee and J. A. Miller, *Sandia National Laboratories, Livermore, California*, and B. Rogg, *University of Cambridge, Cambridge, England, United Kingdom*, and M. D. Smooke, *Yale University, New Haven, Connecticut*, and G. Stahl and J. Warnatz, *Universität Stuttgart, Stuttgart, Germany*

Simulation of Partially Premixed Methane-Air Counterflow-Diffusion Flames and Comparison with Experimental Results 145
F. Behrendt and J. Warnatz, *Universität Stuttgart, Stuttgart, Germany*

Extinction of Premixed Curved Flames Stabilized in a Stagnation Flow 161
K. Asato, H. Nagata, and T. Kawamura, *Gifu University, Gifu, Japan*, and K. Mabuchi, *Sanyo Electric Company, Ltd., Osaka, Japan*, and T. Ban, *Nippondenso Company, Ltd., Kariya, Japan*

Chapter III. Dynamics of Flames and Shear Layers 177

Dynamics of an Unsteady Diffusion Flame: Effects of Heat Release and Viscosity 179
J. L. Ellzey, K. J. Laskey, and E. S. Oran, *Naval Research Laboratory, Washington, DC*

Dynamics of Turbulent Diffusion Flame Surface 193
H. Yamashita, G. Kushida, and T. Takeno, *Nagoya University, Nagoya, Japan*

Structure of the Reaction Zone in a Reacting Mixing Layer 220
A. F. Ghoniem, O. M. Knio, and G. Heidarinejad, *Massachusetts Institute of Technology, Cambridge, Massachusetts*

Flow-Combustion Interactions in a Turbulent Jet 237
A. F. Ghoniem and A. Krishnan, *Massachusetts Institute of Technology, Cambridge, Massachusetts*

Chapter IV. Turbulent Flames 257

Investigation on the Laminar and Turbulent Burning Velocities of Premixed Lean and Rich Flames of CH_4-H_2-Air Mixtures 259
Y. Liu, B. Lenze, and W. Leuckel, *Universität Karlsruhe, Karlsruhe, Germany*

Geometry Effects on Premixed Turbulent Propagating Flames 277
S. S. Ibrahim, R. S. Cant, and K. N. C. Bray, *University of Cambridge, Cambridge, England, United Kingdom*

Multivariate PDF Closure Applied to Oxidation of CO in a Turbulent Flow 288
E. Gutheil, *Technische Hochschule Darmstadt, Darmstadt, Germany*

Modeling Turbulent Reacting Flows: Detailed Chemical Reaction Mechanisms and Sensitivity Analysis 303
H. Bockhorn, *Technische Hochschule Darmstadt, Darmstadt, Germany*

Numerical Simulation of a Premixed Flame in a Turbulent Boundary Layer 317
B. Rogg, *University of Cambridge, Cambridge, England, United Kingdom*

Chapter V. Flame Propagation in Combustion Engines 333

Flame Imaging Studies of Flame Development in a SI Four-Stroke Engine 335
S. C. Bates, *General Motors Research Laboratories, Warren, Michigan*

Flame Propagation Behavior in a Variable Hydrodynamic Constant Volume Combustion Chamber 378
A. Floch and M. Trinité, *Université de Rouen, Rouen, France*, and F. Fisson and T. Kageyama, *Laboratoire d'Energétique et de Détonique, URA 193 au CNRS, ENSMA, Poitiers, France* and C. H. Kwon and A. Pocheau, *Université de Provence, Marseille, France*

Effect of Turbulent Fluid Motion on Low-Temperature Autoignition of Fuel-Air Mixture Under Piston Compression 394
Y. Ohta, S. Kadowaki, and K. Terada, *Nagoya Institute of Technology, Nagoya, Japan*, and H. Takahashi, *Meijo University, Nagoya, Japan*

Experimental Study of N-Butane Autoignition in a Rapid Compression Machine 407
J. D. Gabano, T. Kageyama, and F. Fisson, *Laboratoire d'Energétique et de Détonique, URA 193 au CNRS, ENSMA, Poitiers, France*

Author Index for Volume 131 418

List of Series Volumes 419

Table of Contents for Companion Volume 132

Preface xvii

Chapter I. Combustion of Dust-Air Mixtures 1

Deflagration and Detonation Combustion of Dust Mixtures 3
P. Wolański, *Warsaw University of Technology, Warsaw, Poland*

Experiments on Unconfined Dust-Air Flames 32
A. Makris, N. R. Valerio, and J. H. S. Lee, *McGill University, Montreal, Quebec, Canada*

Critical Dust Layer Thickness for Combustion of Grain Dust 51
K. Lebecki, J. Sliz, and Z. Dyduch, *Experimental Mine "Barbara," Mikolów, Poland*, and P. Wolański, *Warsaw University of Technology, Warsaw, Poland*

Particle Segregation Effects on the Combustion Safety of Dust-Containing Systems 59
A. L. Berlad and V. Tangirala, *University of California, San Diego, La Jolla, California*

Experimental Measurement of the Aerodynamic Entrainability of Dust Deposits 73
E. A. Ural, *Factory Mutual Research Corporation, Norwood, Massachusetts*

Starch Dust Combustion Characteristics in a Closed Spherical Vessel 93
L. Lemos and R. Bouriannes, *Laboratoire d'Energétique et Détonique, URA 193 au CNRS, ENSMA, Poitiers, France*

Determination of Turbulence Parameters in Closed Explosion Vessels 107
Y.-K. Pu, *Chinese Academy of Science, Beijing, China*, and Y.-C. Li, C. W. Kauffman, and L. P. Bernal, *University of Michigan, Ann Arbor, Michigan*

Numerical Study on Gas-Solid Two-Phase Nozzle and Jet Flow 124
A. K. Hayashi, M. Matsuda, and T. Fujiwara, *Nagoya University, Nagoya, Japan*

Chapter II. Droplet Combustion 137

New Apparatus and its Performance for Free Droplet Combustion Under Microgravity 139
H. Hara and S. Kumagai, *Noritz Corporation, Research Laboratory, Tokyo, Japan*

Water-Gas Shift Reaction in Droplet Burning 150
I. K. Puri, P. A. Libby, and F. A. Williams, *University of California, San Diego, La Jolla, California*

Liquid-Waste Incineration in a Parallel-Stream Configuration: Effect of Auxiliary Fuel 164
J. P. Delplanque, R. H. Rangel, and W. A. Sirignano, *University of California, Irvine, California*

Spray Dynamics and Fuel Vapor Distributions in a Spinning Combustion Chamber 187
G. Avital, J. B. Greenberg, Y. Tambour, and Y. M. Timnat, *Technion—Israel Institute of Technology, Haifa, Israel*

Ignition of Liquid Fuel Drops in Shock Waves 205
V. M. Boiko, V. V. Lotov, and A. N. Papyrin, *USSR Academy of Sciences, Novosibirsk, USSR*

Chapter III. Combustion At Solid and Liquid Surfaces 221

Oxidation of a Porous Graphite Cylinder with Airflow Through a Coaxial Hole 223
M. B. Richards and S. S. Penner, *University of California, San Diego, La Jolla, California*

Premixed Flame in a Radiatively Active Porous Medium Under the External Radiant Heating 248
S. W. Baek and J. S. Kim, *Korea Advanced Institute of Science and Technology, Seoul, Korea*

Major Factors Influencing Flame Spreading Over Solid Fuel Layer 261
J. Fangrat, *Building Research Institute, Warsaw, Poland*, and P. Wolański, *Warsaw University of Technology, Warsaw, Poland*

Combustion Kinetic of a Homogeneous Double Base Propellant: Pseudo-Detailed and Global Mechanism 275
F. Faddoul, J. M. Most, and P. Joulain, *Université de Poitiers, Laboratoire de Chimie Physique de la Combustion, Poitiers, France*

Radiative Heat Flux and Energy Balance at the Surface of a Small Scale Kerosene Pool Fire 297
A. Bouhafid and J. P. Vantelon, *Université de Poitiers, Laboratoire de Chimie Physique de la Combustion, Poitiers, France*

Velocity and Temperature Measurements in a Bidimensional Pool Fire: Influence of a Vertical Wall Close to the Fire 314
M. Annarumma, J. M. Most, and P. Joulain, *Université de Poitiers, Laboratoire de Chimie Physique de la Combustion, Poitiers, France*

Chapter IV. Combustion Diagnostics 339

Fluorescence Measurements in Laminar Nonpremixed Flames of CH_4/N_2 341
A. R. Masri, *University of Sydney, New South Wales, Australia*, and R. W. Dibble, *Sandia National Laboratories, Livermore, California*

Speckle Tomography of Unsteady Gasdynamic Objects 352
O. V. Achasov, G. N. Blinkov, N. A. Fomin, O. G. Penyazkov, and D. E. Vitkin, *USSR Academy of Sciences, Minsk, USSR*

Spectroscopic Measurements of Hydroxyl Radical in Axisymmetric Hydrogen-Air Diffusion Flames 360
N. Kunioshi, S. Fukutani, and H. Jinno, *Kyoto University, Yoshida-honmachi, Sakyo-ku, Kyoto, Japan*

Re-Evaluation of Laser Schlieren Data for Acetylene Pyrolysis 372
W. S. Kolln, *Simpson College, Indianola, Iowa*, and S. M. Hwang, K. S. Shin, and W. C. Gardiner, Jr., *University of Texas, Austin Texas*

Author Index for Volume 132 386

List of Series Volumes 387

Table of Contents for Companion Volume 134

Preface xvii

Chapter I. Vapor Cloud Explosions 1

Scaling of Vapor Cloud Explosions After Turbulent Jet Release 3
M. Stock and W. Geiger, *Battelle Europe, Frankfurt, Germany*, and H. Giesbrecht, *BASF AG, Ludwigshafen, Germany*

Effect of Asymmetric Ignition on the Vapor Cloud Spatial Blast 21
C. Desrosier, A. Reboux, and J. Brossard, *Laboratoire de Recherche Universitaire, Bourges, France*

Experimental Investigation Concerning the Influence of Turbulence on the Flame Front Velocity of Fuel Gas-Air Mixture Deflagrations 38
M. Christill and W. Leuckel, *Universität Karlsruhe, Karlsruhe, Germany*

Mechanism of Flame Acceleration Along a Tube With Obstacles 66
Y.-K. Pu and S. Yuan, *Chinese Academy of Science, Beijing, China*, and J. Jarosinski, *Institute of Aeronautics, Warsaw, Poland*, and V. G. Johnston and C. W. Kauffman, *University of Michigan, Ann Arbor, Michigan*

Explosion in a Vented Vessel Connected to a Duct 85
F. Bouhard, B. Veyssiere, *Laboratoire d'Energétique et de Détonique, URA 193 au CNRS, ENSMA, Poitiers, France*, and J.-C. Leyer, *Université de Poitiers, Laboratoire de Chimie Physique de la Combustion, Poitiers, France*, and J. Chaineaux, *CERCHAR, Verneuil-en-Halatte, France*

Modelization and Validation Tests of the Discharge in Air of a Vessel Pressurized by a Flammable Gas 104
J. Chaineaux, G. Mavrothalassitis, and J. Pineau, *CERCHAR, Verneuil-en-Halatte, France*

Applicability of a Chemical-Equilibrium Model to Explosion Products 138
A. A. Borisov, *USSR Academy of Sciences, Moscow, USSR*, and S. A. Gubin and V. A. Shargatov, *Moscow Institute of Physical Engineering, Moscow, USSR*

Inverse Numerical Process for Inlet Conditions Calculus from Experimental Front Determination 154
R. Saurel, J.-C. Loraud, B. Porterie, and M. Larini, *IUSTI/SETT Université de Provence, Marseille, France*

Chapter II. Blast Wave Reflections and Interactions **169**

Reflection of Shock and Explosion Waves from Surfaces Covered with Layers of Polyurethane Foam 171
D. K. Rayevsky, L. G. Gvozdeva, Y. Faresov, *USSR Academy of Sciences, Moscow, USSR*, and J. Brossard and P. Bailly, *University of Orléans, Orléans, France*

Simulating the Impact Made by a Shock Wave on a Body Surrounded by a Layer of Hot or Cold Gas 183
V. N. Lyakhov, D. K. Raevsky, and S. M. Kharitonov, *USSR Academy of Sciences, Moscow, USSR*, and B. I. Zaslavksky, S. Y. Morozkin, and V. R. Shlegel, *VNIIFTRI, Moscow, USSR*

Turbulent Wall Jet in a Mach Reflection Flow 201
A. L. Kuhl, *R & D Associates, Los Angeles, California*, and R. E. Ferguson, K.-Y. Chien, W. Glowacki, and P. Collins, *Naval Surface Warfare Center, Silver Spring, Maryland*, and H. Glaz, *University of Maryland, Silver Spring, Maryland*, and P. Colella, *University of California, Berkeley, California*

Numerical Simulation of the Change in the Supersonic Flow Past a Body Produced by Switching on a Nearby Heat Source 233
T. V. Bazhenova, V. N. Lyakhov, and S. M. Kharitonov, *USSR Academy of Sciences, Moscow, USSR*

Shock Waves Produced by Reflected Detonations 244
J. E. Shepherd, *Rensselaer Polytechnic Institute, Troy, New York*, and A. Teodorczyk, R. Knystautas, and J. H. S. Lee, *McGill University, Montreal, Quebec, Canada*

Spherical Wave Interaction with a Liquid-Air Interface: Analysis of the Holographic Records 265
N. Fargère, *Centre d'Etude du Bouchet, Vert-le-Petit, France*, and C. Brochet, *Laboratoire d'Energétique et de Détonique, URA 193 au CNRS, ENSMA, Poitiers, France*

Formation of Zones with High Particle Concentrations in Dusty Gas 287
V. P. Korobeinikov, *Far-Eastern Branch of USSR Academy of Sciences, Vladivostok, USSR*

Chapter III. Vapor Explosions **293**

Shock Waves from Vapor Explosion in a Shock Tube 295
B. E. Gelfand, S. P. Medvedev, A. N. Polenov, and S. M. Frolov, *USSR Academy of Sciences, Moscow, USSR*

Dynamics of Explosive Interactions Between Molten Tin and Water in Stratified Geometry 307
G. Ciccarelli, D. L. Frost, and C. Zarafonitis, *McGill University, Montreal, Quebec, Canada*

Steam Explosion Studies with Molten Iron-Alumina Generated by Thermite Reactions 326
D. F. Beck, M. Berman, and L. S. Nelson, *Sandia National Laboratories, Albuquerque, New Mexico*

Application of the Integrated Fuel-Coolant Interaction Code to a FITS-Type Pouring Mode Experiment 356
M. F. Young, *Sandia National Laboratories, Albuquerque, New Mexico*

Multiphase Physical Explosion Modeling Using the CULDESAC Code 387
D. F. Fletcher and A. Thyagaraja, *Culham Laboratory, Abingdon, Oxfordshire, United Kingdom*

Author Index for Volume 134 408

List of Series Volumes 409

Preface

The four companion volumes on *Dynamics of Deflagrations and Reactive Systems* and *Dynamics of Detonations and Explosions* present 91 of the 149 papers given at the Twelfth International Colloquium on the Dynamics of Explosions and Reactive Systems (ICDERS) held at the University of Michigan in Ann Arbor during July 1989.

These four volumes are included in the Progress in Astronautics and Aeronautics series published by the American Institute of Aeronautics and Astronautics, Inc. *Dynamics of Deflagrations and Reactive Systems: Flames* (Volume 131) and *Dynamics of Deflagrations and Reactive Systems: Heterogeneous Combustion* (Volume 132) span a broad area, encompassing the processes of coupling the exothermic energy release with the fluid dynamics occurring in any combustion process. *Dynamics of Detonations and Explosions: Detonations* (Volume 133) and *Dynamics of Detonations and Explosions: Explosion Phenomena* (Volume 134) principally address the rate processes of energy deposition in a compressible medium and the concurrent nonsteady flow as it typically occurs in explosion phenomena. The colloquium, in addition to embracing the usual topics of explosions, detonations, shock phenomena, and reactive flow, includes papers that deal primarily with the gasdynamic aspects of nonsteady flow in combustion systems, the fluid mechanic aspects of combustion (with particular emphasis on turbulence), and diagnostic techniques used to study combustion phenomena.

In this volume, *Dynamics of Detonations and Explosions: Detonations*, the papers have been arranged into chapters on gaseous detonations, detonation initiation and transmission, nonideal detonations and boundary effects, and multiphase detonations. Although the brevity of this preface does not permit the editors to do justice to all papers, we offer the following highlights of some of the especially noteworthy contributions.

In Chapter I, Gaseous Detonations, *Bauer et al.* present an extremely useful review of research and publications on detonation waves prior to 1922—many of which have been overlooked previously. The chronological review includes references to "high explosives" dating back to the fifteenth century. In the "discovery phase," the work of Abel (1869), Berthelot (1870–1871), Nobel (1873–1874), Berthelot and Vielle (1878–1883), and the work of Mallard and Le Chatelier (1883) play a primary role. They also show that the Chapman-Jouget (C-J) theory is based on the pioneering work of Michelson (1890), Berthelot (1891), Dixon (1893), Chapman (1899), and Vielle (1900), as well as the independent development by Jouguet (1901–1905). In addition, this chapter contains articles on the detonability of hydrocarbon fuels at ambient pressure, and the detonation characteristics of methane-oxygen-nitrogen mixtures at high initial pressures (100–400 bars).

Chapter II, Detonation Initiation and Transmission, begins with articles on the initiation of detonations in unconfined clouds of gaseous mixtures by *Murray et al.* and *Borisov et al. Frolov and coworkers* describe the initiation of a detonation wave due to multistage self-ignition. *Borisov et al.* report on the critical energy required for direct initiation of detonations in gaseous mixtures, while *Dupré et al.* provide a limit criterion for detonations in circular tubes. This chapter concludes with articles on the transmission of detonation waves by *Desbordes and Lannoy, Dabora et al.*, and *Jones et al.*

Chapter III, Nonideal Detonations and Boundary Effects, investigates the fundamental propagation mechanisms of nonideal or quasidetonations that travel at a fraction of the C-J velocity. *Teodorczyk and coworkers* report on quasidetonations in H_2-O_2 observed in a narrow, two-dimensional channel filled with periodic obstacles. Photographic observations clearly reveal that autoignition by reflected shocks is the primary propagation mechanism for this type of quasidetonation. Depending on the obstacle height and spacing, three different types of shock reflection processes were observed: 1) reflection of the diffracted shock from the bottom wall; 2) normal reflection of a Mach stem from the front face of an obstacle; and 3) reflection of the transverse shock from the top wall. *Frolov and coworkers* present an approximate analytical theory (based on enhanced drag and heat transfer coefficients), to predict the attenuation of detonations in rough-walled tubes. *Thomas and coworkers* report on detonations propagating through tubes with tightly-packed, perforated steel foils. On the macroscopic scale, the tightly-packed foils completely destroyed the natural transverse structure of the detonation. However, flow interactions with the frontal area of the foils generated sufficient gasdynamic heating to cause autoignition that sustained the continued propagation of the wave. *Borisov et al.* describe low velocity (1.3 k/s) gaseous detonations propagating in tubes with thin water layers. The entrainment of droplets (caused by shock spalling of the water surface) quenches the heat release until the reaction wave decouples from the shock front. *Plewinsky and coworkers* report on surface detonation experiments in liquid tetramethyldihydrogen-disiloxane. Low detonation velocities (0.7–0.8 km/s) were measured.

Chapter IV, Multiphase Detonations, presents recent results on detonations observed in droplet clouds and dust clouds. *Benedick et al.* investigated the detonability of fuel-air clouds formed by the explosive dissemination of various liquid fuels into the air. They found that the detonability of such clouds depended not only on the sensitivity of fuel, but also on the amount of fuel that was in the vapor phase at the time of ignition. Very sensitive fuels such as propylene oxide, and low-vapor-pressure fuels such as nitrated hydrocarbons could be detonated quite readily, even in the aerosol form. Insensitive fuels such as hexane could only be detonated in the vapor phase. *Dabora* reports on shock tube tests that were used to determine the lean detonability limit of kerosene sprays in air. *Zhang and*

Groenig report on experiments with cornstarch particles suspended in an oxygen atmosphere. Steady, self-sustained detonations were observed. Also, spinning detonations were found to exist for a wide range of particle concentrations. *Dahab et al.* report on detonation experiments with dust-air mixtures. They found that Egyptian brown coal dust and linen dust were very easy to detonate, while it was practically impossible to detonate flour dust.

The companion volumes, *Dynamics of Deflagrations and Reactive Systems: Flames* (Volume 131), *Dynamics of Deflagrations and Reactive Systems: Heterogeneous Combustion* (Volume 132), and *Dynamics of Detonations and Explosions: Explosion Phenomena* (Volume 134), include papers on ignitions dynamics, diffusion flames and shear effects, dynamics of flames and shear layers, turbulent flames, flame propagation in combustion engines, combustion of dust-air mixtures, droplet combustion, combustion at solid and liquid surfaces, combustion diagnostics, vapor cloud explosions, blast wave reflections and interactions, and vapor explosions.

These four volumes will, we trust, help satisfy the need first articulated in 1966 and will continue the tradition of augmenting our understanding of the dynamics of explosions and reactive systems begun the following year in Brussels with the first colloquium. Subsequent colloquia have been held on a biennial basis: 1969 in Novosibirsk, 1971 in Marseilles, 1973 in La Jolla, 1975 in Bourges, 1977 in Stockholm, 1979 in Göttingen, 1981 in Minsk, 1983 in Poitiers, 1985 in Berkeley, 1987 in Warsaw, and 1989 in Ann Arbor. The colloquium has now achieved the status of a prime international meeting on these topics, and attracts contributions from scientists and engineers throughout the world.

To provide an enduring focal point for the administrative aspects of the ICDERS, the organization was formally incorporated in the state of Washington under the name Institute for Dynamics of Explosions and Reactive Systems (IDERS). Professor J. R. Bowen is serving as the current president. Communications may be sent to

Dean J. R. Bowen
President, IDERS
College of Engineering FH-10
University of Washington
Seattle, Washington, USA 98195

The proceedings of the first six colloquia have appeared as a part of the journal *Acta Astronautica*, or its predecessor, *Astronautica Acta*. With the publication of the Seventh Colloquium, the proceedings now appear as part of the Progress in Astronautics and Aeronautics series published by the American Institute of Aeronautics and Astronautics.

Acknowledgments

The Twelfth Colloquium was held under the auspices of the Department of Aerospace Engineering, University of Michigan at Ann Arbor, July 23–July 28, 1989. Local arrangements were organized by Professors C. W. Kauffman and M. Sichel. Publication of the Proceedings was made possible by grants from the National Science Foundation and the Defense Nuclear Agency of the United States.

Preparations for the Thirteenth Colloquium are now under way. The meeting will be held at Nagoya University, July 28–August 2, 1991. Professor T. Fujiwara of the University's Department of Aeronautical Engineering is chairman of the local organizing committee.

A. L. Kuhl
J.-C. Leyer
A. A. Borisov
W. A. Sirignano

January 1991

Chapter I. Gaseous Detonations

Chronology of Early Research on Detonation Wave

P. A. Bauer*
University of Paris, Paris, France
and
E. K. Dabora†
University of Connecticut, Storrs, Connecticut
and
N. Manson‡
Laboratoire d'Energétique et de Détonique, URA 193 au CNRS, ENSMA, Poitiers, France

The purpose of this paper is to recall past research studies that led to the discovery of the detonation wave and to the description of its basic properties as determined by what is now called the Chapman-Jouguet (CJ) wave and defined by the classical thermo-gasdynamic theory. The chronology of investigations that we present is restricted to publications prior to 1922. We have chosen this date for two reasons:

1) with the passage of time, it appeared that after the lecture given in 1905 by Nernst (29) before the Verein der Deutschen Ingenieure (VDI) (of which we comment again subsequently) and in particular because of the publication in 1917 of "Mécanique des Explosifs" by Jouguet (16c) and the two papers by Becker (2), the discovery stage of the studies was completed. It was then generally agreed that the gasdynamic theory summarized in a satisfactory way the then known properties of the detonation wave;

2) in papers, textbooks, and review articles on combustion and detonation published thereafter, some of the pioneering contributions were not acknowledged at all, and in some cases there were gaps and serious omissions.

We seek to give credit to those who have been perhaps inadvertently slighted, to fill some gaps, and to correct errors of interpretation. In particular, we wish to present the main results, observations, and conclusions of the investigations that have been carried out by the pioneers.

For brevity's sake, we shall not describe the very early, experimental investigations, nor shall we give a detailed account of motivations, even though they were

*Professor.

†Professor.

‡Honor. Dean.

quite revealing of the industrial expansion taking place at that period.

The reader can find details of the techniques and the reasons for their application in the original articles listed in the references, as well as in the publications on one of the authors (N. Manson (21)) containing extended references. One should note that some articles quoted here have not been listed before since they were only recently discovered.

I. Chronology

A. Before Discovery

15th-18th Centuries

First "high explosives" (gold, silver, and mercury fulminates) discovered by Alchemists.

J.R. Glauber (1604-66) discovers potassium nitrate and chlorate. (37, 46a).

1799

E. Howard (14) describes a practical method for the manufacture of fulminate of mercury.

1860-1863

Industrial manufacture of nitroglycerine (discovered in 1846 by A. Sobrero (41) is developed by A. Nobel and his father. (27, 28)

1864

A mercury fulminate ignitor ("detonator") that initiates in a reproducible manner especially violent explosions (i.e. detonation) in nitroglycerine charges is patented by A. Nobel. (31)

1865-1867

F.A. Abel (1a) improves radically the industrial manufacture of gun-cotton discovered by C.F. Schönbein in 1847. (39, 46b)

1867

A. Nobel invents dynamite Kieselguhr. (31)

B. Discovery

1869

F.A. Abel (1b) demonstrates that unconfined charges of gun-cotton, nitroglycerine, dynamite, and Hg fulminate i) burn if ignited by means of a "heat source" (flame or hot-wire); and ii) or detonate if submitted to an impulsive force (hammer blow a Nobel detonator (31)). He assumes, that detonation is transmitted by means of some "synchronous vibrations".

1870-1871

M. Berthelot defines the "strength" of explosives (condensed and gaseous) and emphasizes the role of a mechanical shock in the propagation ("layer to layer") of the detonation. (3a)

L.P. Roux and E. Sarrau (38) point out the difference between explosions of "first order," i.e., detonations, and explosions of "second order," i.e., deflagrations.

1873-1874

First detonation velocity measurement in condensed explosives (gun-cotton, dry or moist; dynamite; and nitroglycerine) by F.A. Abel. (1c)

Using a chronograph built by A. Noble (32), F.A. Abel ascertains that in 10-15-m-long condensed explosive charges the detonation propagates at a uniform velocity whose value depends upon the explosive (e.g.., density and moisture content in gun-cotton).

As pointed out in Nature, October 1873, "... the detonation of gun-cotton travels more rapidly than any other known medium with the exception of light and electricity."

F.A. Abel also measured the velocity in discontinuous cylindrical charges and noted that it depends on the distance between the charges as well as on the roughness of the interior surface of the tube walls connecting the charges.

1878-1883

M. Berthelot and P. Vieille (3b, 4b) ascertain the existence of the detonation wave in explosive gaseous mixtures. Using a Le Boulengé chronograph (4b) (they measured the propagation velocity D of this wave in about

50 mixtures of fuels (H_2, C_2H_4, C_2H_2, etc) and oxidizers (O_2, NO, and N_2O_4), diluted by different amounts of nitrogen. In particular, they conclude that in lead and rubber tubes (40-80-m-long, 1.5-15mm i.d.) the propagation velocity D is uniform and depends on the mixture composition but depends neither on the tube material nor the tube diameter, as long as the latter is not too small.

1882

M. Berthelot (3c) compares the detonation velocity D with the root-mean-square velocity u of the molecules of the burned gases, whose density relative to air is and where the temperature T is supposed to be that of isobaric combustion, i.e.:

$$D = u = 29.354(T/\rho^*)^{1/2}$$

But, as pointed out later by H. Dixon (11), the equality of D and u noted by M. Berthelot is wrong, since the temperature of the detonation products is more likely to be that of isochoric combustion.

1883

E. Mallard and H. Le Chatelier (20) using the first drum-camera, demonstrate that in gaseous explosives a deflagration can transit into a detonation wave. They state that a detonation wave propagates as a sudden adiabatic compression wave that initiates the chemical reaction and that its propagation velocity (supersonic with respect to the fresh gas) is comparable to the sound velocity of the products.

1885

M. Berthelot (3d) reports the results of the measurements of the detonation wave velocity (by means of Le Boulengé chronograph) made since 1878 by French engineers in long, (up to 200 m) 1-6-mm-cylindrical charges of various condensed explosives. In particular, the results indicate an increase in the detonation velocity with the charge density and thus complete and extend those of F.A. Abel. In addition, they led to the invention, of the detonating fuse (23,34,4) ("cordeaux détonants").

II. The CJ Theory

1890

V.A. Michelson (26) applies the Rankine (35) theory and: 1) derives a relation for the detonation pressure in terms of the detonation velocity and the heat of reaction, showing that, in general, two pressures equally are possible; 2) states that, under a certain condition corresponding to the coincidence of the two pressures, there exists a "minimum limit" for the detonation velocity, thus hinting at the Chapman-Jouguet condition; and 3) concludes that a more detailed study of his theory should be made in the future.

1891

In order to elucidate the influence of composition and confinement, M. Berthelot (3e) investigated detonations propagating in methyl nitrate confined in thick, long tubes (i.d. up to 5 mm) of rubber, glass, steel, etc. The unexpected values of the detonation velocity he obtained (1600-2400 m/s-representing, as we know now, a "low velocity detonation") led him to the conclusion that in condensed explosives "detonations obey the same physico-chemical laws as gaseous mixtures, but their characteristics are more complex."

1893

H. Dixon (11a) makes new measurements of detonation velocities in several gaseous mixtures, using a endulum chronograph, that confirm and extend the results of M. Berthelot and P. Vieille. In particular, he demonstrates that the velocity increases with initial pressure (0.25-2 bars) of the H_2-O_2 mixtures, whereas it decreases with initial temperature (from 20°C to 100°C).

In accordance with A. Schuster, he stated that R. Riemann's theory (36) leads to the conclusion that the motion of the detonation wave may not be strictly steady: Jets of heated gas may be projected in the front of the wave into the unburnt gas, with sudden jets becoming the centres of fresh disturbance which coalesce to reform a wave. If the jets are produced in sufficient number, their average effect would be spatially uniform."

1899

D.L. Chapman (8) establishes the relations of the new classic CJ theory of the steady, plane detonation waves in gaseous mixtures. He demonstrates that the lowest value of the velocity with respect to the burned gas, i) is equal to the sound velocity in this gas; ii) if computed for some mixtures, is in agreement with the measured one, and iii) if known by measurements, allows the determination of the heat capacities of some gases at high temperature.

1900

Using the shock tube he invented, P. Vieille (47) demonstrates experimentally the validity of the Hugoniot (15) relation:

$$D_s = a_o \left[1 + (1 + (\frac{\gamma + 1}{2\gamma})(\frac{p - p_o}{p_o})\right]^{1/2}$$

where D_s is the velocity of a shock wave propagating at an overpressure $p-p_o$ in a perfect gas at a pressure p_o in which the sound velocity is a_o and γ is the ratio of specific heats. Moreover, he describes the detonation wave as a discontinuity (in the Hugoniot sense) supported by the chemical reaction induced by this discontinuity, and, in particular, he computes quite correctly for the $2H_2 + O_2$ mixture the pressure (40 bars) of what is currently termed the Zeldovich - Neuman - Döring spike..

1901-1905

E. Jouguet (16) (unaware of V.A. Michelson's and D.L. Chapman's papers, but cognizant of A. Schuster's attempts) demonstrates that the main properties of a detonation wave propagating at constant velocity can be described if one assumes that behind the wave front the velocity of detonation products with respect to this front is equal to the (local) velocity of sound. The detonation velocities that he calculated (in assuming that the products are at thermodynamic equilibrium and using values of the heat capacities of gases at a high temperature that were available at that time, as provided by E. Mallard and E. Le Chatelier (20) for gaseous mixtures) are in good agreement with those measured by M. Berthelot and P. Vieille (4) as well as by H. Dixon. (11)

1905

W. Nernst (29) theorizes that the "anomaleous combustion" (i.e., knocking) in reciprocating internal combustion engines may be a consequence of the buildup of a detonation wave. This assertion becomes the main motivation for future research on detonation waves in gaseous mixtures.

1906

H. Dautriche (10) invents a new method for the measurement of detonation velocities in condensed explosives whereby it became possible to confirm and improve the previously made measurements (especially by C.E. Bichel (5) in 1902 to 1905) using mechanical (Le Boulengé) and electrical (Mettegang-Siemens (25)) chronographs.

1907

Assuming that "burned gases" behind the shock and combustion wave can be treated as a mixture of perfect gases with constant heat capacities, J.L. Crussard (9) specifies the peculiarities of the locus of their possible thermodynamic states on the pressure-specific volume plane, referred to as the "combustion" or "Crussard" adiabat (better known as Hugoniot). He also demonstrates the equivalence of the three properties of the detonation wave stated respectively by D.L. Chapman (minimum velocity) and E. Jouguet (velocity of burned gases with respect to the wave front equal to the local velocity of sound, and minimum entropy of these gases along the "Crussard adiabat").

1907-1917

E. Jouguet and J.L. Crussard (16c) improve and extend the theory of combustion and shock waves to non planar waves and emphasize the role played in the transition of deflagration to detonation by the gas movement ahead of the flame ("chasse préalable") that was in the meantime systematically investigated by J. Taffanel. (42)

1913

J. Taffanel and H. Dautriche (43) attempt to apply the CJ theory to compute the detonation velocity in dynamite, assuming that the equation of state (EOS) of the product is a simplified Van der Waals equation (i.e., Noble-Abel-EOS). (33)

1914-1917

R. Becker (2) using a drum camera, obtains the first chronophotographs of detonation waves in condensed explosives.

1922

R. Becker (2b, 2c) analyzes and discusses the CJ theory and its application to condensed explosives and proposes a new EOS for the products that later becomes a starting point of research on the EOS of gases at very high pressures and temperatures.

III. Comments

1) Among the results presented in both first and second parts, some are worthwhile noticing.

On the first point, let us note that in 1872 P. Champion and H. Pellet (7) attempted to reveal experimentally the genesis of these synchronous vibrations by triggering a detonation in a small charge of iodine azide placed in the focus of a parabolic mirror, with another charge placed at the focus of another parabolic mirror located coaxially at some distance away to act as the initiator. As it turned out later, what they recorded was the effect of a shock wave that was predicted in 1860 by B. Riemann (36) and physically demonstrated in 1877 by E. Mach and J. Sommer. (18) One should note, however, that it was F.A. Abel who was first to measure the velocity of these waves. (1a)

2) Now, let us note that:

i) As early as the 17th century, it had been observed that the "liveliness" of the explosives and their effects on the adjoining media were different depending on whether the explosive was black powder (industrially produced) (37) pressed in a more or less rigid container or a laboratory substance (dangerous to handle) such as gold, mercury, or silver fulminates or, later on (beginning of the 18th century), silver or lead azide and, thereafter, iodine azide. (46a, 21b) This liveliness was already a matter of "detonation" vs "deflagration" as well as combustion velocity of the black powder tested in the open air or confined under different pressures; although no clear definitions of these terms was given. These definitions could be specified only after it became possible to manufacture in a reproducible manner nitroglycerine, and cotton powder as well as dynamite Kieselguhr.

ii) for at least 50 years the pioneering contributions of F.A. Abel (1b, 1c) have not been mentioned in the detonation literature. To our knowledge, only J. Taylor (44) and L. Médard (24) did so, the former with regard to the transmission of the detonation from one charge to another and the latter to call attention to the measurement of the propagation velocity. Before these citations, the latest mention of these investigations was in Refs. 16, 17 and 22.

3) Concerning the discovery, it is worth stating that contrary to what is generally believed, the investigations conducted by M. Berthelot and P. Vieille (4) on the one hand and, on the other, by E. Mallard and H. Le Chatelier (20) did not have the same objectives. The latter were charged by the "Commission to Study means of Preventing Explosions in Firedamps" (a commission created specifically to remedy the numerous fatal catastrophes that were occuring in France and elsewhere) to undertake experimental investigations on the flammability of gaseous mixtures. In contrast, M. Berthelot and P. Vieille embarked on their experiments on their own initiative, (4a) undoubtedly because: i) they were aware of the studies of F.A. Abel on condensed explosives, which sometimes deflagrated or at other times detonated depending on the experimental conditions (e.g., initiator and confinement); and ii) they knew that in gaseous explosives, like in condensed explosives, the flame propagation is due principally to heat transfer by conduction and believed, therefore, that gaseous explosives could detonate.

Let us note also that according to Mallard and Le Chatelier themselves, detonation waves in gaseous mixtures were discovered by Berthelot and Vieille. (20b)

4) Finally as far as theory is concerned, our main remark refers to the pioneering studies of V.A. Michelson are currently known thanks to their rendition in his Complete Works published in 1930 (26b). However, this version differs in certain aspects from the original publications of 1890 and 1893. (26c) The 1893 publication was rewritten and completed by him in 1920, when the opportunity of editing all of his papers arose. These papers were concerned with meteorological problems, except for his thesis and two or three other articles dealing with thermodynamics.

In particular, he reasons that the state of detonation products corresponds to maximum entropy and finds that the density ratio across the wave is

$$\rho_b / \rho_u = (\gamma+1)/\gamma$$

where γ is the ratio of specific heats, assumed to be the same for the burned and unburned gas. He also gives a relation for the pressure in terms of the detonation velocity as

$$p_b = \rho_u D^2/(\gamma+1)$$

As was indicated in Ref. 21b, when we were unaware of the version of (26a), it appears that Michelson did not know of Chapman (8) or Jouguet.

5) Concerning the latter we consider it important to underscore the fact that Ya.B. Zeldovich (48) cites P. Vieille's work (47) and elaborates his theory of the structure of the detonation wave by developing the ideas of Vieille, but taking into account the friction and heat transfer losses as well as the chemical kinetic effects behind the shock front. This, however, is not the case with J. von Neuman (30) nor with W. Döring (12) who obviously did not cite the work of Vieille (47) even though it was mentioned by E. Jouguet (16) and summarized by H. Dryden, F.O. Murnagham, and N.H. Bateman (13) in 1932, (See also Refs. (24a and 24b)).

References

1 Abel, F.A.,
a "Researches on Gun-cotton: on the Manufacture and Composition of Gun-cotton," Philosophical Transactions, Vol. 156, 1866, pp. 268-308.

b "Researches on Gun-cotton. Second Memoir: On the Stability Gun-cotton," Philosophical Transactions, Vol. 157, 1867, pp. 181-253.

c "Contribution to the History of Explosive Agents," Philosophical Transcactions, Vol. 159, 1869, pp. 489-516. Also Comptes-Rendus Académie des Sciences, Paris, France, Vol. 69, 1869, pp. 105-121.

d "Contribution to the History of Explosive Agents: Second Memoir," Philosophical Transactions, Vol. 164, 1874, pp. 337-395. Also Comptes-Rendus Académie des Sciences, Paris, France, Vol. 78, 1874, p. 1227, 1301, 1362, 1432-1436.

2 Becker, R.,
a "Zur Theorie der Detonation," (On the theory of detonations) Zeitschrift fur Elektrochemie, Vol. 23, 1917, pp. 40-49, 93-95, 304-309.

b "Physikalische uber Feste und Gasformige Sprengstoffe," (The Physics of solid and gaseous Explosives), Zeitschrift fur Techn. Phys., 1922, pp. 152-159, 249-256.

c "Stosswelle und Detonation," (Shock Wave and Detonation), Zeitschrift fur Physik, 1922, pp. 321-362.

3 Berthelot, M.,
a "Sur la Force des Poudres et des Matières Explosives," (On the strength of powder and Explosives), Comptes-Rendus Académie des Sciences, Paris, France, Vol. 71, 1870, pp. 619-625, 667-677, 709-728, and "Annales de Chimie et de Physique, 4e Serie, Vol. 23, 1871, pp. 223-273.

b "Sur la Vitesse de Propagation des Phénomènes Explosifs dans les Gaz," (On the Propagation Velocity of Explosive Phenomena in Gases), Comptes-Rendus Académie des Sciences, Paris, France, Vol. 93, 1881, pp. 18-22.

c "Sur l'Onde Explosive," (On Explosive Wave), Comptes-Rendus Académie des Sciences, Paris, France, Vol. 94, 1882, pp. 149-152.

d "Sur la Vitesse de Propagation de la Détonation dans les Matières Explosives Solides et Liquides," (On the Detonation Velocity in Solid and Liquid Explosive Media), Comptes-Rendus Académie des Sciences, Paris, France, Vol. 100, 1885, pp. 314-320.

e "Sur l'Onde Explosive, sur les Données Caractéristiques de la Détonation et sa Vitesse de Propagation dans les Corps Solides et Liquides, et Spécialement dans le Nitrate de Méthyle," (On Explosive Wave, on Detonation Characteristics and its Velocity in Solid and Liquid Media and especially in Methyl Nitrate), Comptes-Rendus Académie des Sciences, Paris, France, Vol. 112, 1891, pp. 16-27.

4 Berthelot, M., and Vieille, P.,
a "Personal Investigations" of the Members of the French "Commission des Substances Explosives," Mémorial des Poudres et Salpêtres, Vol. 1, 1878-1883, p. 462.

b "L'Onde Explosive," (The Explosive Wave), Annales de Chimie et de Physique, 5e Serie, Vol. 28, 1883, p. 289.

5 Bichel, C.E.,
a "Untersuchungsemthoden fur Sprengstoffe," (Investigation Methods for Explosives), Zeitschrift fur Berg und Hutten und Salinenwesen in Preuss. State, 1902, pp. 669-689.

b "Uber Zundung von Schlagwettern durch Detonierende Sprengstoffe," (The Ignition of Permissible Explosives), Gluckauf, Vol. 40, 1904, pp. 1040-1048.

6 Brunswig, H., "Explosifstoffe," Vol. X of Handbuch der Angew, Physikal. Chemie Georg Bredig, 1st ed., 1909, 2nd ed., 1922, J.A. Barth, Leipzig, East Germany.

7 Champion, P., and Pellet, H., "Sur la Théorie de l'Explosion des Composés Détonants," (On the Theory of Explosion of Detonating Media), Comptes-Rendus Académie des Sciences, Paris, France, Vol. 75, 1872, pp. 210-214.

8 Chapman, D.L., "On the Rate of Explosion in Gases," Philosophical Magazine, Vol. 47, 1889, pp. 90-104.

9 Crussard, J.C.,
a "Sur Quelques Propriétés de l'Onde Explosive," (On Some Properties of Explosive Wave), Comptes-Rendus Académie des Sciences, Paris, France, Vol. 144, 1907, pp. 417-420.

b "Ondes de Choc et Onde Explosive," (Shock Wave and Explosive Wave), Bulletin de la Société d'Industrie Minérale (Bulletin of Society of Mineral Industry, 4th Serie, Vol. VI, 1907, pp. 287-30.

10 Dautriche, H., "Sur la Vitesse de Détonation des Explosifs," (On the Detonation Velocity in Explosives), Comptes-Rendus Académie des Sciences, Paris, France, Vol. 143, 1906, pp. 641-664, Vol. 144, 1907, pp. 1030-1032.

11 Dixon, H.,
a "The Rate of Explosion in Gases," Philosophical Transactions A, Vol. 184, 1893, pp. 97-188.

b "The Rate of the Flame in the Explosion of Gases," Philosophical Transactions A, Vol. 200, 1903, pp. 315-351.

12 Döring, W., "Detonationvorgange in Gasen," (Detonation Phenomena in Gases), Ann. der Phys., Leipzig, Vol. 43, 1943, pp. 44-46.

13 Dryden, H., Murnagham, F.D., and Bateman, N.H., "Report of the Commission on Hydrodynamics," Bulletin of the US National Research Council, Vol. 84, 1932.

14 Howard, E., "On a New Mercury Fuliminate," Philosophical Transactions of the Royal Society, Vol. 90, 1800, pp. 204-238 and 1 plate.

15 Hugoniot, H., "Propagation des Mouvements dans les Corps et Spécialement dans les Gaz Parfaits," (Propagation of Movements in Bodies and Specially in Ideal Gases), Journal de l'Ecole Polytechnique, Cahiers 57 et 58, 1887-1889.

16 Jouguet, E.,
a "Sur l'Onde Explosive," (On Explosive Wave), Comptes-Rendus Académie des Sciences, Paris, France, Vol. 140, 1904, p. 1211.

b "Sur la Propagation des Réactions Chimiques dans les Gaz," (On the Propagation of Chemical Reactions in Gases), Journal des

Mathématiques Pures et Appliquées, (J. of Pure and Applied Mathematics), 6th Serie, 1905-1906, Vol. 1, p. 347, Vol. 2, p. 5.

c La Mécanique des Explosifs, (Mechanics of Explosive), O. Doin, Paris, 1917.

17 Kast, H., "Die Brisantzbestiummung und die Messung der Detonationsgeschwindigkeit von Sprengstoffen," (Measurement of the Brisance and the Detonation Velocity of Explosives), Zeitschrift fur das Gesamte Schiess u. Sprengstoffwesen, Vol. 9, n°4, 1913.

18 Mach, E., and Sommer, J., "Uber die Fortpflanzungsgeschwindigkeit von Explosionschallwellen," (On the Propagation of Blast Waves), Sitzung Berichte Akad. Wissensch. Wien, (Reports of the Academy of Sciences of Vienna), Austria, Vol. 75, 1877, pp. 101-130.

19 Mallard, E., "De la Vitesse avec Laquelle se Propage l'Inflammation dans un Mélange d'Air et de Grisou et la Théorie des Lampes de Sureté," (Propagation Velocity in Firedamp Air Mixtures and the Theory of Safety Lamps), Annales des Mines, 7th Serie 8, 1875, pp. 355-381 (see also (21c).

20 Mallard, E., and Le Chatelier, H.,
a "Sur la Vitesse de Propagation de l'Inflammation dans les Mélanges Gazeux Explosifs," (On the propagation Velocity of Inflammation in Gaseous Explosive Mixtures). Comptes-Rendus Académie des Sciences, Paris, France, Vol. 93, 1881, pp. 145-148.

b "Recherches sur la Combustion des Mélanges Gazeux Explosifs," (Research on Combustion of Gaseous Explosives), Annales des Mines, 8th Serie, Vol. 4, 1883, pp. 274-568; 10 PLATES/ VIII-XVII.

21 Manson, N.,
a "Contribution de Paul Vieille à la Connaissance des Détonations et des Ondes de Choc," (Contribution of Paul Vieille to Knowledge of Detonations and Shock Waves), Science et Technologie de l'Armement, Vol. 60, N°232, 1980, pp. 26-59.

b "Historique de la Découverte de l'Onde de Détonation," (History of the Discovery of Detonation Wave), Colloque C4, Journal de Physique, Vol. 48, Supplement to N°9, 1987, pp. 7-37.

c "Some Notes on the First Theories of the Flame Velocity in Gaseous Mixtures," Combustion and Flame, Vol. 71, 1988, pp. 179-187.

d "Notes et Remarques sur la Contribution de M. Berthelot à l'Etude des Matières Explosives," (Notes and Remarks on M. Berthelot's Contribution to the Study of Explosive Materials), Presenté au Colloque : "M. Berthelot, une Vie, une Epoque, un Mythe," (M. Berthelot a Life, a Era, a Legend), Saint-Maur, France, 1988, (Proceeding to be published in 1990).

22 Marshall, A., Explosives, J. and A. Church, London, England, 1917.

23 Massin, P., "Cordeaux Détonants," (Detonating Fuses), French Patent 13F 190073, Apr. 1888.

24 Médard, L.,
a "Les Explosifs Occasionnels," (The Occasional Explosives), Techniques and Documents, Paris, France, 1978.

b "Paul Vieille et son Oeuvre," (Paul Vieille and his work), Science et Technique de l'Armement, Vol. 60, 2nd fasc., N° 232, 1986, pp. 11-23.

25 Mettegang, H., "Messung der Detonationsgeschwindigkeit von Sprengstoffen," (Measurement of Detonation Velocity in Explosives), Vth Int. Cong. fur Angewandte Chem., Vol. 2, Berlin, Germany, 1904, p. 322.

26 Michelson, V.A.,
a "On the Normal Ignition Velocity of Explosive Gaseous Mixtures," Ph.D. Dissertation printed by the Moscow University Printing Service, Moscow, USSR, 1890; Republished in Scientific papers of the Moscow Imperial University on Mathematics and Physics, Vol. 10, 1893, p. 1-93 (with 2 plates).

b Collected Works, Novi Agronom Moskow, USSR, 1930.

27 Miles, F.D., "A History of Research in Nobel Division of the Imperial Chemical Industry," ICI Ltd, Nobel Div., Ardeer, U.K., 1955.

28 Mosenthal, H. de, "The Life-Work of Alfred Nobel," Journal of Society and Chemical Industry, Vol. 18, May 1899, pp. 443-451.

29 Nernst, W., "Physikalische-Chemiche Betrachtungen uber den Verbrennungprogress in den Gasmotoren," (Physical and Chemical Considerations on Combustion in Engines), Zs V.D.I., (Zeitschrift des Vereines der Deutschen Ingenieure), Vol. 49, 1905, pp. 1426-36.

30 Neumann, J., von, "Theory of Stationary Detonation Waves", O.S.R.D. Rept 549, Selected Works, Vol. 6, MacMillan New York, 1942.

31 Nobel, A.B.,
a "Detonators," Brit. Pat. 2359, 1867.

b "Kieselguhr Dynamite," Brit. Pat. 1345, 1867.

c "Blasting Gelatine," Brit. Pat. 4179, 1875.

32 Noble, A., "Artillery and Explosives," John Ringway, London England, 1906.

33 Noble A., Abel, F.A.,
a "Researches on Explosives I: Fired Gunpowder," Philosophical

Transactions of the Royal Society of London, Vol. 165, 1875, pp. 129-237.

b "Researches on Explosives II: Fired Gunpowder," Philosophical Transactions of the Royal Society of London, Vol. 171, 1880, p. 204.

34 Pepin-Lehalleur, J., Traité des Poudres, Explosifs et Artifices," (Treatrise on Propellants, Explosives and Pyrotechnics), Ballieres, Paris, France, 1935.

35 Rankine, W.J., "On the Thermodynamic Theory of Finite Longitudinal Disturbance," Philosophical Transactions, Vol. 160, 1870, pp. 277-288.

36 Riemann, B., "Uber die Fortpflanzung ebener Luftwellen von endlicher Schwingungsweite," (On the propagation of plane air waves of finite amplitude), Sitz. Berichte der Wissenschaftlicher Ges. der Univ. Göttingen, (Proceedings of the Scientific Society, Univ. of Gottingen), Germany, 1860.

37 Romocki, S.J., von, "Geschichte der Explosivstoffe," (History of Explosives). Vol. I: - Geschichte der Sprengstoffchemie, der Sprengstechnik und des Torpedowesen, (History of Explosives, Chemistry of Explosives and Blasting Technics and of Torpedos), Vol. II - Die rauchschwachen Pulver in ihrer Entwicklung, (The Propellants and their Development), R. Oppenheim (Gustav Schmidt), Berlin, Germany, 1895-1896.

38 Roux, L.P., and Sarrau, E., "Expériences sur les effets de la dynamite," (Experiments of the dynamite), Comptes-Rendus Académie des Sciences, Paris, France, Vol. 76, 1873, pp. 1089-1092.

39 Schönbein, C.F., Reports of the Natural Researcher Society, Basel 7, 1846, p. 26; Sitzung Berichte der Naturforscher Geselschaft; U.S. Pat. 4674, 1846.

40 Sebert, et al., "Comptes-Rendus des Travaux de la Commission des Substances Explosives," (Reports of the Explosives Committee), Memorial des Poudres et Salpêtres, (Memorial of Powders and Nitre), Vol. 1, 1878-1883, pp. 434, 451, 481; Vol. 2, 1884-1886, pp. 36, 53, 100.

41 Sobrero, A., "Sur Plusieurs Composés Détonants Produits avec l'Acide Nitrique et le Sucre, la Dextrine, la Lactine, la Marnite, et la Glycérine," (On Several Detonating Compounds obtained from Nitric Acid, Sugar, Dextrine and Glycerine), Comptes-Rendus Académie des Sciences, Paris, France, Vol. 25, 1847, pp. 247-248.

42 Taffanel, J., "Nouvelles Expériences sur les Poussières de Houille et les Moyens de Combattre leurs Explosions," (New Experiments on Coal Dust and the means to Fight their Explosions), 1th Part.: Développement et Arrêt des Coups de Poussière. Théorie des Explosions," (Development of Dust Explosion. Theory of Explosions), Annales des Mines, 11th Ser., Vol. 1, 1912, pp. 259-554.

43 Taffanel, J., and Dautriche, M.,
a "Sur la Propagation de l'Onde Explosive dans les Solides," (On the Propagation of Explosive Wave in Solids), Comptes-Rendus Académie des Sciences, Paris, France, Vol. 125, 1912, pp. 1221.

b "Sur la Détonation de la Dynamite,", Comptes-Rendus Académie des Sciences, Paris, France, Vol. 125, 1912, pp. 2595.

44 Taylor, J., Detonation in Condensed Explosives; Clarendon, Oxford, England, 1952.

45 Upmann, J., and Meyer, E. von, "Traité sur la Poudre, les Corps Explosifs et la Pyrotechnie," (Treatise of Propellants Explosives and the Pyrotechnics). Vieweg u. Sohn, Baunschweig, Germany, 1874, French translation from German by E. Dessortiaux, Dunod, France, 1878.

46 Urbanski, T.,
a Chemistry and Technology of Explosives, Pergamon, Oxford, England, 1964.

b Urbanski, T., "Le Centenaire de la Nitro-cellulose," (Hundred years of the Nitrocellulose), Mémorial de l'Artillerie Française, Vol. 13, 1934, pp. 885-842?

47 Vieille, P., "Rôle des Discontinuités dans la Propagation des Phénomènes Explosifs", (The role of the discontinuities in the Propagation of Explosive Phenomena), Comptes-Rendus Académie des Sciences, Paris, France, Vol. 130, pp. 413-416. Mémorial des Poudres et Salpêtres (Memorial of Powder and Nitre), Paris, Vol. 10, 1899-1900, pp. 177-260.

48 Zeldovich, Ya. B., "Teoria pazprostranenia Detonazii b Gasovikh smesseĭ," (On the Theory of the Propagation of Detonation in Gaseous Systems), Zh. Exp. i Teor. Fiz., Vol. 10, 1940, pp. 542-568. English Translation NACA TN N° 1261, 1950.

Detonability of Hydrocarbon Fuels in Air

H. D. Beeson,* R. D. McClenagan,†
C. V. Bishop,‡ and F. J. Benz§
NASA White Sands Test Facility, Las Cruces, New Mexico
and
W. J. Pitz¶ and C. K. Westbrook¶
Lawrence Livermore National Laboratory, Livermore, California
and
J. H. S. Lee**
McGill University, Montreal, Quebec, Canada

Abstract

The detonation of gas-phase mixtures of n-hexane and JP-4, with oxidizers varying from pure oxygen to air, was studied. Experiments were carried out in a heated detonation tube 150 mm in diameter and 1.75 m long. An exploding bridge wire with 2000 J was used for initiation, except for less sensitive mixtures, when a solid explosive charge was used. Detonation velocities and cell sizes λ were measured as a function of stoichiometry and diluent concentration. A theoretical model was used to calculate the induction length Δ of a one-dimensional Zeldovich-von Neumann-Doring (ZND) detonation using the detailed kinetics for the reaction of the hydrocarbon fuel used. Using a constant of proportionality of 29 (i.e., $\lambda = 29\Delta$), the theoretical prediction for the cell size of n-hexane agrees well with the experimental data. No detailed modeling was carried out for JP-4. The cell

*Senior Research Scientist, currently with NMERI, Albuquerque, NM.
†Project Engineer, Lockheed-ESC.
‡Director, currently with McGean-ROHCO, Cleveland, OH.
§Project Manager, NASA.
¶Physicist, Computational Physics Division.
**Professor, Department of Mechanical Engineering.

size data obtained for n-hexane and JP-4 were found to be slightly less than that of the other lighter hydrocarbons in the same alkane family with the exception of methane (i.e., ethane, propane, n-butane). Critical energy and critical tube diameter are compared for a relative measure of the detonability of the heavy hydrocarbon fuels studied, and it appears from kinetic modeling that the detonation sensitivity increases slightly with increasing carbon number; however, this trend cannot be distinguished experimentally with the error in cell size measurement.

Introduction

Detonations represent a major source of hazard in the use of hydrocarbon fuels in practical combustion systems. The study of detonations also provides insight into the fundamental interactions between the fluid mechanics and chemical kinetic properties of combustion. Detonations have been the subject of many experimental, theoretical, and computer modeling studies, and it is beyond the scope of the present paper to survey this extensive literature. The entire field was reviewed quite recently,[1] and other reviews have appeared.[2]

Most previous detonation studies have dealt with relatively small hydrocarbon-fuel molecules, both in experimental work[3,4] and modeling analyses.[5,6,7] From the kinetic modeling point of view, one important reason for the focus on smaller fuels has been the fact that only very recently have mechanisms for fuels as large as n-pentane[8] and n-octane[9,10] been developed.

From the experimental point of view, a different factor has been responsible for the past concentration on smaller fuels: larger fuels (those with about six or more carbon atoms) exist as liquids under normal conditions of temperature and pressure. Although it is not impossible to produce detonable fuel-oxidizer mixtures for these larger hydrocarbon molecules, their lower volatility certainly complicates experimental studies of their combustion and detonation properties. However, many practical hydrocarbon fuels, particularly those often used for aviation fuels and other propellants, include large fractions of these liquid fuels. Therefore, it is very important to be able to study these fuels, in spite of the difficulty of dealing with them in well-posed experimental conditions.

With liquid fuels such as n-hexane, two distinct types of detonations are very important: those involving aerosols or sprays, and those consisting of purely gas-phase fuel-oxidizer mixtures. Of particular interest are detonations that propagate through a droplet or aerosol spray. Bull[11] reported studies of spray detonations in which n-hexane was the fuel, whereas experiments[12] indicated that, under conditions very similar to those for n-hexane, n-decane would not support a spray detonation. A great deal of work remains to be done to unravel the mysteries of spray detonations, including the identification of the controlling physical and chemical processes in these detonations.

To provide insight into some of those processes, gas-phase detonations can be studied as a form of limiting condition. The purely chemical kinetic factors that influence detonability can be examined in detail by eliminating complicating factors such as the formation and vaporization of a spray, the interaction of a shock wave with arrays of droplets, the mixing of the vaporized fuel with the oxidizer, and chemical ignition in a fuel-oxidizer medium that is not homogeneous. These results can then be integrated into a future, more complex model that includes the additional factors important in a true spray detonation. Without a thorough description of the gas-phase problem, analysis of the spray case would be overly speculative. The present study reports the results of exactly this type of problem: the detonability of gas-phase mixtures of n-hexane, oxygen, and nitrogen, and mixtures of JP-4, oxygen, and nitrogen.

For practical applications, mixtures of fuels with air are of primary importance. The spatial scales for such detonations and the amounts of energy required to initiate such detonations are often quite large, however, making it inconvenient and even dangerous to carry out the fuel-air experiments. It is therefore common to study fuel-oxygen mixtures, for which the relevant scales are much smaller, and systematically dilute these mixtures with increasing amounts of nitrogen until the scales become larger. These results are then used to extrapolate, in terms of further dilution by nitrogen, to mixtures that have a nitrogen/

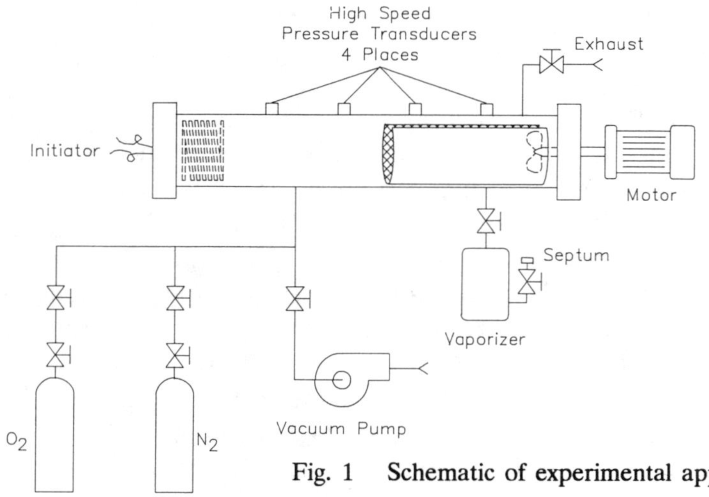

Fig. 1 Schematic of experimental apparatus.

oxygen ratio equal to 3.76, the ratio in normal air. This approach has been used productively in many experimental studies,[3,13,14] and modeling studies have provided further refinements for this extrapolation procedure.[5,7] These approaches are employed in the present study.

Experimental Details

The experiments were carried out in a 150-mm-diam and 1.75-m-long insulated detonation tube (Fig. 1). The tube could be heated above ambient temperatures by electrical tape to maintain the hydrocarbon fuel in the vapor phase. A 1-liter heated vaporizer vessel was used to evaporate the liquid fuel so that the fuel vapor could be introduced into the evacuated detonation tube. The fuel concentration was monitored by partial pressure. The rest of the mixture components (oxygen and nitrogen) were added to the detonation tube, and mixing was accomplished by a propeller fan at the end of the tube.

The initiation of the detonation was achieved with a 2000-J exploding bridge wire for the sensitive mixtures. For the less sensitive, highly dilute mixtures, however, a solid explosive charge of PETN (or various lengths of detonating cord) initiated by a #8 blasting cap was used. A short length of Schelkhin spiral was also placed at the ignition end to insure the formation of the

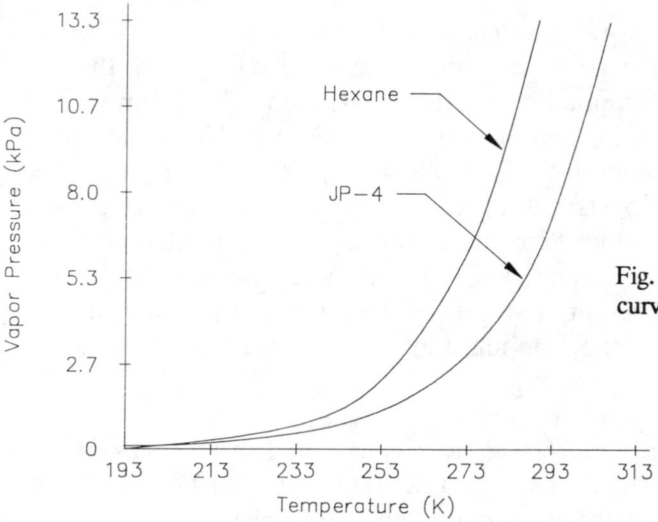

Fig. 2 Vapor pressure curve for JP-4 and hexane.

detonation. For diagnostics, four PCB piezoelectric transducers were used to measure the detonation velocity. Detonation cell sizes were measured by inserting a metallic smoked foil at the end of the detonation tube.

The vapor pressure (as a function of temperature) of the JP-4 used in the experiment and of n-hexane is shown in Fig. 2. For the mixtures studied with the highest fuel concentration, the vapor pressure corresponds to about 8.1 kPa. This vapor pressure corresponds to a temperature of 293 K. Similar results were obtained for the JP-4 liquid fuel used. The maximum initial temperature of the heated detonation tube required to maintain the fuel in the vapor phase was, therefore, about 293 K. This modest initial temperature above ambient played a negligible role in the equilibrium detonation properties and induction kinetics of the mixture.

Theoretical Modeling

The Zeldovich-von Neumann-Doring (ZND) model was employed, in which a detonation (locally) consists of a shock wave traveling at the Chapman-Jouguet (CJ) velocity, followed by a reaction zone. The shock wave compresses and heats the fuel-oxidizer mixture, which then begins to react. In most mixtures, the fuel oxidation consists of a relatively long induction period, during which the temperature and pressure remain nearly constant, followed by a rapid release of chemical energy and temperature increase. For each fuel-oxidizer mixture considered, the CJ conditions were computed using the TIGER code.[15] From the detonation velocity D_{CJ}, the conditions in the von Neumann spike (including the temperature T_1, pressure P_1, and the particle velocity u_1 of the postshock unreacted gases) were computed and used as initial conditions for the chemical kinetics model. In actuality, the shock velocity and other physical properties of the detonation varied within a single detonation cell; therefore, the CJ conditions (and the computed induction times) represented average values.

The reactive mixture volume was assumed to remain constant over the mixture's reaction time, and the induction time was defined in terms of the mixture's temperature history. The mixtures considered in the present study experienced a temperature increase of more than 1000 K, and the induction time was defined as the time of maximum rate of temperature increase. This coincided closely with the time at which the temperature had completed about half of its total increase. This is not, strictly speaking, a true induction period (which is often defined as the time required for a small [1-5%] temperature or pressure increase), but it represents a time scale for the release of a significant amount of energy. In a detonation, it is this macroscopic energy release that reinforces the shock wave and permits a detonation to propagate; therefore, the definition of the characteristic time used here was motivated by the properties of a detonation. In addition to the induction time τ, it is useful to define the induction length $[\Delta - \tau (D_{CJ} - u_1)]$, which represents a characteristic length scale in the postshock unreacted gas mixture.

The computed induction times and lengths defined characteristic time and length scales rather than the precise history of a gas element through the detonation front. The evolution of the reacted gas subsequent to the induction period considered here was dominated by the fluid mechanics of the postinduction expansion of the reaction products. This expansion reduced the pressure and density of the products and altered the kinetic equilibrium, leading eventually to the CJ state. Because virtually all of the reactants had been consumed by this time, the kinetics of this final expansion phase were controlled by relatively slow radical recombination processes. The present model does not attempt to follow that entire relaxation phase, concentrating on the details of the induction kinetics in the von Neumann spike. Shepherd has constructed a model that follows the entire process (combining the fluid mechanics with the reaction kinetics in the case of hydrogen-air mixtures) and found results that were very similar to those obtained using the present simplified approach.[16]

This model of the detonation neglects some potentially significant effects associated with the fluid mechanics parts of the gas history. Variations of density, temperature, and particle velocity in the postshock unreacted mixture were not considered. Multiple shock wave reflections, rarefactions, interactions with confining walls, cellular structure, and related effects were also not treated directly by the present simplified model.

For the kinetic model, the same type of approach employed at present has been used in the past to study detonation parameters for fuels such as hydrogen, methane, ethane, ethylene, methanol, acetylene, and propane.[5-7,17,18] Recent applications have included a wide variety of larger fuels, including ethers, nitrates, and other complex fuels.[19] In the present case of n-hexane, the reaction mechanism was assembled by adding reactions for n-hexane and its immediate products to an existing reaction mechanism for the oxidation of n-pentane.[8] Reactions for the unimolecular decomposition of n-hexane were based on those for n-pentane. Abstraction reaction rates of H atoms from n-hexane were based on the principle that rates of abstraction of primary and secondary H atoms from a hydrocarbon fuel are relatively insensitive to the size and structure of the parent fuel, so that these rates could be estimated from analogous reactions in n-pentane and other comparable fuels. Subsequent reactions of hexyl radicals were assumed to be dominated by the process of β scission, producing smaller olefins and alkyl radicals. In this formulation, the reaction mechanism for the oxidation of n-hexane was viewed as a perturbation of mechanisms for n-pentane and smaller hydrocarbon fuels.

The second fuel used in the present study was JP-4, a complex mixture of hydrocarbon fuels. Modeling was not attempted with this particular fuel, but it is known that its combustion characteristics are quite closely related to those of n-hexane, and n-hexane is a major component in JP-4. Most of the conclusions of the present kinetic analysis of n-hexane would be expected to apply closely to the case of JP-4.

The important kinetic features of the ignition of n-hexane at the high temperatures characteristic of detonation conditions were dominated by the reactions of the H_2-CO-O_2 submechanism. The chain branching reaction,

$$H + O_2 \longrightarrow O + OH \qquad (1)$$

which produces two radicals O and OH for each H atom radical, is particularly important. Those reactions that produce H atoms accelerate the overall rate of ignition because the H atoms then produce two new radical species via reaction (1). Smaller radical species, which lead to H atom production, also accelerate the overall rate of ignition. An example of this is the ethyl radical C_2H_5, which decomposes at elevated temperatures to produce H atoms and ethylene through the reaction

$$C_2H_5 \longrightarrow C_2H_4 + H \qquad (2)$$

In contrast, those reactions that produce methyl CH_3 radicals actually retard the overall rate of ignition because many of those methyl radicals recombine to produce ethane,

$$CH_3 + CH_3 \longrightarrow C_2H_6 \qquad (3)$$

which, because it is relatively stable, does not produce new radical species.

The major reactions consuming n-hexane under the present conditions are its unimolecular decomposition, followed by reactions with OH, H, O, CH_3, and HO_2. The important decomposition reactions are

$$nC_6H_{14} \longrightarrow pC_4H_9 + C_2H_5$$

$$nC_6H_{14} \longrightarrow nC_3H_7 + nC_3H_7$$

The former leads to the production of H atoms from both the p-butyl and ethyl radicals, whereas the second reaction leads to two methyl radicals. The H atom abstraction reactions produce three logically different hexyl radicals, those in which the H atom has been taken from the 1, 2, and 3 sites in the hexane molecule,

```
    1   2   3   3   2   1
1 - C - C - C - C - C - C - 1
    1   2   3   3   2   1
```

where the numbers 1, 2, and 3 indicate logically distinct H atoms in the n-hexane molecule. These sites must be kept distinct, because the subsequent decomposition of the product alkyl radicals leads to different products. The dominant decomposition paths are as follows:

1) $1\text{-}C_6H_{13} \longrightarrow C_2H_4 + 1\text{-}C_4H_9$

$1\text{-}C_4H_9 \longrightarrow C_2H_4 + C_2H_5$

$C_2H_5 \longrightarrow C_2H_4 + H$

2) $2\text{-}C_6H_{13} \longrightarrow C_3H_6 + 1\text{-}C_3H_7$

$1\text{-}C_3H_7 \longrightarrow C_2H_4 + CH_3$

3) $3\text{-}C_6H_{13} \longrightarrow 1\text{-}C_4H_8 + C_2H_5$

$C_2H_5 \longrightarrow C_2H_4 + H$

$3\text{-}C_6H_{13} \longrightarrow 1\text{-}C_5H_{10} + CH_3$

Thus the abstraction of H atoms from the 1 site in n-hexane leads eventually to the production of H atoms and acceleration of the overall rate of ignition, whereas the 2 site leads to a slower rate of ignition, and the 3 site leads to both H atoms and CH_3 radicals. In addition to the thermal decomposition of the alkyl radicals, the $1\text{-}C_6H_{13}$ and $2\text{-}C_6H_{13}$ radicals are interrelated through internal H atom abstraction reactions. The subsequent

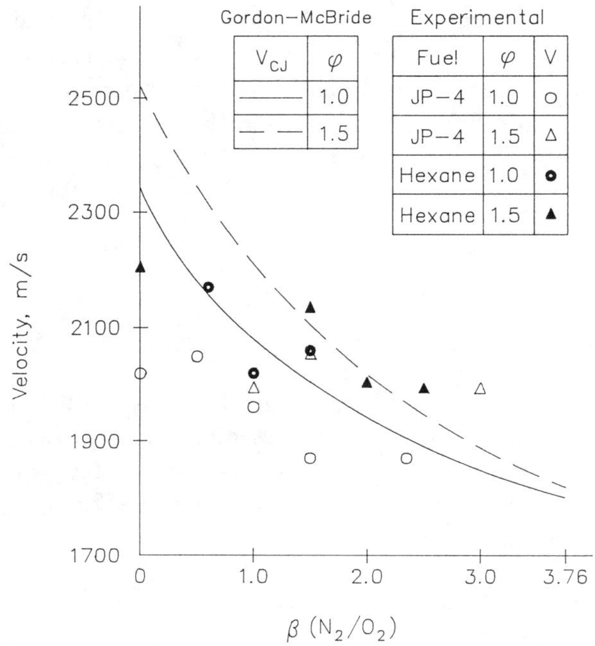

Fig. 3 Theoretical detonation velocity as a function of equivalence ratio for JP-4 and hexane.

reactions of the ethylene, propene, and 1-butene are already well understood from previous modeling studies; therefore, the most important features of the n-hexane mechanism can be presented in terms of the ways in which they eventually impact the mechanisms for smaller fuels and lead to H atom or CH_3 radical production. Minor paths in the n-hexane reaction mechanism are also included (paths that lead to production of hexenes and other intermediate hydrocarbon species), but these play an insignificant part in the induction process.

Results and Discussion

Measurements for the detonation velocity deduced from the time of arrival at the various pressure transducers are shown in Fig. 3. Velocities were obtained for hexane and JP-4 at two fuel compositions of $\phi = 1$ (stoichiometric) and $\phi = 1.5$ and for various nitrogen dilutions β. Also shown are the theoretical curves for the CJ detonation velocity computed using the Gordon-

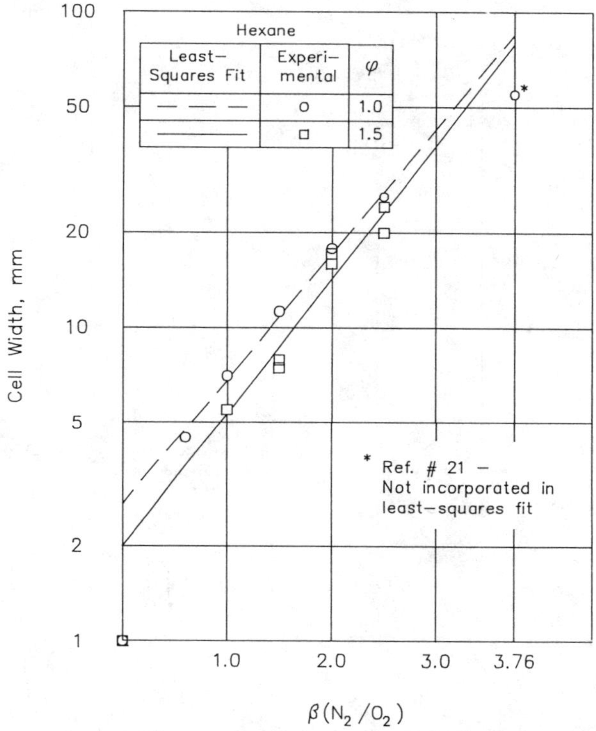

Fig. 4 Experimentally determined detonation cell widths at varying diluent/oxidizer ratios and equivalence ratios for hexane.

McBride code.[20] The experimental results are in accordance with the theoretical CJ values with typical velocity deficits of less than 10% for most cases. For the high-nitrogen diluted mixtures when very large amounts of initiation energy were used, the detonation velocities appear to be slightly overdriven because the detonation tube was only 1.75 m long. Results for both hexane and JP-4 showed insignificant differences between the two fuels, within the experimental error of the present study.

The cell sizes for hexane and JP-4 detonations are shown in Figs. 4 and 5, respectively. The detonation sensitivity (as inferred from the cell size) for both JP-4 and hexane are about the same, indicating that the presence of other hydrocarbons in JP-4 are not important in influencing its induction kinetics.

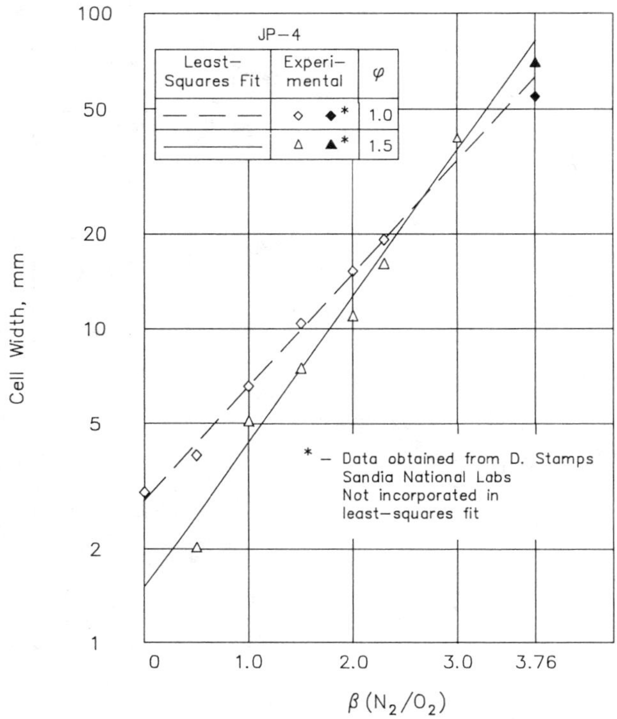

Fig. 5 Experimentally determined detonation cell widths at varying diluent/oxidizer ratios and equivalence for JP-4.

Because of limitations of the present detonation tube (short length and small diameter), the less sensitive mixtures with higher values of β could not be studied. For JP-4, the cell size was measured in support of this study in a larger tube at Sandia National Laboratory. The Sandia heated detonation tube was 43 cm in diameter and 13.1 m in length. The Sandia cell size data for JP-4-air mixtures ($\beta = 3.76$) are also plotted in Fig. 5, and are consistent with the data obtained for more sensitive mixtures obtained in the smaller apparatus. Additionally, recent results reported by Sandia National Laboratories for hexane-air ignitions are plotted in Fig. 4.[21]

For kinetic modeling, computations were carried out using the approach outlined earlier, which has been used in previous studies for other similar hydrocarbon fuels. Only n-hexane was

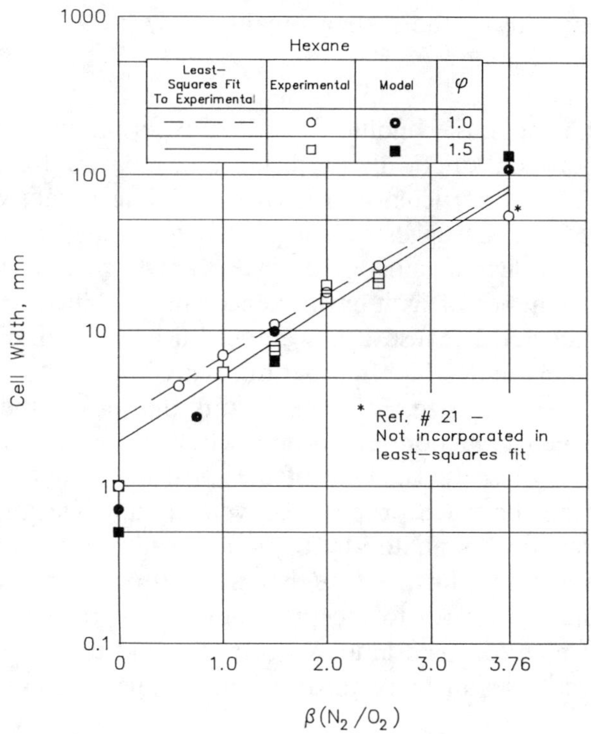

Fig. 6 Comparison of experimentally determined cell-widths results and model-predicted cell widths for hexane.

included in the modeling analysis. Kinetic induction times were computed for stoichiometric n-hexane-oxidizer mixtures and for fuel-rich mixtures with $\phi = 1.5$. The entire range from pure oxygen to air as the oxidizer was included for both equivalence ratios.

As outlined previously, the computed induction length $[\Delta - \tau (D_{CJ} - u_1)]$ has been related to the cell size by a simple constant of proportionality. In earlier work, the majority of the fuel-oxidizer mixtures were best correlated if this constant of proportionality was approximately 29.[5,6] If exactly the same relationship between computed induction time and cell size is used, the model predictions for cell sizes agree quite well with the experimentally measured values over the range of conditions studied. These results are indicated in Fig. 6, in which the computed cell sizes (using the proportionality of $\lambda = 29\Delta$) are indicated as filled symbols.

Over the range of conditions studied experimentally, the agreement between kinetically predicted and measured cell size is very good. This is particularly true in the middle range, where $1.0 < \beta < 2.5$. This suggests strongly that n-hexane, in the gas phase, is very similar to most other hydrocarbon fuels in the way the chemical kinetics of ignition influences detonability. This result was expected because it is well established that the ignition of n-alkane fuels under high temperature conditions is a very weak function of fuel size.[22] Those factors that affect the relationship between induction length and cell size are very much the same for all of these fuels. One of the most important of these factors must be the wave propagation velocity and, because all of these mixtures are dominated by oxygen and nitrogen, this conclusion is consistent with the fact that the hydrocarbon fuel does not contribute very much to the thermodynamic properties of the combustible mixture. The amounts of water and carbon dioxide in the product gases in these mixtures were roughly constant as well.

Of particular interest are the limiting conditions for the cell sizes as $\beta \rightarrow 0$ and as $\beta \rightarrow 3.76$ (normal air). For n-hexane-oxygen mixtures, the predicted cell size is slightly smaller than that observed experimentally, but the two values are quite close together. If the experimental results are extrapolated to $\beta = 0$, the estimate of the cell size is larger than that observed by more than a factor of 2. The kinetic model, therefore, provides an improved basis for extrapolation of the experimental results. This really means that linear extrapolation to the experimental data in Figs. 4 and 5 is not valid, in part because the variations in induction length with nitrogen dilution are very nonlinear. Similar curvature as $\beta \rightarrow 0$ is seen for other hydrocarbon fuels.[5-7] At the other limit, the predicted cell size for stoichiometric n-hexane-air is $\lambda = 111$ mm. When an extrapolation to n-hexane-air conditions is made solely on the basis of the experimental results, the best estimate is $\lambda = 85$ mm for $\phi = 1.0$ and $\lambda = 79$ mm for $\phi = 1.5$, whereas the direct measurement of cell size is reported to be 55 mm.[21] Clearly, the kinetic model provides a significantly larger estimate of the cell size under n-hexane-air conditions.

Several overall observations can be drawn from the present computational results. The constant of proportionality between the computed induction length and the cell size data (≈ 29) is the same as that determined previously for many other hydrocarbon fuels.[5] This indicates that the same type of scaling can be applied for a very wide range of hydrocarbon fuels, including many for which experiments may not have been carried out but kinetic mechanisms may exist. It also indicates that the reaction mechanism and numerical model are generally reliable.

For stoichiometric fuel-air mixtures with hydrocarbon fuels larger than methane or ethane, the predominant species in the reactive mixture are oxygen and nitrogen. This means that physical quantities such as the sound speed and heat capacity of the reactant mixtures will vary only slightly from one hydrocarbon fuel to another. The adiabatic flame temperature and heat of reaction are also very similar for these fuels. Combined with the near constancy of the computed induction times for these fuels,[20] this suggests very strongly that the cell sizes will also be very similar for the same gaseous fuels. Therefore, if major differences are observed experimentally between detonation cell sizes for hydrocarbon-air mixtures (particularly for n-alkane-air mixtures with 4-10 carbon atoms), they are probably attributable to variations in parameters (such as vapor pressure) that affect the ability of the fuel to support stoichiometric gas-phase mixtures. There is very little distinction between the kinetic factors of any n-alkane fuels; therefore, the only remaining conclusion is that differences are due to variability in the abilities of these fuels to provide detonable fuel-oxidizer mixtures.

The cell size data for stoichiometric fuel-air mixtures for the alkane family (i.e., ethane, propane, hexane, and JP-4) are shown in Fig. 7a. With the well-known exception of methane, all of the fuels given show similar cell sizes. Although there is a slight trend in decreasing cell size with increasing carbon number, this trend is well within the experimental error in cell size measurements. From the cell size data, all the dynamic detonation parameters can be estimated (e.g., critical tube diameter, $d_c = 13\lambda$). The critical initiation energy for these fuels is compared in Fig. 7b. With the exception of methane, stoichiometric mixtures

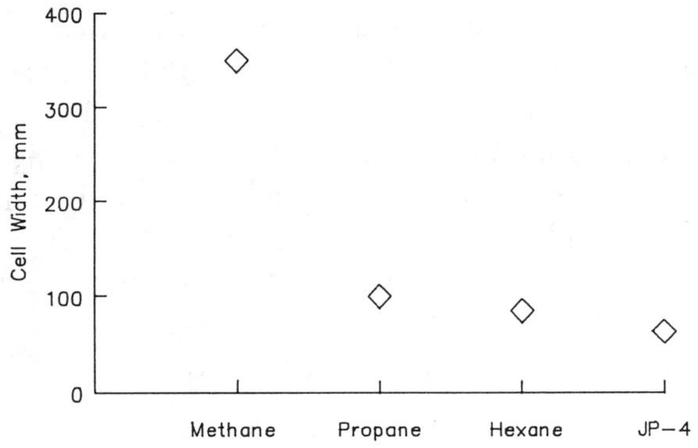

Fig. 7a Cell widths for several hydrocarbon fuels.

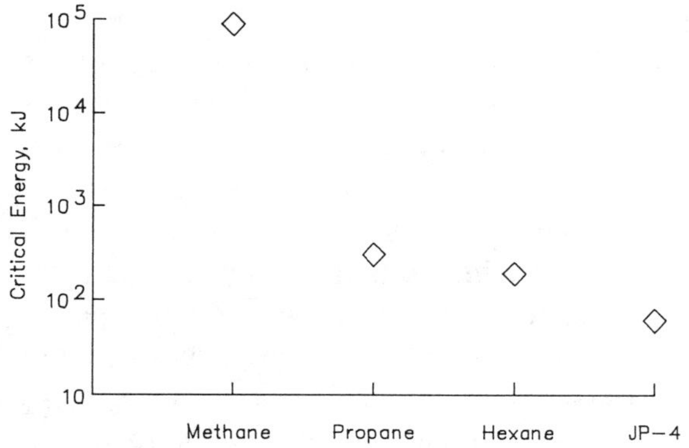

Fig. 7b Critical energy for several hydrocarbon fuels.

of alkanes in air typically require 100 kJ of energy (about 25 g of high explosive) for initiation.

Acknowledgments

The experimental portions of this work were carried out under the auspices of NASA at the Johnson Space Center White Sands Test Facility. The computational portions of this work were carried out under the auspices of the U.S. Department of Energy by the Lawrence Livermore National Laboratory under contract number W-7405-ENG-48. The authors wish to thank Doug Stamps of Sandia National Laboratory for the JP-4/air cell size data.

References

[1]Lee, J. H. S., and Guirao, C. M. (eds), "Fuel-Air Explosions," Proceedings of the International Conference on Fuel-Air Explosions, November 4-6, 1981, Univ. of Waterloo Press, Waterloo, Ontario, 1982, p. 1005.

[2]Vasiliev, A. A., Mitrofanov, V. V., and Topchiyan, M. E., "Detonation Waves in Gases," Fizka Goreniya i Vzryva, Vol. 23, 1987, p. 109.

[3]Matsui, H., and Lee, J. H., "On the Measure of the Relative Detonation Hazards of Gaseous Fuel-Oxygen and Air Mixtures," Seventeenth Symposium (International) on Combustion, The Combustion Institute, Pittsburgh, 1979, p. 1269.

[4]Knystautas, R., Lee, J. H., and Guirao, C. M., "The Critical Tube Diameter for Detonation Failure in Hydrocarbon-Air Mixture," Combustion and Flame, Vol. 48, 1982, p. 63.

[5]Westbrook, C. K., "Chemical Kinetics of Hydrocarbon Oxidation in Gaseous Detonations," Combustion and Flame, Vol. 46, 1982, p. 191.

[6]Westbrook, C. K., The Chemistry of Combustion Processes, ACS Symposium Series, No. 249, edited by T. M. Sloane, The American Chemical Society, Washington, D.C, 1984.

[7]Westbrook, C. K., and Urtiew, P. A., "Chemical Kinetic Prediction of Critical Parameters in Gaseous Detonations," Nineteenth Symposium (International), The Combustion Institute, Pittsburgh, 1983, p. 615.

[8]Westbrook, C. K., Pitz, W. J., Thornton, M. M., and Malte, P. C., "A Kinetic Modeling Study of n-Pentane Oxidation in a Well-Stirred Reactor," Combustion and Flame, Vol. 72, 1988, p. 45.

[9]Axelsson, E. I., Brezinsky, K., Dryer, F. L., Pitz, W. J., and Westbrook, C. K., "Chemical Kinetic Modeling of the Oxidation of Large Alkane Fuels: N-Octane and Iso-Octane," Twenty-First Symposium (International) on Combustion, The Combustion Institute, Pittsburgh, 1988, p. 783.

[10]Westbrook, C. K., Warnatz, J., and Pitz, W. J., "A Detailed Chemical Kinetic Reaction Mechanism for the Oxidation of Iso-Octane and n-Heptane Over an Extended Temperature Range and Its Application to Analysis of Engine Knock," Twenty-Second Symposium (International) on Combustion, The Combustion Institute, Pittsburgh, PA., 1989.

[11]Bull, D. C., "Detonation of Unconfined Fuel Aerosols," Progress in Astronautics and Aeronautics: Gasdynamics of Detonations and Explosions, Vol. 75, edited by J. R. Bowen, N. Manson, A. K. Oppenheim, and R. Soloukhin, AIAA, New York, 1981, p. 48.

[12]Lu, P. L., Slagg, N., and Fishburn, B. D., "Relation of Chemical and Physical Processes in Two-Phase Detonations," Acta Astonautica, Vol. 6, 1979, p. 815.

[13]Bull, D. C., Elsworth, J. E., Hooper, G., and Quinn, C. P., "A Study of Spherical Detonation in Mixtures of Methane and Oxygen Diluted in Nitrogen," Journal of Physics D: Applied Physics, Vol. 9, 1976, p. 1991.

[14]Vandermolen, R., and Nicholls, J. A., "Blast Wave Initiation Energy for the Detonation of Methane-Ethane-Air Mixtures," Combustion Science and Technology, Vol. 21, 1979, p. 75.

[15]Cowperthwaite, M., and Zwisler, W. H., TIGER Computer Program Documentation, Vol. I, Stanford Research Institute, Menlo Park, CA, SRI-Pub-Z106, 1973, p. 263.

[16]Shepherd, J. E., "Chemical Kinetics of Hydrogen-Air-Diluent Detonation," Progress in Astronautics and Aeronautics: Dynamics of Explosions, Vol. 106, edited by J. R. Bowen, J. C. Leyer, and R. I. Soloukhin, AIAA, Washington, DC, 1986, p. 263.

[17]Westbrook, C. K., Pitz, W. J., and Urtiew, P. A., "Chemical Kinetics of Propane Oxidation in Gaseous Detonations," Progress in Astronautics and Aeronautics: Dynamics of Shock Waves, Explosions, and Detonations, Vol. 94, edited by J. R. Bowen, N. Manson, A. K. Oppenheim, and R. I. Soloukhin, AIAA, New York, 1984, p. 151.

[18]Westbrook, C. K., "Hydrogen Oxidation Kinetics in Gaseous Detonations," Combustion Science and Technology, Vol. 29, 1982, p. 65.

[19]Tieszen, S. R., Stamps, D. W., Westbrook, C. K., and Pitz, W. J., "Gaseous Hydrocarbon Detonations," Presented at Western States Section Meeting of The Combustion Institute, 1988.

[20]Gordon, S., and McBride, B. J., "Computer Program for Calculation of Complex Chemical Equilibrium Compositions, Rocket Performance, Incident and Reflected Shocks, and Chapman-Jouguet Detonations," NASA SP-273, March 1976.

[21]Tieszen, S. R., Stamps, D. W., Westbrook, C. K., and Pitz, W. J., "Gaseous Hydrocarbon-Air Detonations," Sandia National Laboratories, Albuquerque, NM, SAND 88-1825J, 1988.

[22]Westbrook, C. K., and Pitz, W. J., "Chemical Kinetics Modeling of the Influence of Molecular Structure on Shock Tube Ignition Delay," Shock Waves and Shock Tubes, edited by D. Bershader and R. K. Hanson, Stanford Univ. Press, Stanford, CA, 1986.

Detonation of Cryogenic Gaseous Hydrogen-Oxygen Mixtures

M. Plaster,* R. D. McClenagan,† and F. J. Benz‡
NASA White Sands Test Facility, Las Cruces, New Mexico
and
J. E. Shepherd§
Rensselaer Polytechnic Institute, Troy, New York
and
J. H. S. Lee¶
McGill University, Montreal, Quebec, Canada

Abstract

A study was performed on the detonation properties of cryogenic mixtures of gaseous hydrogen and oxygen at 100 K initial temperature and various initial pressures and equivalence ratios. Measured detonation pressures and velocities agreed quite well with the computed Chapman- Jouguet (CJ) values based on the ideal-gas equation of state. Numerical computations of idealized, steady-state, one-dimensional Zeldovich-von Neumann-Doring (ZND) detonation structure with detailed chemical reaction kinetics were used to evaluate the reaction-zone thickness and to estimate cell size, critical tube diameter, and critical initiation energy. Predictions of critical tube diameter using calculated reaction-zone thickness agreed reasonably well with experimental data. The kinetic modeling also indicated that the initial pressure had a much smaller effect on reaction-zone length at low initial temperatures (50-100 K) than at room temperature (300 K). This was due to a shift in the reaction mechanism as initial density (and, therefore, final pressure) increased with decreasing initial temperature at a fixed initial pressure.

---

* Senior Scientist, Lockheed-ESC
† Engineer, Lockheed-ESC
‡ Aeronautical Engineer, NASA
§ Professor of Mechanical Engineering
¶ Professor of Mechanical Engineering

Introduction

Liquid hydrogen and oxygen are used in large quantities as propellants in rockets and spacecraft. The accidental mixing and explosion of these propellants is a serious hazard the aerospace industry must understand and avoid. Safety analyses of the hazards in current aerospace systems indicate that several mixing mechanisms are possible. In the case of a tank failure on the launch pad, liquid hydrogen and liquid oxygen would spill onto the ground, and the resulting boiling could cause the formation of a large, cold (~100 K) hydrogen-oxygen cloud. To better understand the explosion hazards associated with cold hydrogen-oxygen, an experimental study was performed that determined the detonation parameters of homogeneous cryogenic mixtures of gaseous hydrogen and oxygen. This paper summarizes that study.

There has been relatively little previous work on this problem. Laderman et al.[1] reported detonation pressures and velocities of hydrogen and oxygen mixtures at low initial pressures of 13 kPa and at temperatures down to 155 K. There were no measurements, however, of the dynamic detonation parameters (i.e., cell size, critical initiation energy), and the experiments were not extended to high initial pressures. Edse and Lawrence[2] measured the induction distance for the transition from deflagration to detonation for conditions of 100 kPa and temperatures down to 123 K in small-diameter tubes. However, they made no other measurements of detonation velocity, pressure, or cell size. Minimum ignition energy and flame propagation studies for cryogenic hydrogen and oxygen have been reported by Hayashi and Fujiwara,[3] Wolfe and Yurczyk,[4] Kaye,[5] and Yates et al.[6] None of these studies included the detonation mode of combustion. Thus, there is a need for a detailed study of both the thermodynamic properties (the CJ velocity and pressure) and the dynamic detonation parameters (reaction-zone length, cell size, and critical tube diameter).

The present study includes the following: experimental measurements of CJ parameters and critical tube diameter, simple thermochemical analyses to assess the role of real- gas effects on the detonation parameters, and kinetic modeling of the reaction-zone length to estimate the detonation length scales and initiation energies. The results of the experiments were compared to the values predicted by the kinetic model.

Experimental Apparatus

The experiments were carried out in the vertical detonation tube shown in Fig. 1. The tube was a stainless steel pipe 2.5 m long and 120 mm i.d. A cooling jacket filled with liquid nitrogen surrounded the tube and lowered the temperature to 100 K. Mixtures were prepared by first evacuating and cooling the detonation tube and then filling the tube with the desired amount of hydrogen and oxygen. The gases (oxygen first, then hydrogen) were introduced into the bottom of the detonation tube, and the ratio was determined by partial pressures of the gases. The gases were left to mix by diffusion. The convection created by the buoyancy of the light hydrogen greatly assisted the mixing process.

Detonations were initiated at the bottom of the tube using either an exploding bridge wire (EBW) that released about 200-300 J of energy, a 200-250-J blasting cap, or a special detonator containing 1.3 g of lead azide that was equivalent to 2500 J of energy. Detonation velocities were measured using ion gauges

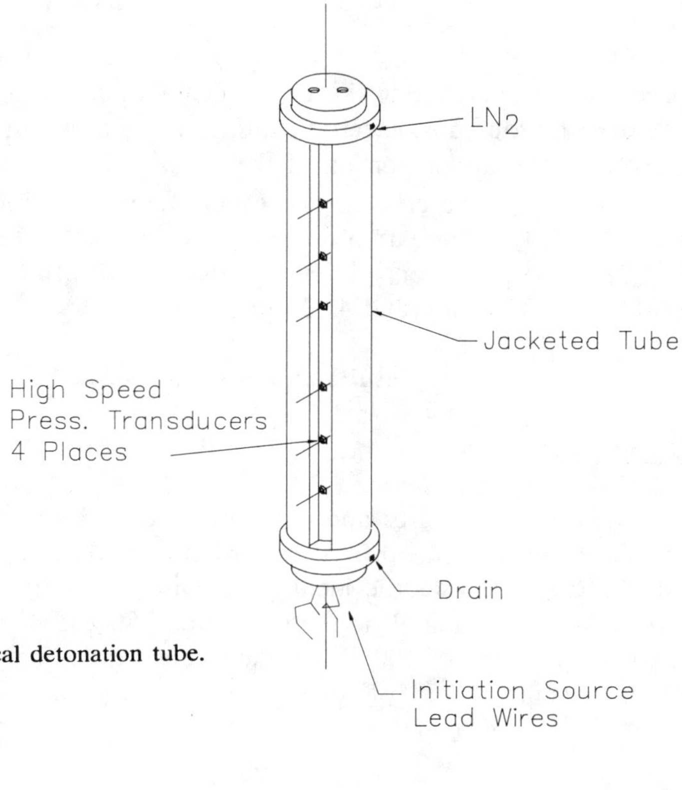

Fig. 1 Vertical detonation tube.

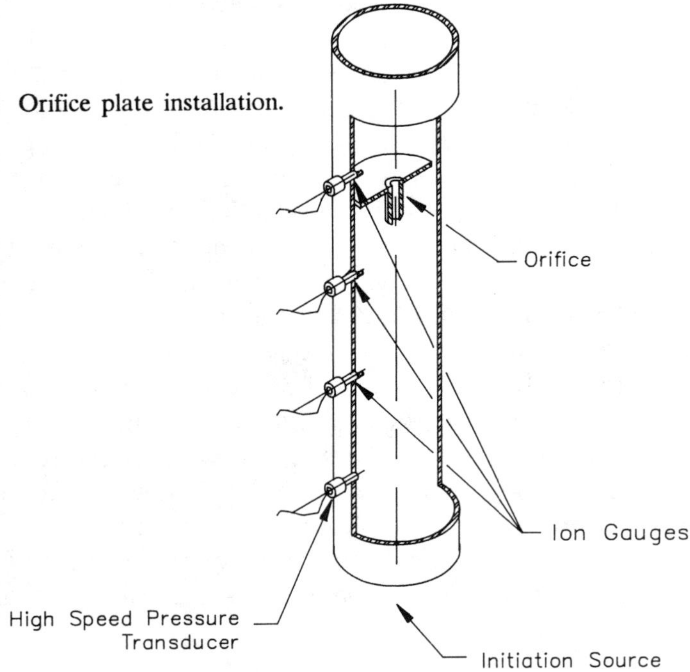

Fig. 2 Orifice plate installation.

spaced 0.3 m apart along the tube. Detonation pressures were measured by four piezoelectric transducers also mounted at 0.3-m intervals in the upper portion of the tube.

For performing critical tube experiments, an orifice plate was inserted into the detonation tube to divide it into two sections (Fig. 2). Three different plates were used, with orifice diameters of 10.8 mm, 25.4 mm, and 41.2 mm.

Results and Discussion

CJ Properties

Measured pressures and detonation velocities were compared with CJ values computed by the Gordon and McBride computer code.[7] This code uses the ideal-gas equation of state. Results were obtained for initial pressures between 86 and 405 kPa at an initial temperature of 100 K for equivalence ratios between 0.3 and 3. Results of a 300 K detonation were obtained for comparison. The calculated and experimental values are given in

Table 1. The measured velocities ranged from 6% lower to 9.8% higher than the computed CJ values. From other research, velocity data are typically slightly lower (1-3%) than the computed CJ values for a carefully controlled experiment.[8] In the relatively short tube used in the present experiment, a steady state may not have been reached, and the deviation from computed CJ values could be due to the influence of the initiation processes. Detonation pressures were measured from the pressure transducer output. A digital Fourier filter technique removed the resonance and the mechanical vibration effects. Measured pressures ranged from 16% lower to 1% higher than the computed CJ values. Experience with other gaseous detonations has shown that accurate pressure measurements are difficult to obtain, and the present level of agreement with the ideal-gas CJ computation is probably satisfactory.

The high initial densities of the cryogenic mixtures used in the present study could cause deviations from an ideal-gas equation of state.[9] The real-gas (compressibility) effect on the computed detonation parameters was examined by repeating the CJ computations using the JCZ3 equation of state and the TIGER thermochemical equilibrium code. The equation of state and the code are used extensively in the computations for condensed explosive detonation properties and have been successfully applied to dense gases by Bauer and Brochet.[8] Detonation velocity vs initial pressure is shown in Fig. 3 for both the ideal-gas and JCZ3 equations of state. At an initial pressure of 400 kPa, the JCZ3

Table 1 Initial conditions, measured and computed detonation pressure and velocity for cryogenic hydrogen-oxygen detonations

Equivalence ratio	Density, kg/m^3	Initial temperature, K	Initial pressure, kPa	Detonation velocity, m/s		Detonation pressure, MPa	
				Calculated	Experimental	Calculated	Experimental
1.00	0.4	300	86	2850	2819	1.7	1.66
1.00	1.5	100	101	2939	3175	5.92	5.10
1.00	2.6	100	188	2972	3115	11.3	9.50
0.33	9.8	100	405	2162	2028	21.35	21.60
0.75	6.8	100	405	2748	3018	24.26	22.80
1.00	5.8	100	405	3011	3008	24.76	24.01
2.00	3.9	100	405	3551	3636	22.90	19.20
3.07	3.0	100	405	3758	3688	19.53	18.70[a]

[a]No detonation using a 200-300-J initiator; detonation achieved with a 2500-J initiator.

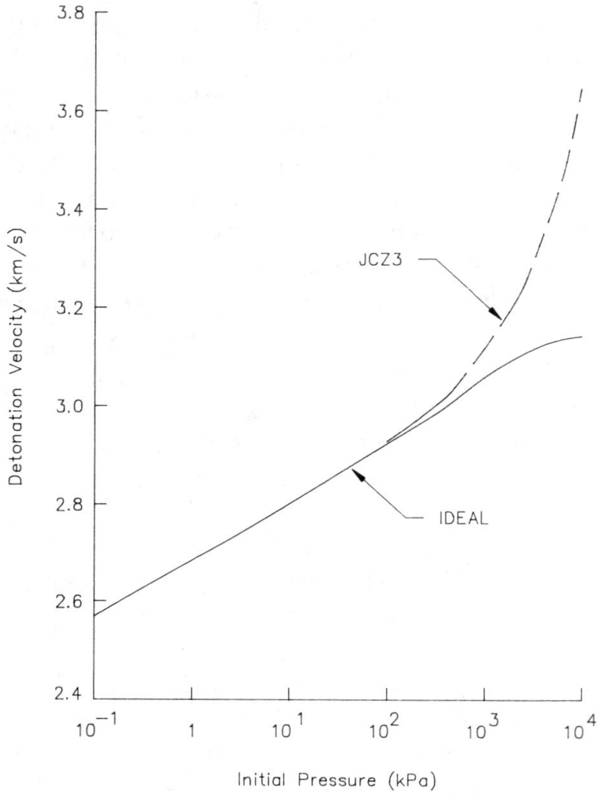

Fig. 3 Detonation velocity vs initial pressure. Chapman-Jouguet computations with ideal-gas and JCZ3 equations of state. Data from WSTF tests. Stoichiometric mixtures of hydrogen-oxygen at an initial temperature of 100 K.

velocity is only 1.3% higher than the ideal-gas value, and this is well within experimental accuracy. At an initial pressure of 10,000 kPa, it is 17% higher. Therefore, no significant real-gas effects would be expected in the present experiments. As shown, the deviations of the measured data from either of the computed values are too large to draw definitive conclusions.

Dynamic Detonation Parameters

It is difficult to directly measure initiation energy and the limits of detonability and, typically, these parameters are inferred from measurements of the cell size λ .[10] In this study, however, cell size could not be measured directly using the standard smoked foil technique because the detonations at high initial density damaged the foils. It was necessary, therefore, to use theoretical modeling to determine the initiation energy and the limits of detonability. Quantitative prediction of these dynamic detonation parameters is among the more difficult problems of computational fluid dynamics; multiple time-scale phenomena, intrinsic instabilities, and three-dimensional flow are a few of the complicating factors. A simplified but useful approach to this problem is to use empirical correlations between ideal detonation properties (e.g., reaction-zone length) and measured dynamic detonation parameters such as cell size, initiation energy, and critical diameter. Reaction-zone structures are computed using an idealized, steady-state, one-dimensional Zeldovich-von Neumann-Doring (ZND) model of detonation with detailed chemical reaction mechanisms and thermochemistry. The principal product of these computations is a characteristic reaction-zone length Δ , usually defined to be the location of maximum energy release relative to the leading shock wave. This approach has been extensively used in many recent studies on hydrocarbon fuel-oxidizer detonations,[11] hydrogen-oxygen-diluent detonations,[12,13] and detonations of monopropellants such as hydrazine.[14]

Reaction-zone lengths were computed using the ZND model, 11 species, and a 22-reaction mechanism for hydrogen oxidation. Details of the reaction-zone computation and the hydrogen-oxidation reaction mechanism are described by Shepherd.[12] The reaction-zone lengths used in the present study are all based on the location of maximum temperature gradient (approximately the location of peak energy deposition rate) in the reaction zone. Results for stoichiometric mixtures with initial pressure of 0.1, 0.3, and 1.0 bar and initial temperatures between 50 and 300 K are shown in Fig. 4. The effect of initial temperature on the reaction-zone length is strongly influenced by the initial pressure. At a pressure of 1 bar, the reaction zone length is almost independent of initial temperature and has nearly the same value (0.06 mm) at 50 K as at 300 K. As the initial pressure decreases, the reaction-zone length strongly decreases with decreasing initial temperature between 300 K and 50 K. At a pressure of 0.1 bar, the reaction-zone length decreases from 0.67 mm at 300-K initial temperature to 0.13 mm at 50-K initial temperature.

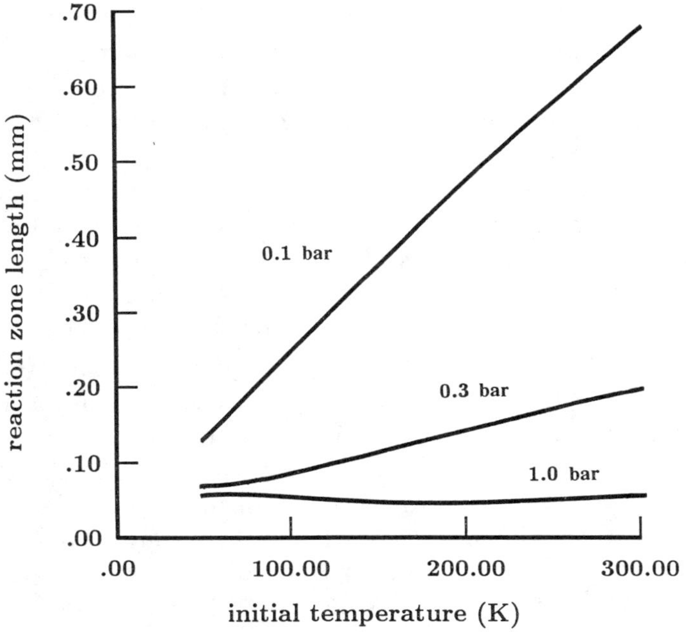

Fig. 4 Reaction zone length vs. initial temperature at three initial pressures (stoichiometric mixtures of hydrogen and oxygen).

The pressure dependence of the temperature effect on the reaction-zone length can be explained by considering the factors that play a role in determining the thermodynamic conditions in the reaction zone and how these factors influence the chemical reaction rates. The reaction-zone length is approximately determined by the product of the postshock velocity and the induction time,

$$\Delta = u_1 t_i \tag{1}$$

where the velocity u_1 is determined by the shock jump conditions for the reactants and the induction time t_i is determined by the characteristic time for reaction. Representing the reaction by an overall Arrhenius rate, the induction time can be written as follows:

$$t_i = a\ \rho_{vn}^{\alpha}\ e^{(E_{eff}/RT_{vn})} \tag{2}$$

where the parameters a, α, and E_{eff} depend on the details of the reaction mechanism. For stoichiometric hydrogen-oxygen reactions,

$\alpha \sim -1$ and $15 \leq E_{eff} \leq 50$ kcal/mole. The subscript vn refers to the von Neumann (VN) or postshock thermodynamic state. The velocity u_1 is only slightly influenced by initial temperature or pressure; all significant variations are due to induction time dependence on the thermodynamic state.

Numerical computations of the CJ detonation properties and von Neumann states as a function of initial conditions demonstrate that the initial density is a more significant parameter than initial temperature or pressure alone. When the initial density is increased by decreasing initial temperature at fixed initial pressure, the postshock density ρ_{vn} will change inversely with initial temperature, but the postshock temperature will decrease only a small amount. Consequently, unless the effective activation energy E_{eff} is rather high, the density term will dominate the induction time expression. For low initial densities (below 0.5 kg/m^3) where $E_{eff} \sim$ 16 kcal/mole, the reaction-zone length is inversely proportional to initial density. As the initial density is increased, the effective activation energy changes, ultimately increasing to about 45 kcal/mole.[12] With a larger activation energy, the exponential term in the induction time expression balances the density term, resulting in a reaction-zone length that is approximately independent of initial density. This change in effective activation energy with pressure and temperature is due to the competition for H atoms by the three-body reaction

$$H + O_2 + M \Longleftrightarrow HO_2 + M \tag{3}$$

relative to the usual chain-branching binary reaction

$$H + O_2 \Longleftrightarrow OH + O \tag{4}$$

The first reaction will dominate the second with increasing initial density. This results in a change in reaction mechanism from the standard chain-branching system involving H, O, and OH, produced via reaction (4), to a slower process involving HO_2, produced via reaction (3), and H_2O_2. Ultimately, the H_2O_2 decomposes to produce OH, and the reaction zone is always terminated by the coupled chain-branching and thermal explosion involving reaction (4). This effect and its significance for diluent and initial-state effects in hydrogen-air detonations is discussed by Shepherd.[12]

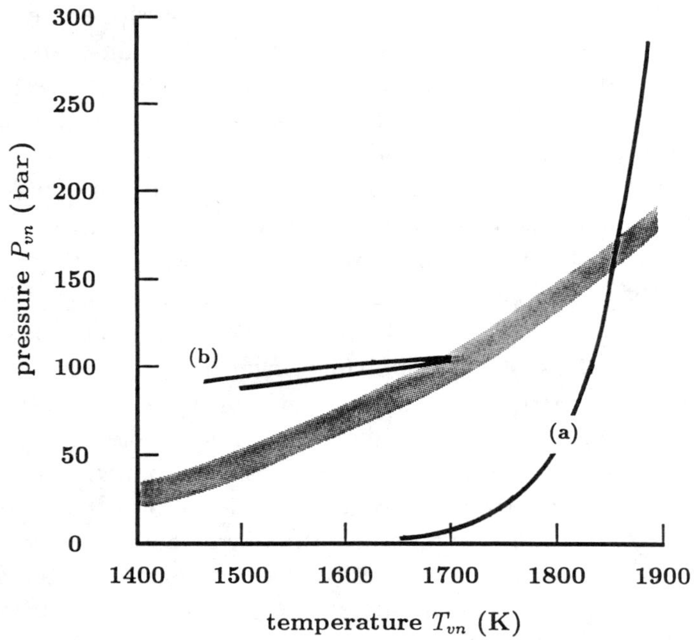

Fig. 5 Computed postshock (VN) pressure-temperature states. Shaded band is extended second limit calculated as in Shepherd.[12] Below the band, the reaction mechanism is the standard H-O-OH branching chain; above the band, the reaction mechanism is the alternative $H\text{-}HO_2\text{-}H_2O_2\text{-}OH$ mechanism discussed in the text. Curve (a) is the locus of p_{vn}-T_{vn} states for stoichiometric hydrogen-oxygen detonations at 300-K initial temperature with pressure varying from 10 to 800 kPa. Curve (b) is the locus of p_{vn}-T_{vn} states for hydrogen-oxygen detonations with 100-kPa initial pressure and 100-K initial temperature; equivalence ratio varies between 0.3 and 3 along this curve.

The location of the crossover between these two mechanisms has been identified as the extension of the standard second explosion limit to the high-pressure and temperature conditions behind the shock wave that precedes the detonation reaction zone.[12] This limit is plotted in Fig. 5, together with the von-Neumann postshock thermodynamic conditions for detonations at initial conditions of 100 K and 100 kPa and at 300 K with variable initial pressure. Note how high pressure and low temperature both shift the state above the extended second limit.

The predominance of initial density and the reaction mechanism shift at higher densities are illustrated in the reaction-zone length vs initial density plot of Fig. 6. In this plot, reaction-zone lengths calculated at two initial temperatures of 100 and 300 K are shown for a range of initial densities that correspond to initial pressures between 5 and 1200 kPa. At low initial densities (less than 0.5 kg/m^3), the reaction-zone length is inversely proportional to density. At moderate values of initial density (between 1 and 10 kg/m^3), the reaction-zone length appears independent of initial density. At initial densities higher than those shown on this figure (greater than 10 kg/m^3), the reaction-zone length is again inversely proportional to initial density. The regime of density independence in Fig. 6 corresponds to von Neumann states near the extended second limit in Fig. 5.

Note that the 100-K curve lies above the 300-K curve in Fig. 6. The effect of increasing the initial temperature (at fixed initial density) is a slight decrease in the reaction-zone length with increasing initial temperature. This is due to the increase in post-shock temperature with increasing initial temperature. As shown

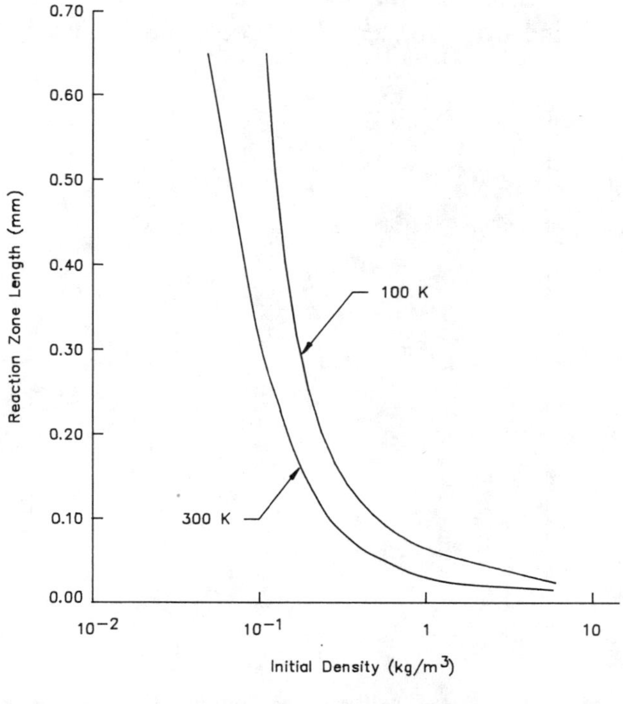

Fig. 6 Calculated reaction-zone length as a function of initial density for initial pressure ranging from 5 to 1200 kPa.

by Eq. (2), if the von Neumann temperature increases while the von Neumann density is fixed, the Arrhenius dependence on temperature will result in decreased reaction time and reaction-zone length.

Real-gas equation-of-state effects have not been considered in the reaction-zone structure computations. Although the effect on the final state and CJ conditions is negligible at an initial pressure of 400 kPa, the effect on the postshock state (where most of the reaction takes place) is much greater. The reactants have much lower critical pressures than the products, and the postshock pressure is two to three times higher than that of the final products. In addition to the effects on the equilibrium constants, the thermodynamic functions, such as heat capacity, will also be affected. The reaction pathways, activation energies, and collision rates (pre-exponential factors) are also affected by pressure; these effects probably become more significant at pressures higher than in the present study.

Computed reaction-zone lengths can be used to estimate cell size λ and critical tube diameter d_c using the empirical relations $\lambda = A\Delta$ [11] and $d_c = 13\lambda$.[15] Westbrook and Urtiew[11] originally

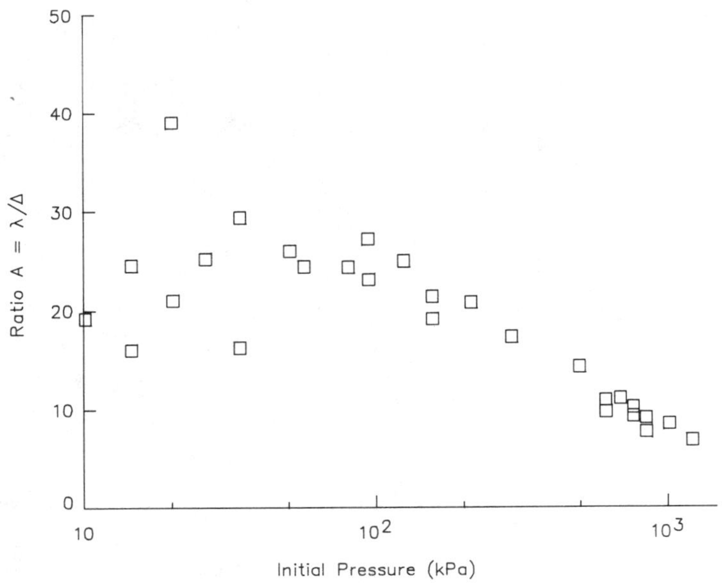

Fig. 7 Ratio of measured cell size[16] to computed reaction-zone length for stoichiometric hydrogen-oxygen detonations at an initial temperature of 300 K and initial pressures between 0.1 and 12 bar.

suggested a value of 29 for A in the hydrogen-oxygen system. Critical evaluation of the existing data reveals that A is not a constant and depends on a large number of factors, such as fuel and diluent type, equivalence ratio, and initial conditions. Variations of A between 10 and 50 are observed in the hydrogen air system for variations in equivalence ratio between 0.3 and 3.[12]

To better determine A for the present system, the 300-K initial temperature detonation cell size data of Manzhalei et al.[16] were re-examined (these data were also analyzed by Westbrook and Urtiew[11]) and compared to computed reaction-zone lengths. The ratio of experimental cell size to computed reaction-zone length is shown in Fig. 7 for initial pressures between 10 and 1200 kPa. Note the large scatter at low pressures and the systematic decrease in A with increasing pressure. Whereas a value of 30 might be an average A below 100 kPa, a value of 10 is more appropriate above 500 kPa.

When choosing a value of A for tests with disparate initial conditions, it is necessary to use data at equal density. The value of A does not depend as much on the initial pressure and temperature as it does on the nature of the reaction mechanism itself. The von Neumann points shown in Fig. 5 for the cryogenic initial conditions indicate that the alternate reaction mechanism based on HO_2 and H_2O_2 dominates, just as in the higher-pressure (500-1000 kPa), room-temperature cases. To evaluate cryogenic hydrogen-oxygen systems at moderate initial pressure, a value of A = 15 is appropriate. This choice leads to predictions of critical tube diameter and initiation energy that are consistent with the present data. Combining the relation $\lambda = 15\Delta$ with the independent relation $d_c = 13\lambda$ yields $d_c = 195\,\Delta$. This estimate is used below in comparing the results of critical tube experiments to the reaction-zone computations.

Critical Tube Measurements

To determine the critical tube diameter, an orifice was inserted in the detonation tube, and the velocity of the shock wave was measured before and after the orifice. The detonation velocity before the orifice plate was compared to the theoretical CJ velocity to determine if a steady detonation was achieved. When steady detonations occurred, the shock velocities before and after

Table 2 Critical tube velocity measurements

Equivalence ratio	Initial temperature, K	Initial pressure, kPa	Calculated detonation velocity,[a] m/s	Experimental detonation velocity, m/s	Experimental detonation velocity after orifice, m/s	Ratio of experimental detonation velocities before and after orifice, %	Detonation propagation
Critical tube diameter: 41.2 mm							
0.25	114	101.5	1946[b]	1990	1405	71	No
2.0	101	109	3489	3544	2655	75	Yes
2.4	101	106	3606	3629	2326	64	No
2.4	106	99.5	3606	3639	2372	65	No
Critical tube diameter: 25.4 mm							
0.25	104	103.2	1946[b]	1798	1332	74	No
0.5	101	102	2377[b]	2424	2777	115	Yes
2.0	102	101.3	3489	3514	3998	114	Yes
3.0	105	101.9	3717	3719	1960	53	No
Critical tube diameter: 10.8 mm							
0.25	105	104.3	1946[b]	2007	1261	63	No
0.5	110	101	2377[b]	2458	1699	69	No
1.0	107	103.1	2926	3043	2506	82	Yes
1.5	108	101	3290[b]	3375	2546	75	Yes
1.54	106	103.8	3290[b]	3377	3604	107	Yes
1.75	106	105	3390[b]	3509	2243	64	No
2.5	102	102.4	3628[b]	3701	2209	60	No
2.9	104	103.5	3702[b]	3740	1579	42	No

[a]Velocity computed by Gordon McBride thermodynamic code. [b]Values estimated by linear interpolation.

the orifice were compared. If the velocity after the orifice was less than 75% of the velocity before the orifice, then the detonation was considered to have failed. When calculating the velocity after the orifice, a correction was made for the change in geometry caused by the orifice plate. Whereas the detonation before the orifice was a plane wave, the shock emerging from the orifice was a hemispheric wave. Thus, the calculation of the velocity after the orifice included the radial distance (hypotenuse) from the center of the baffle plate opening to the ion gauge point located on the detonation tube wall. This method was used because more direct measurements of detonation failure were not possible as a result of the cryogenic conditions and, although the choice of 75% is somewhat arbitrary, the results are meaningful in at least a qualitative sense.

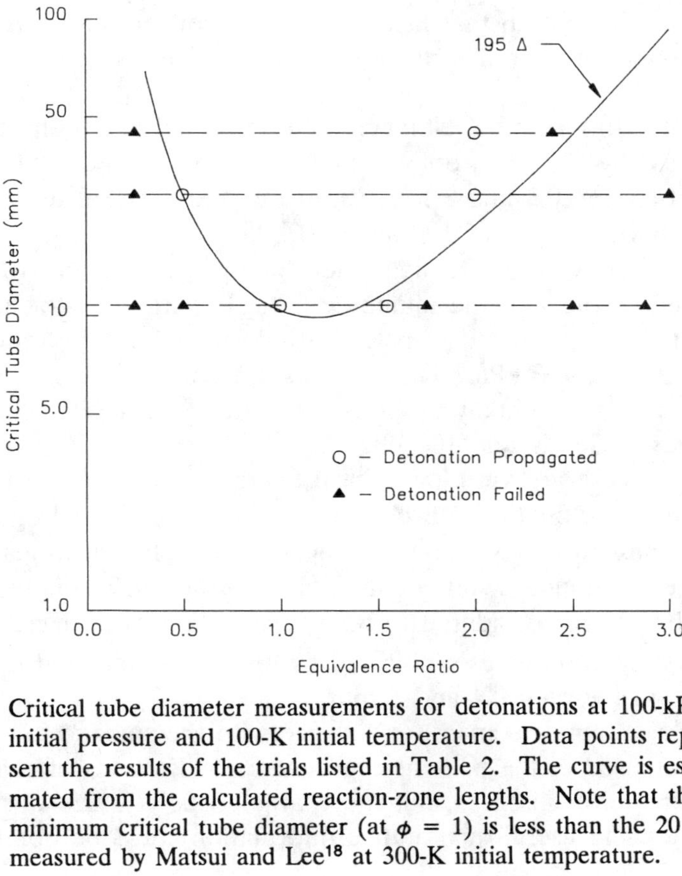

Fig. 8 Critical tube diameter measurements for detonations at 100-kPa initial pressure and 100-K initial temperature. Data points represent the results of the trials listed in Table 2. The curve is estimated from the calculated reaction-zone lengths. Note that the minimum critical tube diameter (at $\phi = 1$) is less than the 20 mm measured by Matsui and Lee[18] at 300-K initial temperature.

Table 2 displays the velocities before and after the orifice with equivalence ratios ranging from 0.25 to 3.0, a temperature of 100 K, and a pressure of 100 kPa. Figure 8 shows a comparison between the measured and calculated critical tube diameter from the kinetic model. Both results are in reasonable agreement.

Calculation of Critical Initiation Energy

Critical initiation energies are typically computed using cell size. Substituting reaction-zone length (using the correlation $\lambda = 15\Delta$ discussed earlier) in the formula developed by Benedick et al.[17] for critical initiation energy of hydrogen-air detonations yields

$$E_c = 1.7 \times 10^6 \rho_0 (U_{CJ})^2 \Delta^3 \quad (5)$$

where E_c is the critical or minimum energy required to initiate a spherical detonation that has a CJ velocity of U_{CJ} and a reaction-zone length of Δ in a mixture with an initial mass density of ρ_0.

Reaction-zone lengths, estimated cell sizes, and initiation energies predicted by Eq. (5) are given in Table 3 at selected equivalence ratios between 0.3 and 3 for mixtures at 100-K initial temperature and initial pressures of 100 and 400 kPa.

For reference, a single data point at 300 K and 100 kPa is included. Note that the minimum critical energy (at stoichiometric conditions) is predicted to be 3 J at 100 K and 100 kPa and 1.4 J at 100 K and 400 kPa. These values are comparable to the 1.6 J obtained experimentally by Matsui and Lee[18] for stoichiometric mixtures at 300 K and 100 kPa.

These are very low critical energies and indicate the ease with which detonation can be initiated in cryogenic hydrogen-oxygen mixtures. As a comparison, stoichiometric hydrogen-air mixtures at standard initial conditions have a critical energy of 4000 J and are considered to be a severe detonation hazard. The ease of initiation is also indicated by the small predicted cell sizes, which are between 0.4 and 5 mm.

An attempt to measure cell size with the smoked-foil technique was used in the present experiments; indications of cellular structure were observed, but the characteristic cell size was so small that no useful quantitative information could be obtained.

Table 3 Calculated ZND reaction-zone lengths Δ, estimated cell size $\lambda = 15\Delta$, and critical initiation energies E_c

ϕ	X_{H_2}	X_{O_2}	Δ, mm	λ, mm	E_c, J
		Initial Conditions: 100 K, 100 kPa			
3.00	0.86	0.14	0.575	8.6	3440
2.50	0.83	0.17	0.239	3.6	263
2.00	0.80	0.20	0.123	1.8	38
1.50	0.75	0.25	0.067	1.0	6.3
1.00	0.66	0.33	0.052	0.78	3.0
0.50	0.50	0.50	0.114	1.70	30
0.308	0.38	0.62	0.410	6.2	1270
		Initial Conditions: 100 K, 400 kPa			
3.07	0.86	0.14	0.30	4.5	1945
2.00	0.80	0.20	0.074	1.1	34
1.00	0.66	0.33	0.025	0.375	1.4
0.75	0.60	0.40	0.039	0.58	5
0.33	0.40	0.60	0.15	2.25	262
		Initial Conditions: 300 K, 100 kPa			
1.00	0.66	0.33	0.054	1.5[a]	1.6[b]

[a]Ref. 17 [b]Ref. 18

Conclusions

Measurements of detonation velocity and pressure in cryogenic mixtures of gaseous hydrogen and oxygen at an initial temperature of 100 K, initial pressures up to 400 kPa, and equivalence ratios between 0.3 and 3 show good agreement with calculations. Computations with both ideal- and real-gas equations of state also showed that nonideal effects were negligible at these conditions. With an equivalence ratio of 1, initial pressure of 400 kPa, and initial temperature of 100 K, the detonation velocity calculated using the JCZ3 equation of state differed from the ideal-gas calculation by only 1%.

Numerical computations of idealized, steady-state, one-dimensional ZND detonation structure with detailed chemical-reaction kinetics were used to evaluate the reaction-zone thickness and to estimate the critical tube diameter. These calculations show that initial density effects overwhelm the individual effects of initial temperature and pressure. These calculations also show good agreement with the experimental data.

Acknowledgments

The authors would like to thank Ed Denzler for his technical assistance in conducting the experiments, Morgan Phillips for performing the mixing study and contributing to the experimental design, Randy Kays for efforts in data reduction, and Stephen Woods for technical assistance.

References

[1]Laderman, A. J., "Detonability of Hydrogen-Oxygen Mixtures in Large Vessels," presented to The Combustion Institute, Western States Section, 1965 Spring Meeting, Hollywood, Ca.

[2]Edse, R., and Lawrence, L. R. Jr., "Detonation Induction Phenomena and Flame Propagation Rates in Low Temperature Hydrogen-Oxygen Mixtures," Combustion and Flame, Vol. 13, Oct. 1969, pp. 479-486.

[3]Hayashi, A. K., and Fujiwara, T., "Oxyhydrogen Ignition and Flame Propagation Under Cryogenic Temperatures," Proceedings of the Symposium on Mechanics for Space Flight, edited by K. Oshima, Tokyo University, Japan, March 1983.

[4]Wolfe, S. W., and Yurczyk, R. F., "A Combustion Test Program to Establish Design Criteria for Liquid-Hydrogen-Fueled Space Vehicles," Advances in Cryogenic Engineering, Vol. 11, 1966.

[5]Kaye, S., "Hazard Studies with Hydrogen and Oxygen in the Liquid and Solid Phases," Advances in Cryogenic Engineering, Vol. 11, 1966.

[6]Yates, G. B., Perl, A. R., Loos, J. F., and Sergeant, R. J., "Spark Ignition Parameters of Cryogenic Hydrogen in Oxygen and Nitrogen Mixtures," Advances in Cryogenic Engineering, Vol. 11, 1966.

[7]Gordon, S., and McBride, B. J., "Computer Program for Calculation of Complex Chemical Equilibrium Compositions, Rocket Performance, Incident and Reflected Shocks, and Chapman-Jouguet Detonations," NASA SP-273, March 1976.

[8]Bauer, P., and Brochet, C., "Properties of Detonation Waves in Hydrocarbon-Oxygen-Nitrogen Mixtures at High Initial Pressures," Progress

in Aeronautics and Astronautics: Shock Waves, Explosions, and Detonations, Vol. 87, edited by J. R. Bowen et al., AIAA, New York, 1983, pp. 408-422.

[9]Astapov, N. S., Nikolaev, Y. A., and Ulyanitskii, V. Y., "Detonation Parameters of Hydrogen-Oxygen and Hydrogen-Air Mixtures at High Initial Density," Fizika Goreniya Vzryva, Vol. 20, No. 1, 1984, pp. 98-105.

[10]Lee, J. H. S., "Dynamic Parameters of Gaseous Detonations," Annual Review of Fluid Mechanics, Vol. 16, 1984, pp. 311-336.

[11]Westbrook, C. K., and Urtiew, P. A., "Chemical Kinetic Prediction of Critical Parameters in Gaseous Detonations," Nineteenth Symposium (International) on Combustion, The Combustion Institute, Pittsburgh, 1982, pp. 615-623.

[12]Shepherd, J. E., "The Chemical Kinetics of Hydrogen-Air-Diluent Detonation," Progress in Aeronautics and Astronautics: Dynamics of Explosions, Vol. 106, edited by J. R. Bowen et al, AIAA, New York, 1986, pp. 263-293.

[13]Shepherd, J. E., Moen, I. O., Murray, S. B., and Thibault, P. A., "Analyses of the Cellular Structure of Detonations," Twenty-First Symposium (International) on Combustion, The Combustion Institute, Pittsburgh, 1988, pp. 1649-1658.

[14]Pedley, M. D., Bishop, C. V., Benz, F. J., Bennett, C. A., McClenagan, R. D., Fenton, D. L., Knystautas, R., Lee, J. H., Peraldi, O., Dupre, G., and Shepherd, J. E., "Hydrazine Vapor Detonations," Progress in Astronautics and Aeronautics: Dynamics of Explosions, Vol. 114, edited by A. L. Kuhl et al., AIAA, New York, 1987, pp. 45-63.

[15]Lee, J. H., Knystautas, R., Guirao, C. M., Benedick, W. A., and Shepherd, J. E., "Hydrogen-Air Detonations," Proceedings of the Second International Workshop on the Impact of Hydrogen on Water Reactor Safety, edited by M. Berman, Sandia National Laboratories, Albuquerque, NM, SAND82-2456, Oct. 1982.

[16]Manzhalei, V. I., Mitrofanov, V. V., and Subbotin, V. A., "Measurement of Inhomogeneities of a Detonation Front in Gas Mixtures at Elevated Pressures," Fizika Greniya i Vzryva, Vol. 10, 1974, pp. 102-110.

[17]Benedick, W. B., Guirao, C. M., Knystautas, R., and Lee, J. H., "Critical Charge for the Direct Initiation of Detonation in Gaseous Fuel-Air Mixtures," Progress in Aeronautics and Astronautics: Dynamics of Explosions, Vol. 106, AIAA, New York, 1986, pp. 181-202.

[18]Matsui, H., and Lee, J. H. S., "On the Measure of the Relative Detonation Hazards of Gaseous Fuel-Oxygen and Air Mixtures," 17th Symposium (Intl.) on Combustion, The Combustion Institute, Pittsburgh, 1979, pp. 1269-1280.

Detonation Characteristics of Gaseous Methane-Oxygen-Nitrogen Mixtures at Extremely Elevated Initial Pressures

P. Bauer,* H. N. Presles,† and M. Dunand‡
Laboratoire d'Energétique et de Détonique, URA 193 au CNRS, ENSMA, Poitiers, France

Abstract

In order to provide data on the detonation velocities of hydrocarbon-oxygen-nitrogen mixtures at an initial pressure beyond 100 bars, a new setup was installed. It is based on the use of tubes that are torn after each shot; that is, in the same way as experiments performed in high explosives. A new initiation technique was used and results were obtained in methane-oxygen-nitrogen mixtures. It was observed that a good stability of the detonation velocity over occurs the range of initial pressures from 100 - to 400 bars. This parameter varies in a linear way vs initial pressure. The influence of the diameter of the tube is discussed. A comparison with a priori calculations based on JCZ3 and Percus Yevick Equations of State is given. These results were obtained with QUATUOR Code.

Introduction

To date, a great number of data have been provided concerning the detonation properties of gaseous explosive mixtures at a high initial pressure, e.g.,.[1] However, the pressure was kept below a certain threshold (i.e., 100 - 150 bars). This was due to the type of confinement, namely either a 20 mm internal diameter (i.d.) or 55 mm i.d. tube that could withstand, at most, a detonation pressure of 7 kbars or 3 kbars,

* Professor
† Research Director
‡ Engineer

respectively.[2] In addition, the use of a detonator for direct initiation turned out to be inappropriate at the upper level of initial pressure, for the high-pressure environment was inhibiting its detonation. The comparison between experimental detonation velocities and calculated ones showed the ability of either Boltzmann or Percus Yevick Equations of State (EOS.)[3] to predict experimental values. However, in order to improve the knowledge on the validity of these EOS., data were required at a more elevated initial pressure. For this reason and, in order to obtain detonation pressures of the same order as that of condensed explosives, the present study was undertaken. Its purpose is to provide new data concerning the detonation properties of gaseous explosives at an initial pressure reaching 400 bars.

Preliminary experiments showed the feasibility of such an investigation[4] since we could measure the detonation velocity in a C_2H_4-Air mixture at a maximum pressure of 320 bars.

Experimental Device

The new experimental setup we used is no longer based on a detonation tube designed to withstand the detonation pressure. In contrast, the thickness of the tubes we used only allows one to hold the fresh mixtures. The tubes are torn after each shot. This type of investigation is the same as the ones used in cases where high explosives are handled. The initiation of the mixtures is obtained by means of a layer of solid explosive pasted around the tube, over 10 cm or so, at one of its ends (Fig. 1). The detonation of this outer explosive is expected to generate a Mach stem in the central core, yielding an overdriven detonation in the inner explosive.[5] Therefore, in the present case, one may expect a satisfying initiation of the detonation. However, its velocity has to slow down to the

Fig. 1 Sketch of the detonation tubes.

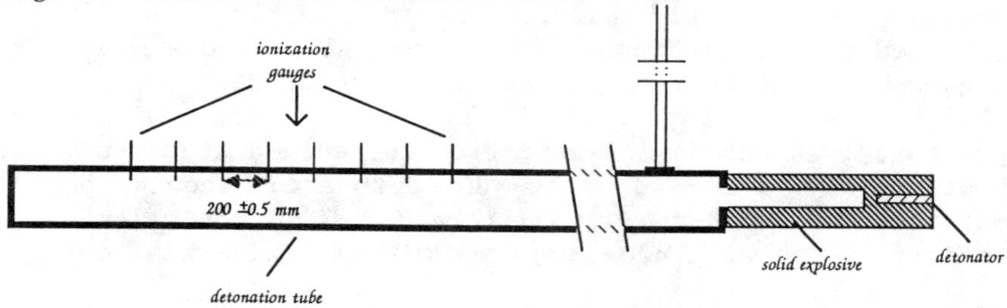

Chapman-Jouguet (C.J.) state after few 10s of cm of propagation. This point will be discussed later in this paper.

Detonation velocities were measured using ionization pickups that required special care in order to avoid any leakage. Four succeeding 20-cm-long sets of measurements were installed along 2-m-long tubes. Several i.d. tubes were used, namely, 12, 16, 21, and 26 mm. A careful check of the detonation stability could thus be made.

Experimental Results and Discussion

Detonation Velocity Profile

Various mixtures of methane-oxygen-nitrogen were studied at initial pressures in the range of 100-400 bars. All measurements were made at a temperature of 283 ± 2 K. The detonation was very stable since the mean discrepancy between all sets of measurement was less than 1%. Although an overdriven state was generated at the initiation point, we could observe that the self-sustained state was obtained very rapidly, i.e., within the 40 cm of propagation between the ignition point and the first set of measurements. The corresponding results are plotted on Fig. 2 for 1) mixture A, $CH_4 + 2.01\ O_2 + 3.92\ N_2$; and 2) mixture B, $CH_4 + 1.85\ O_2 + 5.09\ N_2$.

They obviously stand on a line as a function of initial pressure. As a reference we have plotted on the same diagram the results obtained in the same type of mixture but at a much lower initial pressure, i.e., less than 70 bars. These results were obtained in the 55-mm-i.d. and 4.5-m-long tube.[2]

Particular Case of Methane-Air

To date, it has been impossible to generate a self-sustained steady detonation in such a mixture and using high-purity methane, i.e., in a tube at an initial pressure up to 100 bars. However, we legitimately expected that an increase of the initial pressure would enhance the onset of the detonation. The drastic drop of the critical diameter as a function of the initial pressure[6] was the basic assumption for such a statement. However, we observed that no detonation could be obtained in a methane-air mixture (with an equivalence ratio of 1.15) at a pressure of 400 bars. This trial was made in a 16-mm-i.d tube. The present technique did not allow one to conduct such an experiment in a greater i.d. tube due to the fragility of the ionization gauges. Such a result, showing the very poor ability of methane-air mixtures to detonate, was very unlikely to have happened. In such an i.d. tube, the highest inert amount, i.e., $\beta = n_{N2}/n_{O2}$, where a detonation could be obtained was $\beta = 3.21$. This had been observed in a 6.5-m -long tube through a deflagration to detonation transition process at initial pressures in the range of 5-50 bars.[7]

Influence of the Tube Diameter

It had been shown[8] that the tube diameter no longer plays a limiting role once the initial pressure of the mixture is increased. However, in the present case the

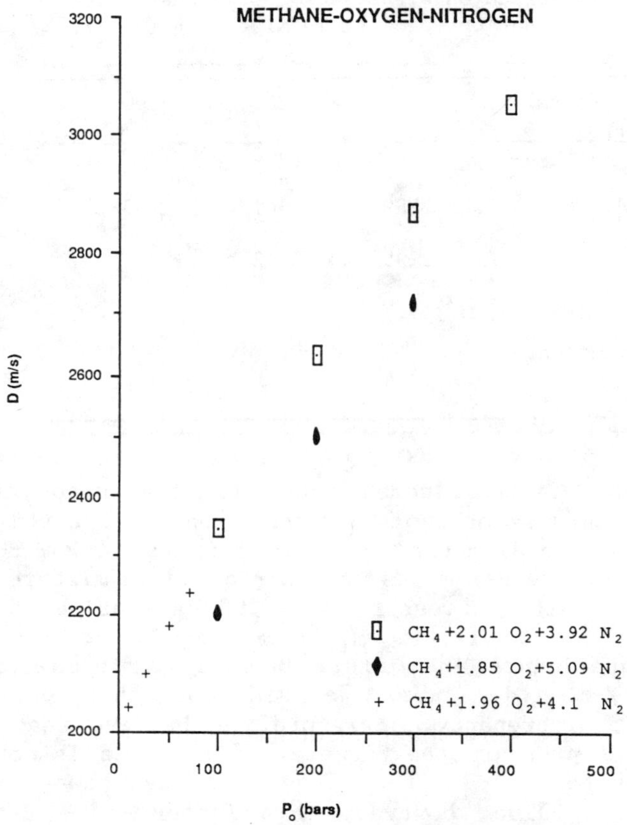

Fig. 2 Detonation velocities vs intial pressure.

Table 1 Experimental (16-mm-i.d. tube) and a priori calculation based on either JCZ3 or Percus Yevick EOS. of the detonation velocity in mixture A (β = 1.95)

Po (bars)		100	200	300	400
D_{exp} (m/s)		2346	2625	2889	3070
D_{calc} (m/s)	JCZ3	2378	2642	2873	3097
	PY	2399	2742	3036	3275
Relative deviation (%)	JCZ3	1.4	0.6	0.5	0.9
	PY	2.2	4.5	5.1	7

Table 2 Experimental (16-mm-i.d. tube) and a priori calculation based on either JCZ3 or Percus Yevick EOS of the detonation velocity in mixture B (β = 2.75)

P_o (bars)		100	200	300
D_{exp} (m/s)		2220	2500	2740
D_{calc} (m/s)	JCZ3	2289	2541	2767
	PY	2295	2612	2883
Relative deviation (%)	JCZ3	3.1	1.6	1
	PY	3.4	4.5	5.2

opening of the confinement due to the detonation pressure generates an expansion wave and we expected an influence of the diameter of the load to occur. For this reason, several experiments were performed in mixture A at the same initial pressure p_o = 200 bars using three tube diameters, namely, 16 mm, 21 mm, and 26 mm. We made similar experiments in mixture B using a 16-mm-i.d., then a 12-mm-i.d. tube. We did not observe any significant discrepancy. Therefore, we believe that at such initial pressure the diameter of the load is not a major parameter, at least as long as we keep away from its critical value. However, this latter value drops drastically at this level of pressure.[9]

Comparison with Calculations

So far most of the a priori calculations up to p_o = 100 bars of the detonation velocity exhibited the validity of the Percus Yevick EOS. and, to some extent that of the much more simple Boltzmann EOS.. This was observed in the case of most hydrocarbon-oxygen-nitrogen mixtures. However, no available data could show whether the range of validity could be extended to a significantly greater initial pressure. The results of calculations provided by QUATUOR code in the present case are reported in Table 1 (mixture A) and Table 2 (mixture B). They show the validity of another EOS., namely, the JCZ3 EOS.[10] This was somewhat expected since the detonation pressure obtained in the present case is of the same order as that of high explosives, which is the field of application of this EOS. The maximum deviation is, thus, less than 2%. The deviation is greater in the case of mixture B at the lowest initial pressure, i.e., 100 bars. This might be explained by the greater value of the inert ratio, which gets closer to that corresponding to propagation limits.

Summary and Conclusion

Following preliminary experiments performed in a C_2H_4-Air mixture at an initial pressure of 320 bars, the investigation was pursued on how methane-oxygen-nitrogen mixtures would behave at a pressure up to 400 bars. Three mixtures were studied based on methane oxygen diluted in various proportions of nitrogen. We observed a stable detonation propagating along the tubes. Although the initiation was obtained through an overdriven state generated by the use of a high explosive, a CJ state could be obtained after a 40-cm-long propagation. The velocities vary in a linear way as a function of the initial pressure and no significant diameter effect was observed. The calculated values based on the JCZ3 EOS. are in fairly good agreement with experimental data. Here again, we failed to generate a detonation in a high-purity methane-air mixture.

References

[1]Bauer, P., "Contribution à l'Etude de la Détonation de Mélanges Explosifs Gazeux à Pression Initiale Elevée," Doctorat es Sciences Thesis, Univ. of Poitiers, Poitiers, France, 1985.

[2]Presles, H.N., Bauer, P., Heuzé, O., and Brochet, C., "Investigation on Detonation of Gaseous Explosive Mixtures at Very High Initial Pressures," Combustion, Science and Technology, Vol. 43, 1985, pp. 315-320.

[3]Heuzé, O., Bauer, P., Presles, H.N., and Brochet, C., "Equations of State for Detonation Products and Their Incorporation into the QUATUOR Code," Proceedings of the 8th International Symposium on Detonation, Naval Surface Weapon Center, White Oak, Silver Spring, MD, 1985, pp. 762-769.

[4]Bauer, P., Presles, H.N., and Dunand, D., "Premières Etudes Expérimentales sur la Détonation dans les Mélanges Explosifs Gazeux à Pression Initiale Supérieure à 100 Bars," Comptes Rendus de l'Académie des Sciences de Paris, France, Vol. 308, 1989, pp. 1405-1408.

[5]Sellam, M., Presles, H.N., Brochet, C., and Chéret R., "Characterization of Strong Detonation Waves in Nitromethane," Proceedings of the 8th International Symposium on Detonation, Naval Surface Weapon Center, White Oak, Silver Spring, MD, 1985, pp. 425-428.

[6]Bauer, P., Brochet, C., and Presles, H.N., "The Influence of Initial Pressure on Critical Diameters of Gaseous Explosion Mixtures," Dynamics of Shock Waves, Explosions and Detonations, Vol. 94, Edited by J.R. Bowen, N. Manson, A.K. Oppenheim, and R.I. Soloukhin, Progress in Astronautics and Aeronautics, AIAA, 1984, pp. 118-119.

[7]Bauer, P., Brochet, C., and Presles, H.N., "Detonation Study of Gaseous Mixtures at Initial Pressures Reaching 10 MPa," Archivum Combustionis, Vol.4, No. 3, 1984, pp. 191-195.

[8]Bauer, P., Vidal, P., Manson, N., and O. Heuzé, "Applicability of the Inverse Method for the Determination of CJ Parameters for Gaseous Mixtures at Elevated Pressures," Dynamics of Explosions, Vol. 114, Edited by A.L. Kuhl, J.R. Bowen, J.C. Leyer, and A. Borisov, Progress in Astronautics and Aeronautics, AIAA, 1988, pp. 64-76.

[9]Bauer, P., Presles, H.N., Heuzé, O., and Brochet, C., "Measurement of Cell Lengths in the Detonation Front of Hydrocarbon Oxygen and Nitrogen Mixtures at Elevated Initial Pressures," Combustion and Flame, Vol. 64, No. 1, 1986, pp. 113-123.

[10]Cowperthwaite M., and Zwisler W.H., "The JCZ Equation of State for Detonation Products and Their Incorporation into the TIGER Code," Proceedings of the 6th International Symposium on Detonation, Naval Surface Weapon Center, White Oak, Silver Spring, MD, 1976, pp. 162-172.

Correlation between Chemical Kinetics and Detonation Structure for Gaseous Explosive Systems

C. Paillard,* G. Dupré,† Al Aiteh,‡ and S. Youssefi‡

National Centre of Scientific Research and University, Orléans, France

Abstract

The thermal decomposition of small energetic molecules has been carried out in order to establish correlations between chemical kinetics and detonation structure and to determine if the linear relationship (established for fuel/air detonation or hydrazine decomposition) between two characteristic lengths, namely, the induction distance of the explosive reaction and the cell size of detonation, could be generalized to other explosive systems. The originality of our method, in which a shock tube is coupled to various means of diagnostics, is to investigate experimentally both the kinetics of the explosive reaction and the propagation of detonation by merely varying initial conditions and shock strength. The means of diagnostics include the laser schlieren technique to measure vibrational relaxation times, time-of-flight mass spectrometry, absorption and emission spectroscopies to determine the reaction rates from chemical species profiles behind shock waves. Pressure transducers are used to measure induction times, and smoked foils to record the structure of detonation.

Two gases with totally different mechanisms were chosen to illustrate the present study: first, chlorine dioxide, for which the mechanism is that of a degenerate chain branching reaction and the induction time has been formulated by a simple relationship; and second, hydrogen azide, for which a more complex expression of the induction

---

*Professor, University of Orleans.

† Chargee de Recherche, Research Center on the Chemistry of Combustion and High Temperatures (CRCCHT, CNRS).

‡ Graduate Student, Research Center on the Chemistry of Combustion and High Temperatures (CRCCHT, CNRS).

time deduced from the adiabatic thermal explosion theory has been established. These relationships are used to estimate the induction distance according to the Zeldovich, von Neumann, and Döring (ZND) model in the detonation wave for both gases. Correlations found between induction distance and detonation cell size are discussed.

Introduction

During the past few years a number of studies in the area of gaseous detonations were devoted to establish a link between chemical kinetics and detonation structure, with the aim of finding a detonability limit criterion for a given confinement. Our own studies were focused on small explosive molecules such as covalent azides HN_3 (ΔH_f° = 300 kJ mol^{-1}), N_3Cl (ΔH_f° = 390 kJ mol^{-1}), nitrogen trichloride NCl_3 (ΔH_f° = 230 kJ mol^{-1}), chlorine dioxide ClO_2 (ΔH_f° = 104.6 kJ mol^{-1}), or hydrazine N_2H_4 (ΔH_f° = 92 kJ mol^{-1}). They consisted of a detailed analysis of the pyrolysis mechanism at high temperature and in a study of the formation and structure of the detonation behind a shock wave.

After a short description of the experimental arrangement, the results obtained with the explosive systems ClO_2 and HN_3, which have totally different types of kinetic mechanisms, are presented.

Experimental Details

The pyrolysis of unstable molecules in an argon bath and the study of the detonation of the undiluted or slightly diluted gases in argon were carried out in a shock tube. The explosive gas is submitted to a shock wave that brings it almost instantly to the temperature and pressure required to initiate the reaction.

The shock tubes have been described elsewhere.[1] They are equipped with piezoelectric transducers for pressure measurements behind the shock wave as well as for velocity measurement. Different techniques of analysis are coupled to the shock tube: time-of-flight mass spectrometry (TOFMS), absorption and emission spectroscopies, and laser schlieren technique based on the determination of gas density gradients,[2] this last technique being used for determining the vibrational relaxation time of the gas submitted to the shock wave.

Decomposition rates were measured by following either the infrared emission (at 4.67 μm for HN_3 and 8.89 μm for ClO_2) or the ultraviolet absorption at 360 nm for ClO_2. The identification and evolution of intermediate species and products were deduced from the mass spectrometry and emission spectroscopy data.

The structure of the detonation wave was recorded via the classical recording technique of the trajectories of the triple points printed on a soot-coated wall.

Results

The kinetic behavior of chlorine dioxide OClO strongly differs from the one observed with hydrogen azide HN_3. Chlorine dioxide is characterized by a rather low enthalpy of formation[3] (101.5 kJ mol^{-1}) corresponding to a low von Neumann temperature, of the order of 1200 K for a detonation in pure ClO_2. The fast rate of its explosive decomposition is preceded by a long induction period, typical of a degenerate chain reaction.[4] From a previous study on ClO_2 flame characteristics, the overall activation energy was found to be very small, of the order of 40 kJ mol^{-1}.[5]

The formation of HN_3 is much more endothermic (ΔH_f° = 300 kJ mol^{-1}).[6] For undiluted HN_3, von Neumann temperatures are above 2000 K, and the overall activation energy for the reaction of decomposition is of the order of 125 kJ mol^{-1}.[7]

The induction period preceding the explosive reaction was estimated assuming the ZND detonation model to be applicable.

Chlorine Dioxide

It was previously shown from measurements of vibrational relaxation times[8] that the thermodynamic temperature is rapidly reached after the shock has passed. The relationship between shock pressure P (in Pa), temperature T (in K), and the vibrational relaxation time of undiluted chlorine dioxide τ_v (in μs) approximately follows the Landau-Teller law:

$$\tau_v = (10^5/P)\exp(-3.4 + 17.3\ T^{-1/3}) \qquad (1)$$

This relationship, established in a restricted range of pressure and temperature, was used to estimate the vibrational relaxation times under von Neumann conditions.

The decomposition of chlorine dioxide was studied and its complex mechanism modelized for shock temperatures lower than 1450 K.[8] The estimation of the autoignition delay τ_i is easier using the following expression deduced from ClO_2 absorption signals behind the reflected shock wave[1] than with τ_i calculated from a kinetic model:

$$\log \tau_i = 3475/T - 0.92 \log(x\ P/T) - 5.80 \tag{2}$$

where x is the ClO_2 molar fraction. τ_i, T, and P are expressed in s, K, and Pa respectively. This expression gives an estimation of τ_i within ± 30% accuracy for a large range of ClO_2 molar fraction (0.05-1), pressure (8-250 kPa), and temperature (650-2000 K). The autoignition delay can be expressed as

$$\tau_i = K\ [ClO_2]^m\ [Ar]^n \exp(E/RT) \tag{3}$$

with $E = 66.5\ kJ\ mol^{-1}$, $m \simeq 1$, and $n \simeq 0$ in the range of initial conditions considered.

Detonation velocities and cellular structures were studied with regard to ClO_2 initial pressures P_1, close to the detonation limit pressure. The cellular structures, although they show substructures, are sufficiently regular to permit an estimation of the mean value of the detonation cell size λ. These values are given in Table 1 for different initial pressures. In the same table we give the corresponding values of the ratio λ/Δ_i, in which Δ_i representing the induction distance is related to the autoignition delay by

$$\Delta_i = \tau_i\ (V_D - v_2) \tag{4}$$

where V_D is the detonation velocity and v_2 is the particle velocity with respect to the detonation front. τ_i is calculated from equation (2) assuming von Neumann temperature T_{VN} and pressure P_{VN}. The value of λ/Δ_i, which is about 6, is lower than the value 29 obtained with hydrazine[9] and with fuel/air mixtures.[10]

Table 1 Detonation velocity and cell size for different ClO_2 initial pressures; initial temperature = 293 K, tube diameter = 38 mm.

P_1, kPa	P_{VN}/P_1	T_{VN}, K	V_D, m/s	V_{CJ}, m/s	λ, mm	λ/Δ_i
13.3	40.8	1277	1250	1286	18	5.2
12.3	40.6	1324	1242	1284	17	5.5
6.7	39.4	1292	1220	1265	33	5.2
5.6	39.2	1286	1219	1261	54	6.9
4.3	38.6	1280	1215	1253	54	6.5
3.6	38.4	1278	1190	1248	60	6.0

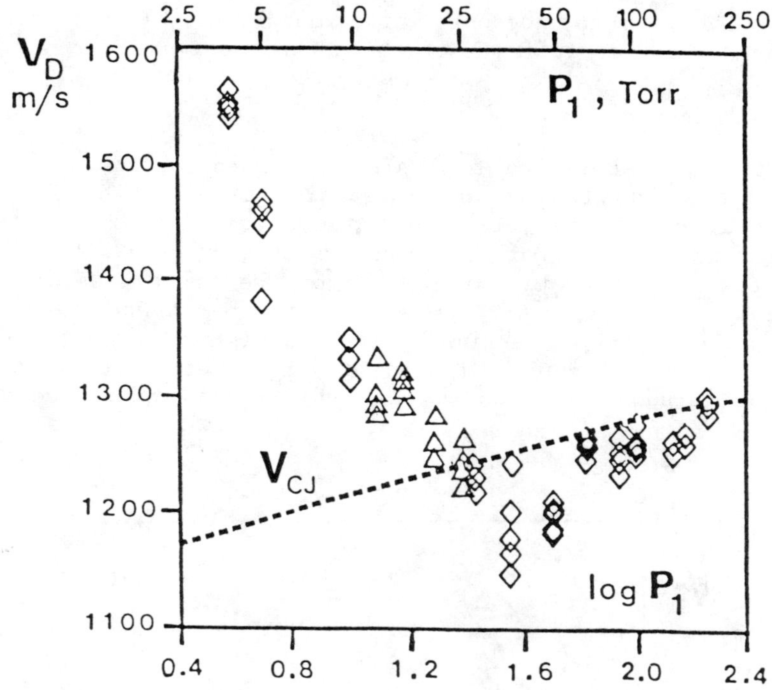

Fig. 1 Influence of ClO_2 initial pressure P_1 on detonation velocity V_D.
◇: detonation with a cellular structure; △: spinning detonation.

The ignition energy must be fairly moderate if significant values of λ are required. Figure 1 shows that detonation cells can be observed at initial pressures much lower than the limit pressure of detonation. The intensity of the initiating shock was very important in this case, provided the pressure ratio P_4/P_1 across the diaphragm was larger than 400. It follows that the temperature behind the initiating shock wave was higher than the von Neumann temperature and the wave velocity was larger than the Chapman-Jouguet velocity.

For initial pressures of ClO_2 larger than 2.67 kPa and a detonation tube of 22-mm i.d., a stable, self-sustained detonation is obtained for a $P_4/P_1 \simeq 50$ and an initiating shock temperature about 560 K (Fig. 2).

As estimated by equation (1), the ClO_2 vibrational relaxation times τ_v behind a detonation front are lower than the autoignition delays by at least one order of magnitude. It follows that vibrational processes need not to be taken into account while modeling the induction distance observed in front of the detonation wave.

Hydrogen Azide

From the detailed analysis of the pyrolysis of HN_3 highly diluted with argon, a reaction mechanism has been proposed and the kinetic model validated for temperatures lower than 1450 K and dilutions greater than 99%.[11-13] In this case, the initiation step is of the first order with respect to HN_3 and Ar. As a consequence, these results cannot be used to estimate the autoignition delay of undiluted or slightly diluted HN_3 prior to detonation. Previous studies have shown that this delay, expressed in

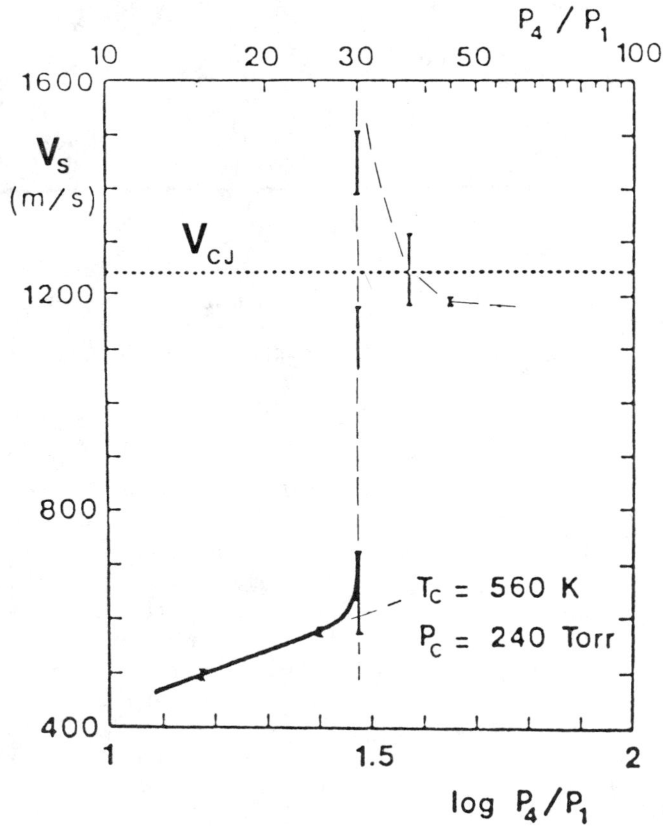

Fig. 2 Influence of the pressure ratio P_4/P_1 on shock velocity V_S for undiluted ClO_2. Initial pressure P_1 = 3.33 kPa, initial temperature T_1 = 293 K. T_c and P_c are the "coupling" temperature and pressure.

terms of temperature and HN_3 and Ar concentrations,[14] was valid in a restricted range of pressure, temperature, and dilution.[13]

In order to link the autoignition delay of undiluted or slightly diluted HN_3 to shock conditions, the model of the adiabatic thermal explosion described by Sokolik[15] was chosen and modified to take into account the following assumptions:

1) A rate of the overall reaction: x HN_3 + (1-x) Ar ⟶ Products expressed in an Arrhenius form as

$$- d[HN_3]/dt = A\ [HN_3]^m\ [Ar]^n\ \exp(- E/RT) \quad (5)$$

where A is the pre-exponential constant, m and n are the partial orders, and E is the activation energy.

2) A rate of energy release at the early stages of the reaction written as

$$- \Delta H\ (d\theta/dt) = \overline{C_p}\ (dT/dt) \quad (6)$$

in which ΔH is the reaction enthalpy, $\overline{C_p}$ the mean calorific capacity at constant pressure, T the temperature, t the time, and θ the HN_3 consumed fraction:

$$\theta = ([HN_3]_o - [HN_3])/[HN_3]_o$$

$$d\theta/dt = - (1/[HN_3]_o)\ (d[HN_3]/dt) \quad (7)$$

3) A reactant concentration varying very little during the induction period and, therefore:

$$dt = (1/K)\ \exp(E/RT)\ dT \quad (8)$$

with

$$K = - (A\ \Delta H/\overline{C_p})\ (xP/RT)^{m-1})\ ((1-x)\ P/RT)^n \approx \text{constant} \quad (9)$$

Writing now T_2 and P_2 respectively for the values of temperature and pressure immediately behind the incident shock front, we have

$$t = (1/K) \int_{T_2}^{T} \exp(E/RT)\ dT \quad (10)$$

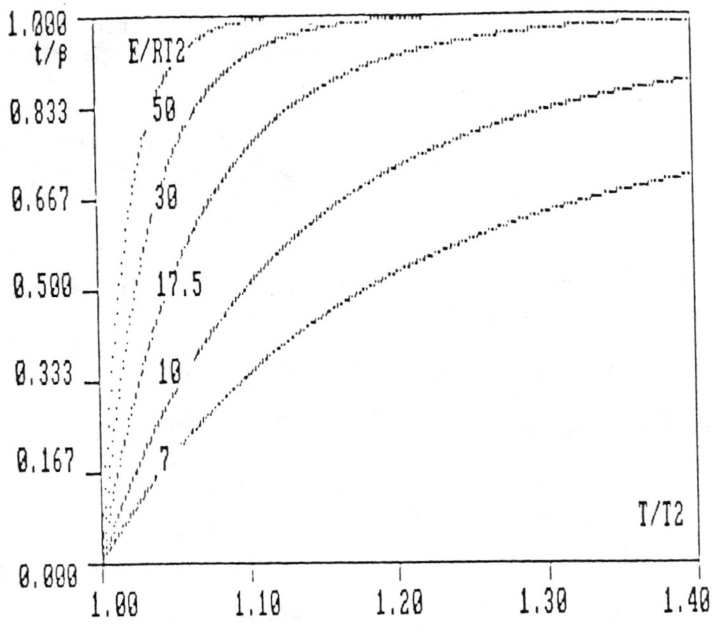

Fig. 3 Variation of relative temperature T/T_2 vs relative times t/β for different values of E/RT_2, according to the adiabatic thermal explosion theory.

with $T = T_2$ at $t = 0$. With the assumption that $E/RT_2 >> 1$, the integration of the above expression can be written as

$$t/\beta = 1-(T/T_2)^2 \ \alpha \ ((T_2/T)-1) \ \exp(E/RT_2) \qquad (11)$$

with $\beta = (RT_2^2/EK) \ [1+(2RT_2/E)+\ldots] \ \exp(E/RT_2)$

and $\alpha = 1-(2RT/E)((T/T_2)-1)+\ldots$

The representation of $t/\beta = f(T/T_2)$ is given in Fig. 3 for different values of E/RT_2. For a slight increase of the ratio T/T_2, t/β tends the faster towards 1 as E/RT_2 increases. When $t/\beta \longrightarrow 1$ (or $\tau_i \longrightarrow \beta$), the ignition delay τ_i can be defined by

$$\tau_i \simeq (RT_2^2/EK) \ \exp(E/RT_2) \qquad (12)$$

Experimentally, τ_i is determined behind the reflected shock wave from the fast pressure rise measured at the tube end.

If these detailed assumptions are valid, a relation of the following form must be verified:

$$\ln \tau_i = E/RT + \ln(R/AE) + 2 \ln T + \ln(\overline{C_p}/\Delta H) + (1-m) \ln(xP/RT) - n \ln((1-x)P/RT) \quad (13)$$

Experimental values of τ_i measured as a function of the molar fraction x of HN_3 at temperature T and pressure P allow us to calculate Y for different values of m and n from

$$Y = \ln \tau_i - 2 \ln T + \ln(\Delta H/\overline{C_p})) + (m-1) \ln (xP/RT) + n \ln((1-x)P/RT) \quad (14)$$

The relationship Y = E/RT + ln (R/AE) must be verified for appropriate values of m and n. For systems with x > 0.2, the best correlation of Y = f(1/T) is obtained with m = 2 and n = 0 (correlation factor = 0.97), which then yields the curve Y = f(1/T) shown in Fig. 4.

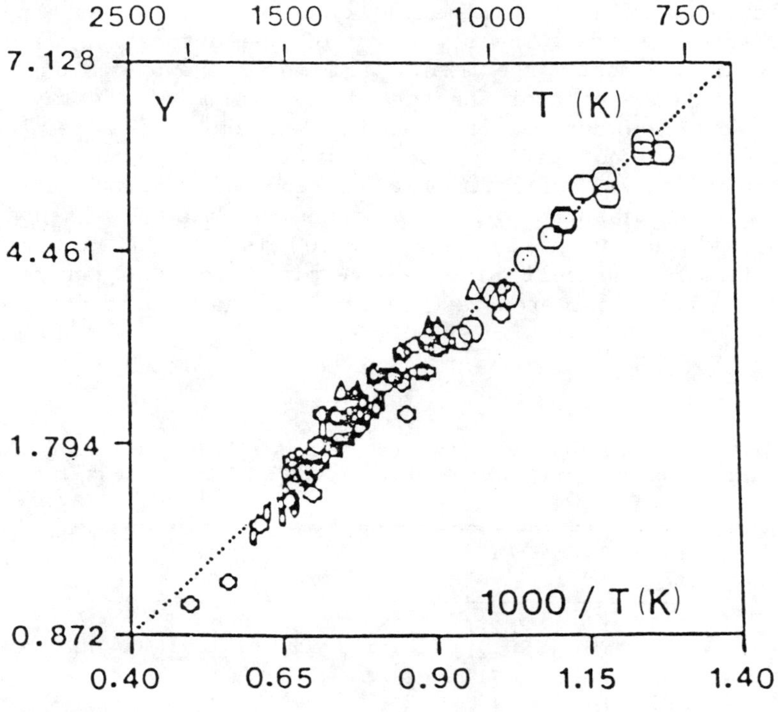

Fig. 4 Variation of the expression Y vs 1/T for HN_3/Ar mixtures with HN_3 molar fraction ranging from 0.2 to 1.

$Y = \ln \tau_i - 2 \ln T + \ln(\Delta H/\overline{C_p}) + (m-1) \ln(xP/RT) + n \ln((1-x)P/RT)$

The numerical expression of the autoignition delay is the following:

$$\log \tau_i = 4650/T - 10.04 + \log[(T^3/xP)(\overline{C_p}/\Delta H)] \tag{15}$$

This relationship is deduced from experiments carried out with $0.2 < x < 1$, $700 < T < 1300$, and $20 < P < 200$, τ_i, T, and P being expressed in s, K, and kPa respectively. The activation energy of the overall reaction is about 90 kJ mol^{-1} and the pre-exponential factor $A = 10^6$ mol^{-1} m^3 s^{-1}. These values are much lower than those obtained from the experiments carried out with considerably diluted HN_3 in argon. In the latter case, the activation energy is about 150 kJ mol^{-1}.

Detonation velocities and the size of the wall-printed detonation cells were determined in tubes of various diameters for different initial pressures down to the detonability limit.[16] The velocity V_D measured independently of the wall effect was found to be systematically higher (by 6% at 1.33 kPa) than the Chapman-Jouguet velocity V_{CJ} calculated using the value 300 kJ mol^{-1} for the enthalpy of formation of HN_3.[6] The structure of the detonation wave is regular in most experiments (Fig. 5). However, one can observe the effect of the tube diameter and of the layer of soot both on detonation velocity and on cell size.[16] The presence of soot in the tube can modify the gas composition in proportions sufficient to cause a discrepancy between velocities measured in the presence and in the absence of soot, which can be as large as 400 ms^{-1} for $P_1 = 0.2$ kPa (Table 2). The cell size λ determined in such conditions cannot be considered as typical of undiluted hydrogen azide.

Table 2 Detonation velocity and cell size for different HN_3 initial pressures; initial temperature = 293 K, tube diameter = 22 mm.

P_1, kPa	P_{VN}/P_1	T_{VN}, K	V_D, m/s	V_{CJ}, m/s	λ, mm	λ/Δ_i
2.67	114.1	2270	2757	2632	0.15	0.5
1.33	110.5	2218	2717	2594	0.5-1	1.3
0.33	104	2120	2427	2520	2.5-3.5	1.6
0.21*	102.4	2094	2526	2498	-	-
0.20	102	2092	2113	2496	7-10	1.9
0.17	101	2075	2106	2488	8-12	1.8

*values found in the absence of soot.

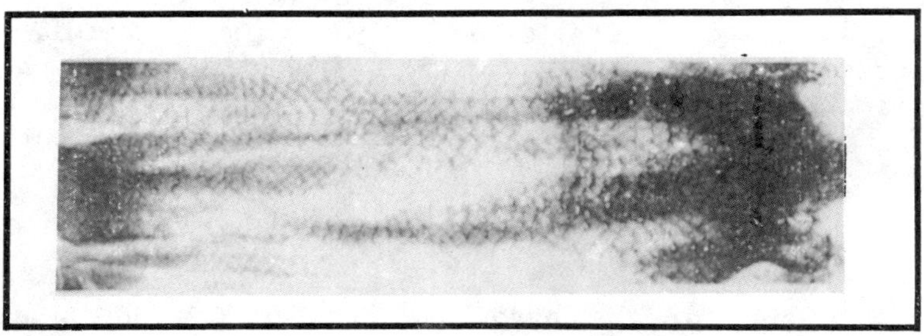

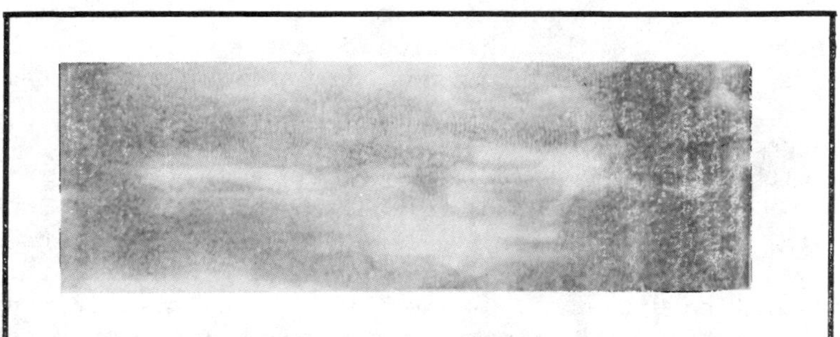

Fig. 5 Cell aspect for a detonation wave propagating in hydrogen azide. Initial pressure: P_1 ; HN_3 molar fraction in argon: x ; tube diameter: 22 mm.

Above 1 kPa, this discrepancy becomes negligible and the estimation of λ can be considered to be trustworthy. The ratios λ/Δ_i, compared to those obtained with hydrocarbon/air mixtures,[10] are small. The values of the induction period for HN_3 can also be criticized, as they are deduced from a relationship issued from experiments carried out at temperatures much lower than von Neumann temperatures. Validation of λ/Δ_i ratios given in Table 2 would require complementary data on the behavior of undiluted HN_3 at temperatures higher than 2000 K, since previous studies of the HN_3 decomposition in an argon bath have shown that the reaction rate depends on the presence of the diluent.[11,12] Therefore, the corresponding modeling cannot be used to calculate the induction period in von Neumann conditions. The low value of this ratio λ/Δ_i is probably due to the change of the reaction mechanism between 1300-2000 K when the main initiation step yields essentially HN_3 ($a^1\Delta$) instead of HN_3 ($X^3\Sigma^-$). In this case the reaction is initiated by a step at a higher activation energy; therefore, its rate increases more rapidly with temperature. From data given by Kajimoto et al.,[11] we can estimate that induction distance is reduced by a factor 5 at 2200 K with respect to that deduced from the relation (15).

Conclusion

The study of the pyrolysis of ClO_2 and HN_3 behind a shock wave leads to a simple formulation of the auto-ignition delays. In the case of ClO_2 the delays were determined at temperatures close to the temperatures calculated assuming a ZND model for the detonation wave. The relationship obtained from kinetic measurements and used to calculate the induction distance behind the shock wave is justified. This is not the case with HN_3, because von Neumann temperatures are much higher than those corresponding to the range of kinetic studies.

With both explosive systems, detonation structure is rather regular, more regular with HN_3, for which E/RT_2 = 4.9, than with ClO_2, for which E/RT_2 = 6.5. In both cases a mean value of cell size λ could be determined with respect to initial conditions. Some problems arise, however, when a characteristic cell size has to be found for a given chemical system, among which are the influence of initiating shock intensity, of wall effects, and of a layer of soot that modifies the gas composition at very low pressure. With sufficient care a constant value of λ/Δ_i for a given explosive mixture, useful for the estimation of detonation limits in a given confinement, can be found.

Acknowledgment

This work was sponsored by the Ministère de la Défense, Délégation Générale à l'Armement, Direction des Recherches, Etudes et Techniques under Contract No. 87/157.

[1]Paillard, C., Youssefi, S., Al Aiteh, H., Charpentier, N., and Dupré, G., "The Explosive Decomposition of Chlorine Dioxide," Journal de Chimie Physique, Vol. 84, No. 1, 1987, pp. 41-47.

[2]Kiefer, J.H., "The Laser-Schlieren Technique in Shock Tube Kinetics," Shock Waves in Chemistry, edited by A. Lifshitz, Marcel Dekker, New York, 1981, pp. 219-277.

[3]Paillard, C., Al Aiteh, H., and Dupré, G., "A Study of Gaseous Explosive Systems: Formation, Structure and Limits of Detonation," Proceedings of the Joint Meeting of the French and Italian Sections of the Combustion Institute, edited by CUEN s.r.l., Naples, Italy, 1987, pp. 6-23.

[4]Paillard, C., Youssefi, S., and Dupré, G., "The Explosive Decomposition of Chlorine Dioxide Behind Shock Waves," in "Dynamics of Reactive Systems, Part II: Modeling and Heterogeneous Combustion," Progress in Astronautics and Aeronautics, edited by J.R. Bowen, J.-C. Leyer, and R.I. Soloukhin, AIAA, Vol. 105, New York, 1986, pp. 394-406.

[5]Laffitte, P., Combourieu, J., Hajal, I., Ben Caïd, M., and Moreau, R., "Charcteristics of Chlorine Dioxide Decomposition Flames at Reduced Pressures," 11th Symposium (International) on Combustion, The Combustion Institute, Pittsburgh, PA, 1967, pp. 941-950.

[6]Gray, P., and Waddington, T.C., "Thermochemistry and Reactivity of the Azides," I Thermochemistry of the Inorganic Azides, Proceedings of the Royal Society, A235, 1956, pp. 106-119.

[7]Laffitte, P., Hajal, I., and Combourieu, J., "The Decomposition Flame of Hydrogen Azide," 10th Symposium (International) on Combustion, The Combustion Institute, Pittsburgh, PA, 1965, pp. 79-85.

[8]Paillard, C., Dupré, G., Youssefi-Stitou, S., and Charpentier, N., "Mechanism of the Pyrolysis of Chlorine Dioxide," in "Shock Tubes and Waves," Proceedings of the 16th International Symposium on Shock Tubes and Waves, edited by H. Grönig, Weinheim, Federal Republic of Germany, 1987, pp. 459-465.

[9]Pedley, M.D., et al., "Hydrazine Vapor Detonations," Progress in Astronautics and Aeronautics, edited by A.L. Kuhl, J.R. Bowen, J.-C. Leyer, and A. Borisov, AIAA, Vol. 114, Washington, DC, 1988, pp. 45-63.

[10]Westbrook, C.K., and Urtiew, P.A., "Chemical Kinetic Prediction of Critical Parameters in Gaseous Detonations," 19th Symposium

(International) on Combustion, The Combustion Institute, Pittsburgh, PA, 1982, pp. 615-623.

[11]Kajimoto, O., Yamamoto, T., and Fueno, T., "Kinetic Studies of the Thermal Decomposition of Hydrazoïc Acid in Shock Waves," Journal of Physical Chemistry, Vol. 83, No. 4, 1979, 429-434.

[12]Dupré, G., Paillard, C., Combourieu, J., Fomin, N.A., and Soloukhin, R.I., "Decomposition of Hydrogen Azide in Shock Waves," in "Shock Tubes and Waves," Proceedings of the 13th International Symposium on Shock Tubes and Waves, edited by C.E. Treanor and J.G. Hall, Albany, NY, 1982, pp. 626-634.

[13]Paillard, C., Dupré, G., and Combourieu, J., "Pyrolysis of Gaseous Endothermic Compounds by the Shock Tube Method," Journal de Chimie Physique, Vol. 82, No. 5, 1985, pp. 489-497.

[14]Zaslonko, I.S., Kogarko, S.M., and Mozzhukhin, E.V., "Induction Period in Exothermic Decomposition Reactions with Energetic Acceleration," Fizika Goreniya i Vrzyva, Vol. 9, No. 3, 1973, 339-344.

[15]Sokolik, A.S., Self-Ignition, Flame and Detonation in Gases, Izdatel'stvo Akademii Nauk SSR, Moscow, 1960, translated from Russian by N. Kaner, edited by R. Hardin, Jerusalem, Israel, 1963.

[16]Paillard, C., Dupré, G., Lisbet, R., and Combourieu, J., "Influence of the Structure of a detonation wave on the Heat Transfer Rates at the wall," Proceedings of the Berthelot Vieille Mallard Le Châtelier International Colloquium, Vol. II, edited by the French Section of the Combustion Institute, Orléans, France, 1981, pp. 449-454.

Influence of Tetrafluoromethane on Hydrogen-Oxygen-Argon Detonations

E. Nzeyimana* and P. J. Van Tiggelen†
Université Catholique de Louvain, Louvain-la-Neuve, Belgium

Abstract

A study of the CF_4-H_2-O_2-Ar detonations is presented. Experiments have been carried out in a square tube (3.2 cm x 3.2 cm) at constant pressure (200 Torr) and constant dilution (50%) but with variable equivalence ratio (ϕ = 1.2, 1.0, and 0.8). A variable amount of CF_4, up to 25%, replaces the argon in the total mixture. Detonation velocities are measured with an accuracy of .5%, and soot imprints of the cellular structure are recorded. A nonmonotonous influence of CF_4 on D is noticed. With 5% CF_4 added, the velocity reaches a maximum, decreases slowly up to 15% CF_4, and decreases rapidly thereafter. No detonation is observed for CF_4 quantities larger than 25%. The cell length varies accordingly all throughout the investigated range of composition. Such a behavior contrasts the one noticed for $CO/H_2/O_2/Ar$ mixtures, where fluorocarbons added in traces were acting as inhibitors, but it is similar to the influence of freons on detonations in hydrocarbon mixtures.The Chapman-Jouguet (C-J) velocity of the investigated mixtures has been computed by taking into account the dissociation of products in radicals. The D_{CJ} remains constant up to 17% of CF_4, and at a higher percentage of CF_4 the D_{CJ} decreases sharply. Chemical mechanism arguments will be presented to suggest an

*Research Chemist, Laboratoire de Physico-Chimie de la Combustion

†Professor, Laboratoire de Physico-Chimie de la Physico-Chimie de la Combustion.

explanation of the behavior of the different systems.

Nomenclature

D = measured detonation velocity
L = cell length
M = Mach number
P_o = initial pressure
t_{car} = characteristic time of a cell
T_{VN} = von Neumann temperature
n = number of runs
α = percentage of diluent in fresh gases mixture
σ = standard deviation of D
ϕ = equivalence ratio of initial gaseous mixture

Subscript

CJ = Chapman-Jouguet state
St = stoichiometric condition.

Introduction

Inhibition phenomena of combustion have been studied extensively. The inhibition can be done by either thermal (thermal diluents) or chemical processes (chemical inhibitors). Recent reviews of the inhibiting influence of fluorocarbon compounds on flames and detonations have been reported by F.N. da Cruz et al.,[1] J. Vandooren et al.,[2] M. Vandermeiren et al.,[3,4] J.C. Libouton et al.,[5] and A. Jacques.[6]

For halocarbons, which are inhibitors for flames, shock tube studies indicate that the behavior depends on the nature of the fuel. For instance, recent studies by Hidaka et al.[7,8] have demonstrated that the influence of an additive, such as CF_3Cl, on the induction periods depends on the nature of the fuel used. For H_2/O_2 and C_2H_4-O_2 mixtures there is an increase of the induction time when CF_3Cl is added, although the opposite is true for CH_4-O_2 mixtures.

For detonations in $CO/H_2/O_2/Ar$ mixtures, Libouton et al.,[5] Dormal et al.[9] have proven that the addition of traces of halogenated compounds inhibits detonation by largely increasing the cell length of the detonation structure. They noticed also that CH_4, CH_3Br suppress the regularity of the cell spacing without modifying the cell length very much in a $H_2/O_2/Ar$ system. Furthermore, the addition of CF_4, up to 15% of the total fuel concentration, in either $CO/O_2/H_2$ or H_2/O_2 mixtures does not change the regularity of the cell length markedly.

Moen and al.[10] have investigated the influence of halogen compounds on the critical tube diameter in C_2H_4/O_2 and H_2/O_2 mixtures. They concluded that in mixtures with great quantities of H_2 the halogenated compounds (CF_3Br or CF_4) induce a slight promoting effect of the detonation. Recently, Vandermeiren et al.[3,4] reported on the role of CF_3Br traces on the behavior of acetylene-oxygen detonations in well-sustained regimes, as well as during the incipient stage. Those results exhibit a slight increase (1.5%) of the detonation velocities when traces of CF_3Br are added.

The present work is devoted to an experimental investigation of the role of CF_4 on $H_2/O_2/Ar$ detonations. The choice of CF_4 as an additive has been made in order to simplify the chemical kinetics of an eventual inhibition action. Detonation velocity and cell length have been measured.

Experimental Technique

A detailed description of the experimental setup has been given previously by Libouton[5] and Vandermeiren.[11] The rectangular cross section (9.2 cm x 3.2 cm) of the detonation tube has been modified by introducing two rods of aluminium (3.0 cm x 3.0 cm) along the tube in order to get a square cross section of 3.2 cm x 3.2 cm. The mixing was achieved during the filling of the tube, and the composition of the gas flow was controlled precisely by employing choked flow conditions through nozzles for the individual gas. Such a technique guarantees a better homogeneity of the mixture, and an accurate and reproducible composition of the gaseous mixture is obtained.

The initiation of detonation is achieved by detonating a H_2/O_2 mixture in the driver section, which bursts a mylar film (diaphragm) and then ignites the test section after passage through an orifice (1 cm in diameter). The structure of detonation is recorded on a glass plate covered by soot. The plate is located 4 m after the diaphragm. The detonation velocity has been measured by six ionizations gauges located on each side of the glass plate. The signal of the ionization gauges is recorded on a Kikusui Memoscope, allowing measurements of time interval to an accuracy of better than 1%.

Experimental Results and Discussion

The experimental conditions for the investigated $H_2/O_2/Ar/CF_4$ system are as follows: stoichiometric H_2/O_2 mixtures (ϕ = 1) with a constant dilution at the initial pressure of 200 Torr. The diluent was either pure argon or argon-tetrafluoromethane mixtures. The quantity of CF_4 in the fresh gases mixtures is varied in such a way that the Ar-CF_4 fraction always remains equal to 50% of the total content (α = 0.5). Some measurements have been performed also on rich (ϕ = 1.2) and lean (ϕ = 0.8) mixtures. The equivalence ratio $\phi = [(H_2)/(O_2)]/[(H_2)/(O_2)]_{st}$. From the data, it appears that a slight increase of the detonation velocity occurs when the diluent is modified from pure argon (D=1791 m/s) to mixtures containing small amounts of CF_4. The maximum detonation velocity is reached with a mixture containing about 5% of CF_4 (D=1844 m/s). Table 1 provides the average detonation

Table 1 $H_2/O_2/Ar/CF_4$ (ϕ=1; α=0.5; P_o=200 Torr)

CF_4,%	D,m/s	n	σ,m/s
0	1791	6	5
2	1806	3	11
3	1821	3	6
5	1844	6	8
7.5	1840	3	4
10	1833	4	2
15	1805	3	4
20	1756	2	17
25	1597	3	6
30	--	3	- -

velocities for increasing amounts of CF_4. A number of experiments with the same composition and standard deviations are mentioned. When over 15% of CF_4 is added, the detonation speed drops rapidly. For mixtures containing more than 25% the detonation is not even ignited (Fig. 1).

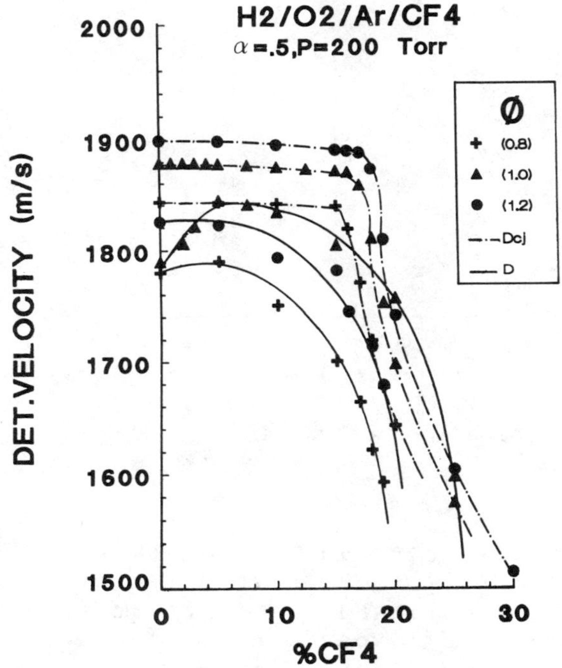

Fig. 1 Comparison of values of D_{CJ} and D in H_2-O_2-Ar-CF_4 mixtures vs the CF_4 percentage.

Table 2 $H_2/O_2/Ar/CF_4$ (α=0.5; P_o=200 Torr)

ϕ = 0.8 (+)				ϕ = 1.2 (•)			
CF4,%	D,m/s	n	σ,m/s	CF4,%	D,m/s	n	σ,m/s
0	1780	3	6	0	1825	4	2
5	1790	3	7	5	1823	5	6
10	1750	6	6	10	1794	6	10
15	1700	4	18	15	1781	5	7
17	1663	4	1	16	1744	4	1
18	1621	2	15	18	1715	3	5
				19	1678	4	16

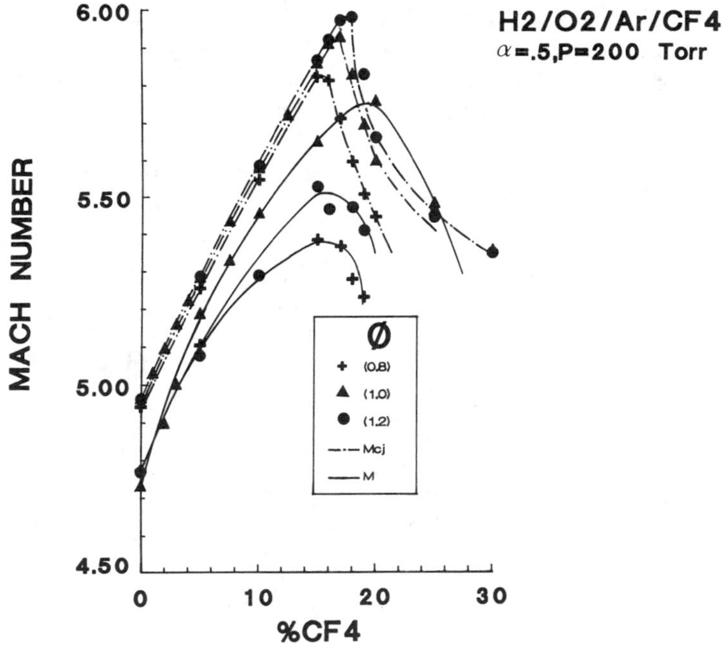

Fig. 2 Comparison of values of M_{CJ} and M in H_2-O_2-Ar-CF_4 mixtures vs the CF_4 percentage.

It should be remembered that, to a first approximation, the detonation velocity is related to the heat released per unit mass. But during these series of experiments, the average molar mass of the fresh gaseous mixture varies substantially when CF_4 is replacing Ar in the original mixture. It is worthwhile to use detonation Mach numbers M, which allow one to take into account the mass variation of the initial mixture in order to compare the influence of a CF4 addition on hydrogen-oxygen detonations. The corresponding Mach numbers M of the experimental data are plotted on Fig. 2. The maximum of M is reached with 20% of CF_4. However, no detonation velocity maxima are observed for nonstoichiometric mixtures (Table 2 and Fig. 1), and the maximum in Mach number is shifted to a lesser CF_4 content (16%).

C-J detonation speeds (D_{CJ}) have been computed by taking into account 17 species (H_2, O_2, Ar, CF_4, C_g, CF, CF_2, CF_2O, CF_3, CO, CO_2, F,

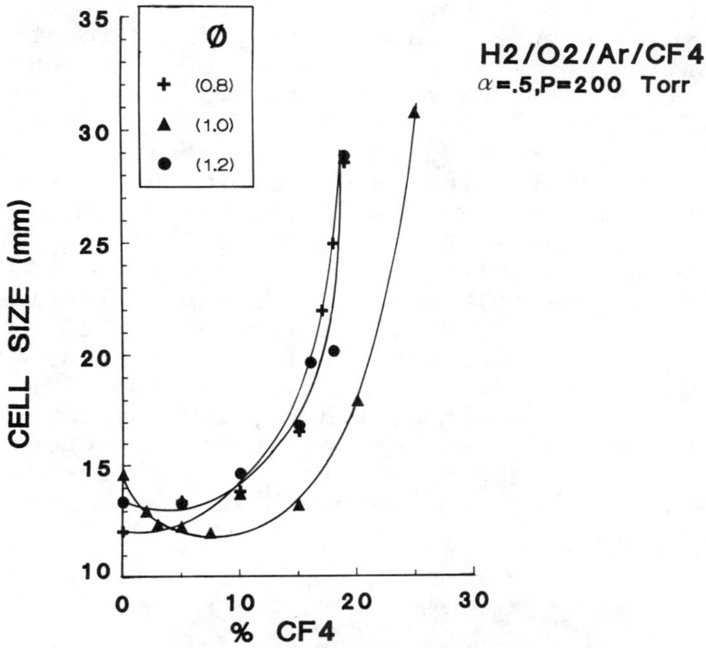

Fig. 3 Cell length vs CF_4 percentage at variable equivalent ratio (ϕ), constant pressure (200 Torr) in H_2-O_2-Ar-CF_4 mixtures.

H, HF, H_2O, O, and OH). From the calculation it appears that D_{CJ} remains almost constant up to 17% of CF_4 (Fig. 1) and drops rapidly thereafter. The maximum C-J Mach number (5.91) is reached for 17% of CF_4 in the stoichiometric mixture (Fig. 2).

Cell lengths L of the detonation, as recorded from the soot imprints, have been studied extensively in the past[12] and more recently in investigations of the detonation limits[13] and the onset[14]. Libouton et al.[5,15] have suggested a relationship between detonation structure and the chemical kinetics of the heat release process by using the characteristic time (t_{car} = L/D) connected to the induction time. For $H_2/O_2/Ar/CF_4$ mixtures, there is first a slight decrease of the cell length from 0 to 7,5% CF_4 (L=15->12 mm). From 10% and greater of CF_4 in the mixture, L increases steadily (Fig. 3) up to 30 mm, which corresponds to the limiting case in our tube geometry. However, the regularity of the cellular structure is diminishing for detonations propagating in mixtures with a CF_4 percentage higher than 10%. According to Libouton et al.'s classification,[16] the highly regular structure (excellent) becomes slightly irregular (poor). Figure 4 exhibits the modification of the cell regularity for mixtures with increasing amount of tetrafluoromethane (0, 5, and 25% CF_4). Such changes in the structure regularity could be related to the role played by some elementary processes in the chemical mechanism of heat release, more specifically, decomposition reactions with a large activation energy.[16]

From the experimental data, one may conclude that CF_4 exhibits a noninhibiting character for detonations propagating in a hydrogen-oxygen mixture. This can be ascribed to two different

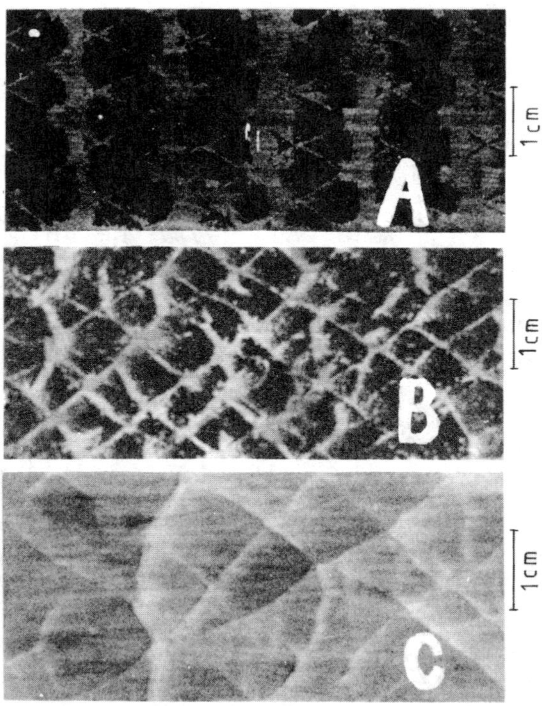

Fig. 4 Typical soot records exhibiting the evolution of the regularity of cell structures with variable amounts of CF_4 in the initial mixture: A) 0% CF_4; B) 5% CF_4; and C) 25% CF_4.

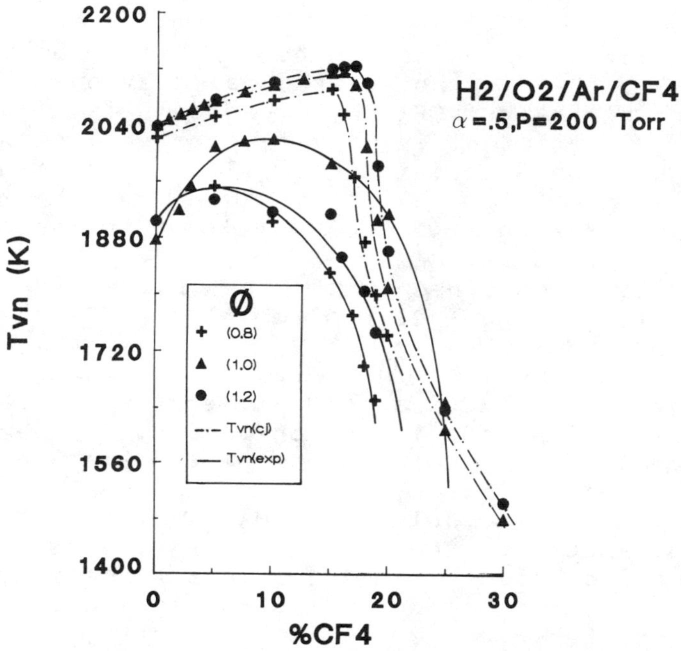

Fig. 5 Experimental $T_{VN(exp)}$ and calculated $T_{VN(CJ)}$ temperatures vs the CF_4 percentage.

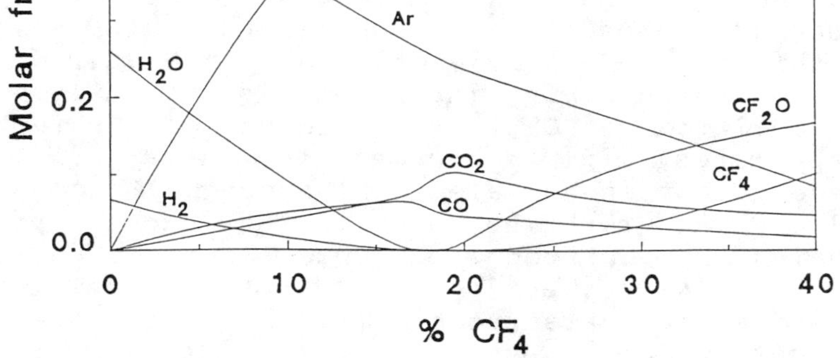

Fig. 6 Equilibrium composition at CJ conditions in H_2-O_2-Ar-CF_4 mixtures at 200 Torr initial pressure (ϕ = 1, α = 0.5) and with variable CF_4 content.

actions:

1)As an opposite to the inhibiting behavior of CF_3Br, CF_4 is not a radical scavenger. The primary attack of CF_4 by H radicals is, indeed, not very efficient in the temperature range of 2000 K characteristic of the von Neumann temperature (Fig. 5).It is due to the high activation energy (42 kcal/mole[17] of the process

$H + CF_4 \rightarrow CF_3 + HF$

This fact, combined with the extremely high activation energy (122 kcal/mole) for CF_4 thermal decomposition:

$CF_4 + M \rightarrow CF_3 + F + M$

as measured by Modica and Sillers,[18] impede CF_4 consumption in the induction period. CF_4 remains a thermal diluent and does not compete with the branching reaction.

2) When the bulk of a H_2/O_2 mixture starts to react during the heat release pulse, the temperature raises and CF_4 starts to be consumed significantly, producing CF_3 radicals that scavenge the H and OH radicals, thereby shifting the equilibrium toward a reduction of radical concentration during the heat release pulse. As a consequence, the rate of increase of temperature during the pulse is faster with CF_4, inducing pressure waves in the flowfield, which, in turn, improve the coupling between the leading shock wave and the reaction zone. A promoting effect is therefore noticeable in detonation velocity as well as in cell length.

One should bear in mind that when CF_4 is present in the mixture the hydrogen is mostly consumed to form HF molecules. The hydrogen fluoride production corresponds to very exothermic processes, noticeable in Fig. 2 where the C-J Mach number increases significantly with the addition of CF_4. If the amount of CF_4 in the fresh gases mixture becomes too large (>17%), there is not enough elemental hydrogen to consume all of the fluorine atoms. The excess of F remains as perfluoro compounds (C_xF_y, CF_2O,...) in the mixture, acting as thermal diluents, and it increases catastrophically the heat losses of the detonation. Such a limiting effect of F content is obvious from Fig. 6, where the equilibrium composition at C-J condition is plotted vs the CF_4 percentage in the initial mixture. With 20-25% of CF_4 in the initial mixture, a composition limit for detonation is reached.

Acknowledgments

The authors acknowledge the help of M. Lefebvre for some computations as well as the financial support of the Fonds de la Recherche Fondamentale et Collective (F.R.F.C.), Belgium, under contract N° 2.9006.88. E. Nzeyimana is also indebted to the Agence Générale pour la Coopération au Développement (A.G.C.D.), Belgium, for a research doctoral fellowship.

References

[1]da Cruz, F.N., Vandooren, J., and Van Tiggelen, P.J., "Comparaison of the Inhibiting Effect of Some Halogenomethane Compounds on Flame Propagation, "Bulletin des Sociétés Chimiques Belges, Vol.97, N°11-12, 1988, pp. 1011-1030.(in french)

[2]Vandooren, J., da Cruz, F.N., and Van Tiggelen, P.J., "The Inhibiting Effect of CF_3H on the Structure of a Stoichiometric $H_2/CO/O_2/Ar$ Flame," 22nd Symposium (International) on Combustion, The Combustion Institute, Pittsburgh, PA, 1988, pp. 1587-1595.

[3]Vandermeiren, M., and Van Tiggelen, P.J., "Structure of Gas Phase Detonations in Acetylene Mixtures -- Role of an Inhibitor, "Proceedings of the International Symposium on Intense Dynamic Loading and its Effects, Beijing, China, 1986, pp. 219-224.

[4]Vandermeiren, M., and Van Tiggelen, P.J., "Role of an Inhibitor on the Onset of Gas Detonations in Acetylene Mixtures," Progress in Astronautics and Aeronautics: Dynamics of Explosions, Vol. 114, edited by A.L. Kuhl, J.R. Bowen, J.C. Leyer, and A. Borisov, AIAA, New York, 1987, pp. 186-201.

[5]Libouton, J.C., Dormal, M., and Van Tiggelen, P.J., "The Role of Chemical Kinetics on Structure of Detonations Waves," 15th Symposium (International) on Combustion, The Combustion Institute, Pittsburgh, PA, 1975, pp. 79-86.

[6]Jacques, A.,Effect of Chemical Composition on the Stucture of Detonation Waves,Mémoire de Licence, Catholic University of Louvain,Belgium,1988. (in french)

[7]Hidaka, Y., Kanano, H., and Suga, M., "Inhibition of H_2-O_2. Reaction by CF_3Cl, "Combustion and Flame, Vol. 59, N° 1, 1985, pp. 93-95.

[8]Hidaka, Y., and Suga, M., "Additive Effect of CF_3Cl on OH*, CH* and C_2* Emissions: Shock Tube Study with C_2H_4-O_2-CF_3Cl and CH_4-O_2-CF_3Cl Mixtures,"Combustion and Flame, Vol. 60, N° 2, 1985, pp. 233-238.

[9]Dormal, M., Libouton, J.C., and Van Tiggelen, P.J., "Evolution of Induction Time in Detonation Cell," Acta Astronautica, Vol. 6, July-August 1979, pp. 875-884.

[10]Moen, I.O., Ward, S.A., Thibaut, P.A., Lee, J.H., Knystautas, R., Dean, T., and Westbrook, C.K., "The Influence of Diluents and Inhibitors on Detonations," 20th Symposium (International) on Combustion, The Combustion Institute, Pittsburgh, PA, 1984, pp. 1717-1725.

[11]Vandermeiren, M., and Van Tiggelen, P.J., "Cellular Structure of Detonation in Acetylene-Oxygen Mixtures, "Progress in Astronautics and Aeronautics: Dynamics of Shock Waves, Explosions and Detonations, Vol. 94, edited by J.R. Bowen, N. Manson, A.K. Oppenheim, and R.I. Soloukhin, AIAA, New York, 1985, pp. 104-111.

[12]Strehlow, R.A., Fundamentals of Combustion, International Textbook, Scranton, Pennsylvania, PA, 1968.

[13]Moen, I.O., Donato, D., Knystautas, R., and Lee, J.H., "The Influence of Confinement of the Propagation of Detonations near the Detonability Limits," 18th Symposium (International) on Combustion, The Combustion Institute, Pittsburgh, PA, 1981, pp. 1615-1622.

[14]Murray, S.B., and Lee, J.H., "On the Transformation of Planar Detonation to Cylindrical Detonation," Combustion and Flame, Vol. 52, N°3, 1983, pp. 269-289.

[15]Libouton, J.C., and Van Tiggelen, P.J., "Influence of composition of Gas Mixtures on the Structure of Detonation Waves," Acta Astronautica, Vol. 3, September-October, 1976, pp. 759-769.

[16]Libouton, J.C., Jacques, A. and Van Tiggelen, P.J., "Kinetic, Structure and Sustain of Detonation Waves," Actes du Colloque International Berthelot-Vieille-Mallard-Le Chatelier, Vol. II, 1981, pp. 437-442.

[17]Kondratiev, V.N., "Rate Constants of Gas Reactions," U.S. Department of Commerce National Bureau of Standards, Washington, DC, 1972.

[18]Modica, A.P., and Sillers, S.J., "Experimental and Theoritical Kinetics of High Temperature Fluorocarbon Chemistry," The Journal of Chemical Physics, Vol. 48, N°.7, 1968, pp. 3283-3288.

Chapter II. Detonation: Initiation and Transmission

Initiation of Hydrogen-Air Detonations by Turbulent Fluorine-Air Jets

S. B. Murray* and I. O. Moen*
Defense Research Establishment Suffield, Defense Technologies Division, Ralston, Canada
and
P. A. Thibault†
Combustion Dynamics Ltd., Medicine Hat, Canada
and
R. Knystautas‡ and J. H. S. Lee‡
McGill University, Montreal, Quebec, Canada
and
A. Sulmistras §
Concordia University, Montreal, Quebec, Canada

Abstract

This study reports on a series of large-scale experiments conducted at the Defence Research Establishment Suffield (DRES) in which highly reactive mixtures of fluorine-air were jetted into stoichiometric hydrogen-air to initiate gaseous detonation. The experimental apparatus consisted of a cylindrical chamber, charged with high-pressure fluorine-air, connected to a large cylindrical plastic bag containing a hydrogen-air mixture. The end of the chamber inside the bag was capped by a thin brass diaphragm, which was ruptured to allow rapid venting of the contents. High-speed cinematography and measured pressure histories inside the bag have shown that initiation of detonation is possible for a narrow range of fluorine concentration. Numerical calculations of an inert transient gas jet have shown that the onset of detonation corresponds to a brief time interval in which a well-defined toroidal vortex exists about three chamber diameters down axis from the exit. The vortex is characterized by decreasing fluorine concentra-

* Defence Research Establishment Suffield, Defence Technologies Division, Ralston, Canada.

† Combustion Dynamics Ltd., Medicine Hat, Canada.

‡ McGill University, Department of Mechanical Engineering, Montreal, Canada.

§ Concordia University, Department of Mechanical Engineering, Montreal, Canada during this work; now at Eyes of Science Ltd., Toronto, Canada.

tions and temperatures as the core is approached. It is hypothesized that these distributions are responsible for a spatial gradient of induction time that leads to ignition and shock wave amplification via a coherent release of energy. It is further postulated that the toroidal explosion results in a shock wave that converges on the jet axis, giving rise to an overdriven detonative Mach stem and, eventually, a spherical detonation wave.

Introduction

In conventional blast initiation of detonation, free radicals for the oxidation processes are brought about by thermal dissociation in the wake of a strong shock wave generated by a powerful energy source. Successful initiation depends on both the shock strength and duration, with the minimum values of these parameters depending on the sensitivity of the combustible mixture. If the initiation source is too weak, chemical reactions can still take place. However, autoignition of the mixture may occur too late for the liberated energy to be of use in supporting the leading shock. If detonation is to occur under such circumstances, some means of shock wave amplification, leading to transition from deflagration to detonation (DDT), must come into play.

An important clue in identifying the critical conditions for the onset of detonation can be drawn from observations about initiation in the wake of a reflected shock wave from the end wall of a tube. In this scenario, the fluid particles are heated initially by the incident wave and heated further by the reflected wave. After an induction time, the particles ignite. Although the induction time is the same for all particles in the wake of the reflected wave, ignition occurs in a definite time sequence. The lamina of gas immediately adjacent to the end wall, having been processed first, will be the first to explode. The resulting weak shock wave propagates into the neighboring lamina, which, having been processed slightly later in time, will itself be on the verge of exploding. The resulting higher-strength shock wave generated by this second explosion propagates into yet a third lamina, where the process is repeated. Although it is not clear whether the shock entering a given lamina actually triggers the explosion or simply arrives there at the precise moment the explosion takes place, it is nonetheless this continuous time sequence of energy release that provides the mechanism for shock wave amplification. In order for amplification to occur, the sequence must be such that the chemical energy release at time t makes an effective contribution to the shock wave produced by the energy release at times less than t. Thus, the phenomenon is one of "shock wave amplification by coherent energy release," or SWACER.[1] This concept suggests that, given a certain amount of available chemical energy, the optimal means of generating a strong shock wave is not to release it instantaneously and uniformly over a region.

Various means of arranging the appropriate temporal and spatial energy release sequence have been examined. Zeldovich and colleagues[2] carried out a numerical study of detonation in nonuniformly preheated gas mixtures. For the case of a mild temperature gradient, the pressure rise in the test volume was uniform and substantially less than the detonation pressure. In the other extreme of a steep temperature gradient, the shock wave and reaction zone were seen to decouple, leading to a deflagration. Between these two limits there existed a range of gradients for which the onset of detonation was observed.

The SWACER concept was first proposed as the mechanism responsible for the photochemical initiation of H_2-Cl_2 mixtures.[1,3] In this study, the energy release sequence was determined by the gradient in chlorine atom concentration produced by the photodissociation of Cl_2 by a flash lamp. Owing to the absorption of light by the gas, the Cl concentration decreased in the direction of the light beam, resulting in a sequence of energy release determined by the dependence of induction time on the Cl concentration. For low flash lamp intensities (steep Cl concentration gradients) no detonation was formed, whereas for very high intensities (leading to uniform irradiation of the volume) the process approached that of constant volume combustion. Between these two extremes a range of intensities was identified for which detonation was possible.

The experimental observation that rapid turbulent mixing between combustion products and unburned explosive mixture can lead to detonation provides further support for the SWACER mechanism. In a study by Knystautas et al.,[4] such mixing within large turbulent eddies led to both a temperature gradient and a free-radical concentration gradient. For a large enough eddy and an appropriate turbulent mixing time with respect to the chemical kinetic time scales, detonation was seen to result. The same mechanism was likely operative in the recent investigations by Moen,[5] Mackay et al.,[6] and Ungut and Shuff.[7] These authors reported transition to detonation near the exit of a tube following entrainment of hot combustion products into the starting ring vortex ahead of the flame.

Experiments carried out by Lee and co-workers[8] have shown that the conditions for the onset of detonation can also be realized in the turbulent mixing region generated by opposing reactive gas jets, one containing propane and the other containing a fluorine-oxygen mixture. In these experiments the delay to ignition was observed to depend on the amount of fluorine present. The chemistry of both the F_2-C_3H_8-O_2 and F_2-C_4H_{10}-O_2 systems has been studied in detail by von Elbe.[9] The study reported by Urtiew et al.[10] was similar, except that the time to the onset of detonation was controlled by the use of an inhibitor rather than a sensitizer. Tetrafluorohydrazine and silane, which normally react in a nearly instantaneous fashion, were employed in these experiments. However, by using a cis-2-butene inhibitor the reaction was delayed to allow turbulent mixing within a volume exceed-

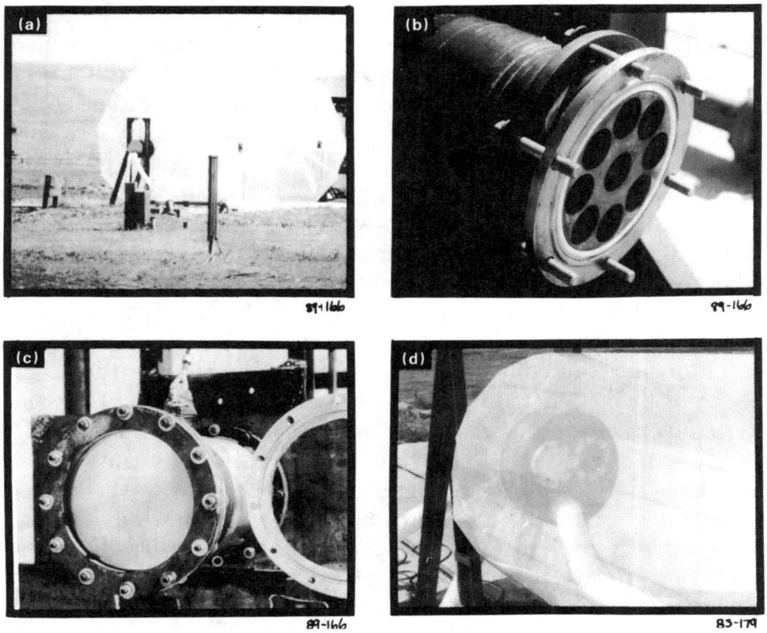

Fig. 1 Experimental apparatus: a) general configuration; b) injection chamber with turbulence grid; c) diaphragm before and after rupture; and d) end-on camera housing.

ing the critical detonable volume for the mixture. Ignition was seen to occur in a localized region of inhibitor deficiency, followed by shock wave amplification through the region of induction-time gradient.

All of these studies that have led to initiation of detonation by induced chemical sensitization have involved relatively sensitive fuel-oxidizer systems. Although attempts have been made to initiate less sensitive fuel-air mixtures (e.g., Tulis,[11] von Elbe and McHale,[12] and Sayles[13]), there is little evidence to suggest that self-sustained detonation has actually been achieved (albeit significant overpressures have been measured). This study reports on a series of field experiments in which high-pressure fluorine-air was jetted from a chamber into large plastic bags containing hydrogen-air mixtures. Small-scale turbulence within the jet structure was induced by a grid plate in the exit plane of the chamber. Successful initiation of detonation was observed for a relatively narrow range of fluorine concentration and nearly invariant mixing parameters.

Experimental Details

The experiments were performed at the DRES large-scale, fuel-air explosives testing facility, described elsewhere in detail.[14,15] As shown in Fig. 1, the experimental configuration consisted of a high-pressure injection chamber connected to a large cylindrical plastic bag. The chamber, measuring 150 mm in diameter and 300 mm in length, was capped at one end by a thin (0.43-mm thick) brass diaphragm. A fluorine-air mixture was prepared in the chamber by the method of partial pressures, with a resultant overpressure between 1.38-1.96 MPa (i.e., $14.8 \leq \Delta P/P_o \leq 21.0$). Rapid venting of this mixture was achieved by piercing the diaphragm with a four-ribbed arrowhead driven pneumatically along the internal axis of the chamber. Small-scale turbulence in the venting gases was promoted by a grid plate installed in the exit plane of the chamber. The plate contained a central circular hole of 38-mm diam surrounded by a series of eight such holes spaced azimuthally apart by 45 deg. This design provided a venting area equal to 58% of the chamber cross-sectional area.

Initial experiments were conducted in plastic bags of 0.90-m nominal diam. This was increased to approximately 2 m for many of the later tests to ensure that the boundaries were not influencing the outcome. The bag length was typically 4-6 m. In most tests, the hydrogen concentration in the bag was obtained by the method of partial volumes. This was accomplished by first measuring the volume of the bag inflated with air. Following evacuation of the bag, the required volume of hydrogen was introduced using a calibrated rotameter. The bag was subsequently topped up with air and the constituents mixed by a sparkless fan. In a few of the tests the fuel concentration was determined by infrared (IR) analysis by adding a small quantity of hydrocarbon tracer (~1% CH_4 or C_3H_8 by volume) to the hydrogen supply.

Two diagnostic techniques were employed. Pressure transducers (Piezoelectronics) were positioned in an axial array along the periphery of the bag to measure pressure histories and wave velocities. In addition, three high-speed cinematographic cameras (Photec, Fastax, and Hycam) with 1-kHz timing mark generators were employed. One camera [~5,000 frames per second (fps)] was placed in a protective housing at the end of the bag opposite the high-pressure chamber so that it was looking along the axis of the jet. A second camera (~12,000 fps) was positioned at the side of the bag looking normal to the jet axis. For many of the tests the third camera (~6,000 fps) was also situated looking normal to the axis, but it was focussed specifically on the region near the chamber exit. Occasionally, this camera (~12,000 fps) was oriented 30 deg off axis looking obliquely into the chamber exit.

Experimental Results

Selected frames from a high-speed cinematographic record (Test No. 8419), showing successful initiation of detonation in stoichiometric hydrogen-air, are shown in Fig. 2. The injection chamber was charged ($\Delta P/P_o = 21.0$) with a mixture of 25% F_2 and 75% air by volume.

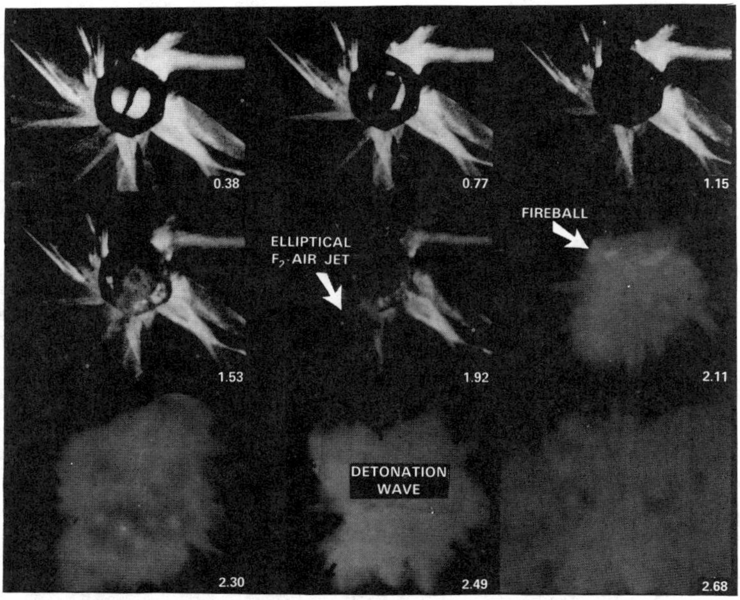

Fig. 2 Frames from end-on cinematographic record showing initiation of detonation in stoichiometric H_2-air (25% F_2 and P = 1.96 MPa in chamber). Time is shown in milliseconds.

The sequence shows the diaphragm opening in two pieces, achieving a "fully open" state at about 1 ms. The emerging F_2-air jet possesses an elliptical cross section as a direct result of the diaphragm rupturing in this manner. The diaphragm petals appear to hinge at the clamped boundary, allowing the mixture to exit the chamber in a relatively clean fashion. There is no sign of ignition until approximately 2.1 ms after initial perforation, at which time a large and intense central fireball suddenly appears. Within two frames of this occurrence (0.4 ms later), a self-sustained detonation wave is observed. Detonation kernels appear to emerge from the fireball in directions aligned with the nearly vertical diaphragm tear. This may be due to the elliptical distribution of fluorine or to more intense turbulent mixing near the ends of the tear.

The measured velocity of propagation and the maximum detonation pressure in the bag were 1963 m/s and 34 bar, respectively. The velocity was deduced from the side-on cinematographic record and represents an average between the time that detonation is first observed and its time of arrival at the end of the bag. The computed velocity is in excellent agreement with the Chapman-Jouguet (C-J) velocity of 1968 m/s for this mixture. The maximum pressure is approximately twice the C-J value. Since the maximum was measured by a ground-level transducer positioned either 0.4 m or 1.4 m down axis from the chamber exit, it is likely that the wave strikes the transducer face at some angle approaching 90 deg, resulting in a pressure close to the

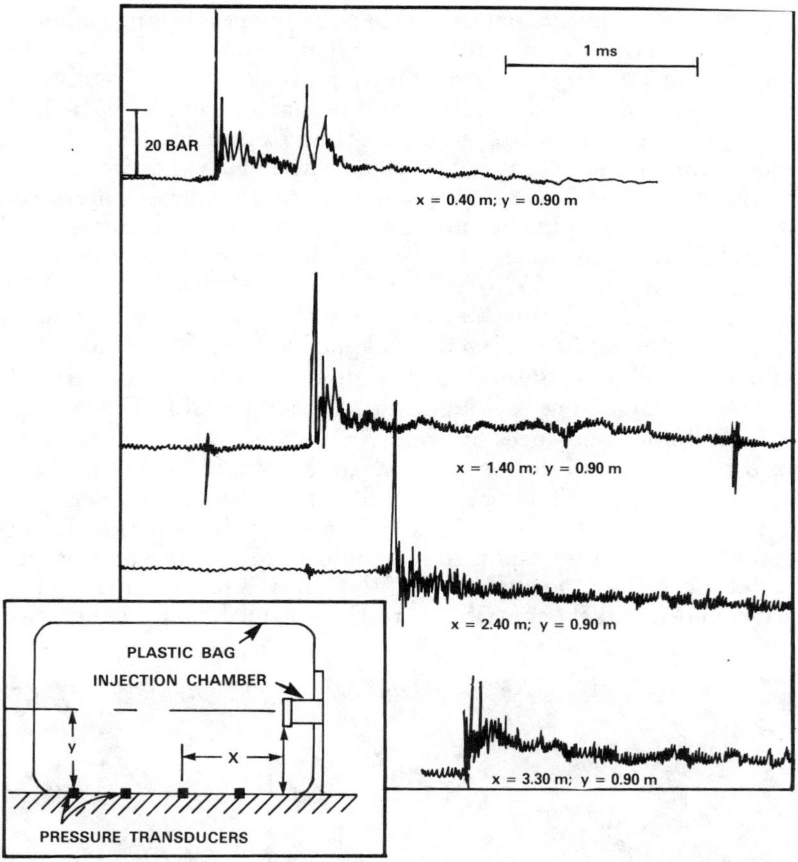

Fig. 3 Ground-level pressure histories confirming initiation of detonation in stoichiometric H_2-air (25% F_2 and P = 1.96 MPa in chamber).

reflected detonation pressure being measured. This conclusion is supported by the detailed pressure histories shown in Fig. 3. The peak pressure decreases with increasing distance from the chamber exit and approaches the C-J pressure at the far end of the bag.

The initiation phenomenon appears to be a function of both the fluorine concentration and the details of the diaphragm breakup. A cinematographic sequence (Test No. 8420) corresponding to a slightly lower initial fluorine concentration (23%) and identical initial pressure is shown in Fig. 4. In this experiment, a ring of ignition centers forms around the periphery of the chamber exit at approximately 1.6 ms after initial perforation. This could be the result of local turbulent mixing in the wake of the accelerating diaphragm petals or in the free shear layer between the emerging jet gases and quiescent surroundings. The diaphragm breakup is somewhat different in this experiment in that the tips of the diaphragm petals curl back as the petals hinge about the clamped boundary. In general, this was seen to be the case for tests in which three or more petals formed during breakup. It is possible that this leads to more extensive mixing and perhaps earlier ignition during the jetting process. However, based on close examination of numerous cinematographic records, it is not clear whether an ignition ring was truly absent in Test No. 8419 or whether such a ring simply formed between cinematographic frames. Despite this apparent difference the ensuing phenomenon in Fig. 4 is essentially the same as that shown in Fig. 2. A large central fireball first appears. Within about 0.4 ms of ignition a pair of detonation kernels emerge from the fireball. As before, these propagate in directions aligned with the diaphragm tears. The upper kernel attenuates, while a detonation wave evolves from the lower one. Oblique views of these tests are compared in Fig. 5. These show very similar sequences of events.

In order for weak ignition to escalate to detonation in these tests, some mechanism for shock wave amplification must have been operative. It is postulated that rapid entrainment of fluorine into the turbulent jet structure is responsible for ignition following an induction delay determined by the chemistry of the fluorine-fuel system. In addition, the fluorine likely contributes to the establishment of the spatial

Fig. 4 Frames from end-on cinematographic record showing initiation of detonation in stoichiometric H_2-air (23% F_2 and P = 1.96 MPa in chamber). Time is shown in milliseconds.

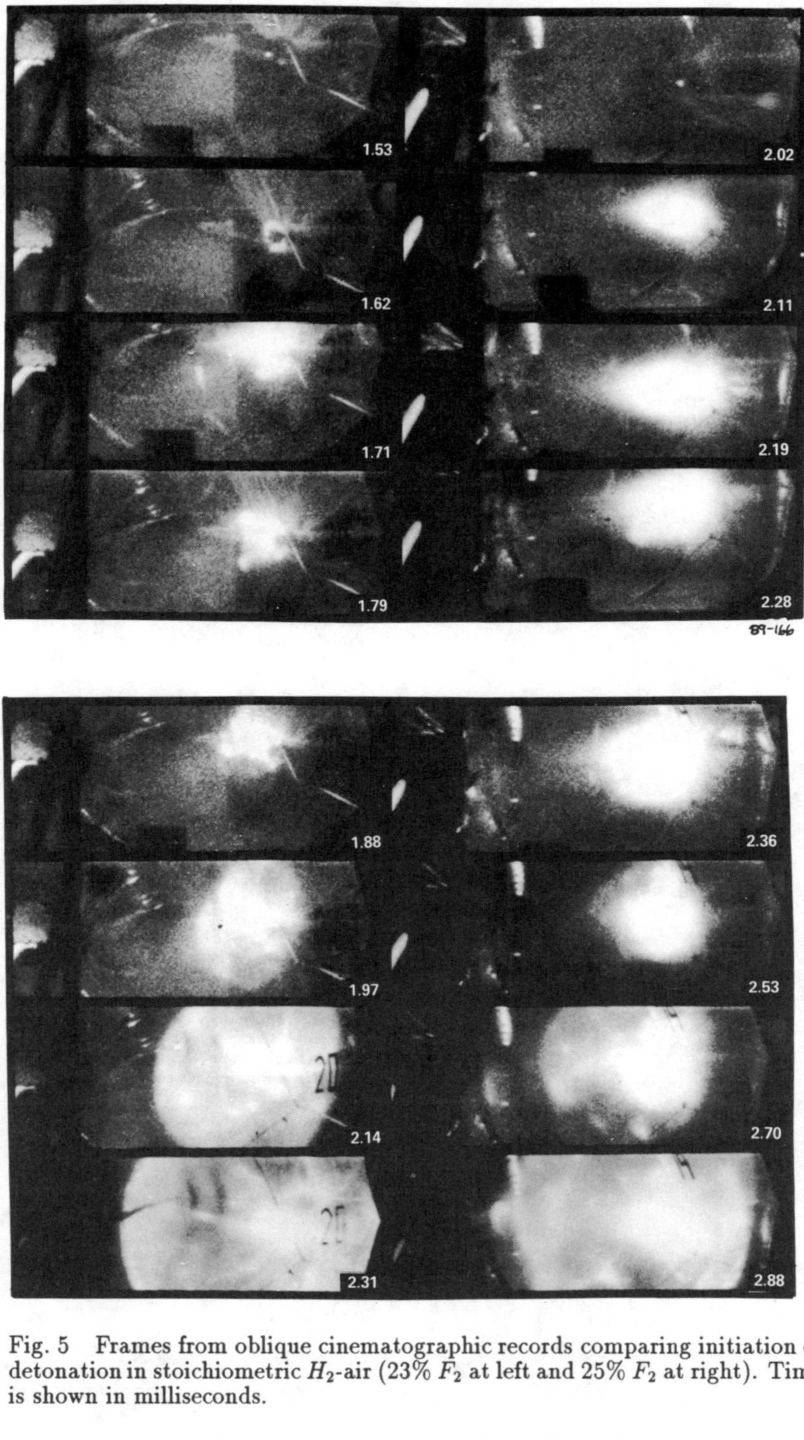

Fig. 5 Frames from oblique cinematographic records comparing initiation of detonation in stoichiometric H_2-air (23% F_2 at left and 25% F_2 at right). Time is shown in milliseconds.

Table 1 Turbulent jet initiation of hydrogen-air mixtures (tests employing turbulence grid in exit plane).

Test No.	Amb. Pres. mbar	Amb. Temp. C	Chamber F_2 %	Chamber Pres. MPa	Bag H_2 %	Bag Diam. m	Outcome
8418	920	26	20.0	1.93	30	1.88	**Deflagration:** Ring of ignition centers formed around periphery of exit. Two explosion nuclei developed adjacent to ends of diaphragm tear but attenuated quickly. Violent explosion observed in bag.
8306	920	30	20.0	1.38	30	0.92	**Detonation:** F_2-air jetted through small hole in diaphragm (~3 ms) prior to rupture. Two flame kernels developed in diametrically opposed regions well displaced from exit in radial direction. Rapid flame acceleration through precursor jet region led to onset of detonation. V_{ave} = 2032 m/s. No pressure data obtained.
8424	931	31	20.4	1.93	30	2.00	**Detonation:** Ring of ignition centers formed around periphery of exit. Two explosion nuclei developed adjacent to ends of diaphragm tear. Lower nucleus attenuated. Detonation evolved from upper nucleus. V_{ave} = 1854 m/s. P_{max} = 26.6 bar.
8420	923	31	23.0	1.93	30	1.96	**Detonation:** Ring of ignition centers formed around periphery of exit. Two explosion nuclei developed adjacent to ends of diaphragm tears. Upper nucleus attenuated. Detonation evolved from lower nucleus. V_{ave} = 2176 m/s. P_{max} = 17.2 bar.
8663	920	14	24.1	1.96	29	0.97	**Deflagration:** Several ignition centers formed around periphery of exit. Two large intense flame kernels developed near axis downstream of diaphragm. Interaction between kernels led to enhanced burning but failed to bring about onset of detonation.
8765	929	17	24.1	1.96	29	1.94	**Detonation:** No cinematographic data obtained. V_{ave} = 2216 m/s. P_{max} = 17.1 bar.
8307	923	26	25.0	1.38	43	1.80	**Detonation:** Numerous ignition centers formed across exit plane with diaphragm only half open. Pair of intense flame kernels developed in vicinity of two lower diaphragm flaps. Detonation ensued near instantaneously. V_{ave} = 2077 m/s. P_{max} = 25.3 bar.
8419	936	25	25.0	1.93	30	1.87	**Detonation:** Diaphragm opened in two halves resulting in jet of elliptical cross section. Ignition centers did not appear. Large central flame ball formed suddenly following time delay. Detonation emerged from top of flame ball in direction aligned with diaphragm tear. V_{ave} = 1963 m/s. P_{max} = 33.8 bar.
8416	921	33	25.0	1.93	30	1.88	**Detonation:** Twist in bag resulted in exit covered by layer of polyethylene prior to diaphragm rupture. Ring of ignition centers formed around 80% of exit periphery. Explosion nucleus developed adjacent to end of diaphragm tear but attenuated. Intense flame kernel appeared in region adjacent to end of second diaphragm tear. Detonation formed well away from exit (radially) in vicinity aligned with diaphragm flap. V_{ave} = 2135 m/s. P_{max} = 33.8 bar.

Table 1 continued on next page

Table 1 (continued)

Test No.	Amb. Pres. mbar	Amb. Temp. C	Chamber F_2 %	Chamber Pres. MPa	Bag H_2 %	Bag Diam. m	Outcome
8303	922	30	25.0	1.38	27[a]	0.90	**Deflagration:** Diaphragm opened in two halves yielding jet of elliptical cross section. Ignition centers did not form. Vigorous exothermic reaction occurred well downstream of exit following slight time delay. Combustion phenomena barely visible in cinematographic records. $P_{max} = 1.6$ bar.
8304	924	26	25.0	1.38	23[b]	0.90	**Deflagration:** Isolated ignition centers formed around periphery of exit. Vigorous exothermic reaction occurred although minimal evidence of flame propagation visible in cinematographic records. $P_{max} = 0.8$ bar.
8664	919	15	26.8	1.96	30	0.97	**Deflagration:** Ring of ignition centers formed around periphery of exit. Several explosion nuclei evolved and intensified through interaction but failed to result in transition to detonation. Violent explosion observed in bag.
8421	931	29	27.0	1.93	30	1.96	**Detonation:** Group of ignition centers formed along small fraction of exit periphery. Sudden formation of large central flame ball generated significant transmitted shock wave. As flame ball attenuated, shock wave reflected from hump in bag to impinge on remnant of jet. This interaction induced transition to detonation. $V_{ave} = 1922$ m/s. $P_{max} = 18.5$ bar.
8302	922	27	30.0	1.38	36[c]	0.90	**Deflagration:** Isolated ignition centers formed around exit periphery but attenuated rapidly. Mild exothermic reaction ensued. Combustion phenomena barely visible in cinematographic records. No pressure data obtained.
8417	920	21	30.0	1.93	30	1.88	**Deflagration:** Ring of ignition centers formed around exit periphery with diaphragm only half open. Annular flame kernel initially grew in size while diminishing in luminosity. Subsequently, large annular flame appeared with centers of intense reaction. Explosion nuclei did not develop. Violent burning observed in bag.
8301	925	20	35.0	1.38	38[d]	0.90	**Deflagration:** Mild exothermic reaction observed. No cinematographic or pressure data obtained.

[a] CH_4 tracer present (1% of total mixture).

[b] C_3H_8 tracer present (0.83% of total mixture).

[c] CH_4 tracer present (1.3% of total mixture).

[d] CH_4 tracer present (1.4% of total mixture).

induction-time gradient required for SWACER to occur. Specifically, it would be possible for a weak shock wave to accelerate in a direction of decreasing F_2 concentration. Other purely gasdynamic factors could also play a role. For example, the temperature field within the various shock and vortex elements constituting the jet would have a strong influence on the induction-time gradient. Provided the scale of the turbulent structure is large enough (e.g., on the order of the critical tube diameter for the surrounding fuel-air mixture) and that sufficient amplification takes place to generate a shock of C-J proportions, initiation of detonation in the surrounding hydrogen-air would occur as the shock wave breaks out of the fluorine-sensitized region.

A detailed summary of the test results is presented in Table 1. Since interpretation of the events from cinematographic records is rather subjective, an explanation of notation is provided below:

1) **Ignition center:** Small, slowly expanding, orange patches; ragged outline; typical of first visual sign of combustion; scale $\ll D_o$ where D_o is the chamber diameter.

2) **Flame kernel:** Small, more rapidly expanding, bright orange or yellow patches; ragged outline; protruding tongues; scale $\leq D_o$.

3) **Fireball:** Similar to flame kernel, but much larger and characterized by less ragged outline; relatively slow rate of growth; scale $\gg D_o$.

4) **Explosion nucleus:** Small, rapidly growing, brilliant yellow or white patches; relatively smooth outline; evidence of precursor compression waves or shock collisions; scale $\leq D_o$.

5) **Detonation kernel:** Intermediate, rapidly expanding, brilliant blue or white volume; smooth, curved outline; sometimes evidence of cellular structure; no precursor wave activity; usually evolves from explosion nucleus; scale $\geq D_o$.

6) **Detonation wave:** Similar to detonation kernel but considerably larger; cellular structure often apparent; growth rate is uniform in all directions and consistent with C-J velocity; scale $\gg D_o$.

The results in Table 1 indicate that initiation of detonation in stoichiometric hydrogen-air is possible for F_2 concentrations ranging between 20-25%. Of the two tests employing 20% F_2, one yielded detonation whereas the other resulted in deflagration. Four experiments involving 25% F_2 produced two detonations and two deflagrations. A fifth test employing 25% F_2 and slightly rich ($\phi = 1.4$) hydrogen-air also yielded detonation. The deflagrations at the upper F_2 concentration limit were observed for slightly lean ($\phi = 0.77$ and 0.91) H_2-air mixtures. Lean mixtures are characterized by a lower effective mixture sensitivity and hence a larger critical volume requirement for initiation. In addition, any increase in chemical kinetic time could lead to a mismatch between the chemical kinetic and gasdynamic time scales that yielded successful shock wave amplification in stoichiometric mixtures.

There are two anomalies in the results of Table 1. The first is the failure to initiate detonation of stoichiometric H_2-air with 24.1% F_2 in the injection chamber (Test No. 8663). Although the reason for this outcome is not clear, possible explanations include: 1) inadequate mixing of the test gas in the bag; 2) an unnoticed shift in the zero position of the gauge used to prepare the F_2-air mixture (this gauge was subjected to heavy blast loading and had to be rezeroed occasionally); and 3) slower kinetics resulting from the lower ambient temperature (14 C) at the time of this test.

The second apparent anomaly (Test No. 8421) is that a fluorine concentration of 27% resulted in successful initiation, whereas concentrations in the range between 25-26.8% did not. The reason for this

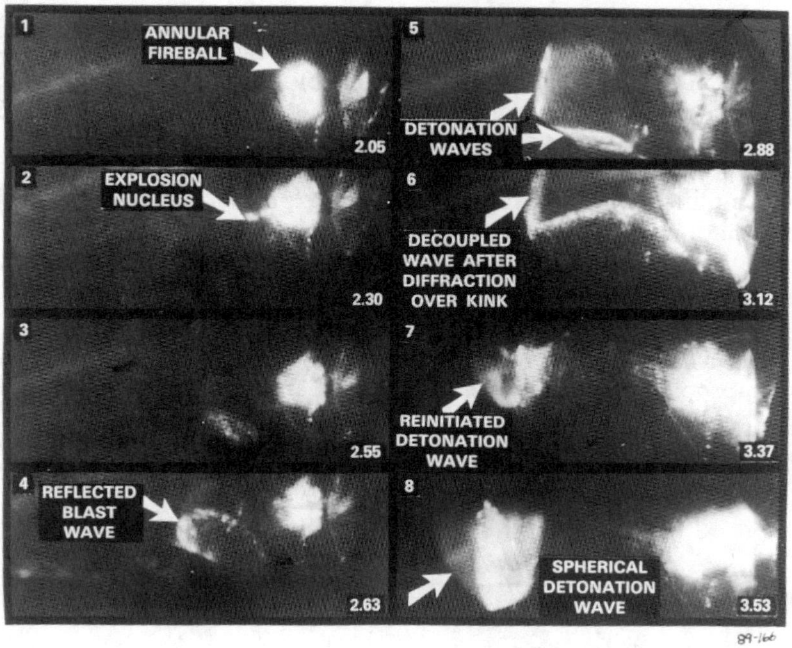

Fig. 6 Frames from side-on cinematographic record showing transition in wake of shock wave reflected from plastic bag boundary (27% F_2 and $P = 1.93$ MPa in chamber). Time is shown in milliseconds.

is clear from the cinematographic sequence shown in Fig. 6. The use of an overly long bag, in conjunction with the buoyancy of the hydrogen fuel, resulted in the bag kinking near the midpoint. Although the emerging F_2-air jet failed to initiate detonation, a strong blast wave was generated. Transition eventually took place in the wake of this wave following its reflection from the bag wall. This cinematographic sequence emphasizes how fluid/boundary interactions can lead to an overstatement of the detonability limits for unconfined mixtures. This may explain why Knystautas and Lee[16] observed initiation for F_2 concentrations as low as 18% and as high as 28% during feasibility tests in their laboratory-scale apparatus.

A graphical summary of the events leading to detonation of stoichiometric hydrogen-air is presented in Fig. 7. The elapsed time from initial perforation of the diaphragm to the first visual signs of ignition and detonation, respectively, are plotted as a function of F_2 concentration in the chamber. All results are for an initial chamber overpressure $\Delta P/P_o$ of 21. Only the data from those cinematographic records yielding unambiguous timing of events are included in the graphical summary. A consistent diaphragm opening time of just over 1 ms is apparent. It is also evident that the delay to ignition is very sensitive to the amount of fluorine in the chamber. For a concentration near 20%, ignition takes place at about the time the diaphragm achieves a "fully open" state. The delay to ignition increases with increasing F_2 concentration and reaches a maximum of about 2.1 ms for 25.5% F_2. This trend reverses for further increases in fluorine concentration. Although it is not clear why this is so, it would appear that the shock wave amplification mechanism responsible for initiation of detonation along the lower branch of the ignition curve is not present along the upper branch. Since the gasdynamics of the jet vary negligibly over this small range of F_2 concentration, failure to initiate must be a consequence of changes in chemistry alone. In view of the fact that the delay to ignition decreases for F_2 concentrations above 25.5%, it is likely that an inappropriate induction-time gradient, and not the induction time itself, is responsible for SWACER ceasing to be successful.

Figure 7 also shows that amplification and transition to detonation are quite rapid once ignition occurs. The amplification time ranges

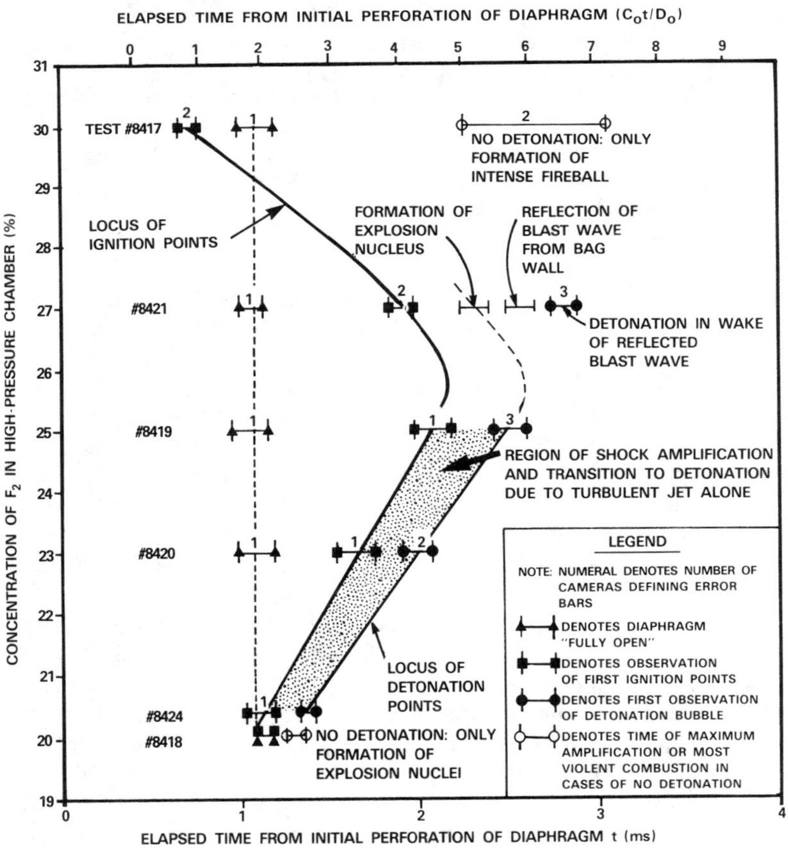

Fig. 7 Graphical summary of events leading to ignition, shock wave amplification and transition to detonation as revealed by cinematographic records.

from about 0.23 ms near the lower F_2 concentration limit to about 0.45 ms at the upper limit. This decrease in chemical kinetic rate with increasing fluorine is consistent with the observations about the ignition delay time. In the absence of detailed information about the velocity profile during amplification, it is only possible to make a crude estimate of the amplification distances. This can be done by assuming that the initial compressive disturbance propagates at sonic velocity in the hydrogen-fluorine-air mixture, and that the resultant velocity of the amplified wave is $\sim V_{C-J}$ for stoichiometric hydrogen-air. This gives a mean velocity of 0.6 V_{C-J}. In conjunction with the times above, the estimated amplification distances are 0.27 m and 0.53 m at the lower and upper F_2 concentration limits, respectively. These compare well with the characteristic transverse dimension of the detonation kernels that appear suddenly in the cinematographic sequences of Figs. 2 and 4, and they are not much larger than the critical tube diameter of 0.2 m for detonation transmission in stoichiometric H_2-air.

Some clues about the failure to initiate detonation outside the F_2 concentration limits are provided by the cinematographic sequences shown in Fig. 8. These compare the jetting process for the cases of 20% F_2 (Test No. 8418) and 30% F_2 (Test No. 8417). The sequence

Table 2 Turbulent jet initiation of hydrogen-air mixtures (tests without turbulence grid in exit plane).

Test No.	Amb. Pres. mbar	Amb. Temp. C	Chamber F_2 %	Chamber Pres. MPa	Bag H_2 %	Bag Diam. m	Outcome
8771	931	28	12.0	1.96	32	1.94	**No Ignition:** Minimal heat release over several seconds evidenced by swelling and wrinkling of bag. No indication of ignition phenomena in cinematographic records. No measurable pressure rise recorded.
8662	922	8	24.0	1.96	27	0.97	**Deflagration:** Series of isolated ignition centers formed around periphery of exit but attenuated rapidly. Combustion phenomena barely visible in cinematographic records. Mild exothermic reaction observed in bag.
8661[a]	923	9	24.0	1.96	30	0.97	**Deflagration:** Series of isolated ignition centers formed around periphery of exit but attenuated rapidly. Combustion phenomena barely visible in cinematographic records. Mild exothermic reaction observed in bag.
8766	927	18	24.0	1.96	32	1.94	**Deflagration:** Diaphragm opened in two halves. Ignition centers first formed in vicinity adjacent to end of diaphragm slit. Subsequently, two arcs of ignition centers developed along periphery of diaphragm opening. These initially grew in size and luminosity but then attenuated. Mild exothermic reaction observed in bag. P_{max} = 1.9 bar.
8764	922	20	24.0	1.96	33	1.94	**Deflagration:** Diaphragm opened in two halves. Series of ignition centers in vicinity adjacent to end of diaphragm slit gave rise to intense fire ball. Simultaneously, two arcs of ignition centers developed along periphery of diaphragm opening. Violent burning observed in bag. P_{max} = 3.7 bar.

[a]An orifice plate having a hole diameter of 102 mm was installed in the exit plane.

corresponding to 20% fluorine shows explosion nuclei developing at either end of the diaphragm tear. However, these attenuate in luminosity after growing to a characteristic transverse dimension of only ~0.15 m. In addition, a dim luminous front appears to propagate outward from the explosion nuclei once they begin to attenuate. This is likely evidence of a mild reaction behind the decaying transmitted shock wave. These observations suggest that the critical volume requirement for initiation of H_2-air may not have been satisfied at the moment a sufficiently amplified shock wave broke out of the jet structure.

The sequence corresponding to 30% fluorine shows early ignition and provides no indication of explosion nuclei or transmitted shock waves, but rather it shows a turbulent fireball that appears suddenly and then expands at a relatively slow rate in comparison with detonation-like time scales. It is possible that the high fluorine concentration resulted in the absence of an induction-time gradient and hence a volumetric explosion in a nearly uniform mixture.

In order to elucidate the importance of small-scale turbulence in the jet-initiation phenomenon, a series of tests was conducted in which the grid plate was removed from the exit plane of the chamber. The test results, summarized in Table 2, show that initiation was not possible without the plate present. This observation emphasizes that small-scale turbulence is essential for the mixing processes leading to a high rate of enery release and hence the conditions for shock wave amplification. Since the phenomena of interest occur quickly in comparison with the characteristic venting time of the chamber, it cannot be argued that removal of the grid plate altered the gasdynamic time scale sufficiently to cause a mismatch between the essential chemical kinetic and gasdynamic processes. Thus, the failure to initiate must be due to the absence of small-scale turbulence alone. This conclusion is substantiated by the outcome of Test No. 8661, in which an orifice plate having a single 102-mm-diam hole was installed in the chamber exit in place of the turbulence grid. This hole diameter gives an open area (46% of the total) only modestly smaller than does the grid plate. Thus, the two configurations have similar venting rates but very different turbulence characteristics. Hence, failure to initiate is most probably due to the absence of small-scale turbulence. Such turbulence is necessary for the rapid mixing between reactive chemical species. In the absence of such turbulence, chemical reactions could only occur at the interface between large pockets of fuel-air and F_2-air during the entrainment processes.

In an attempt to shed more light on the nature of the phenomenon, a third series of tests was carried out using three variations of the original grid plate. The plate was modified by blocking different combinations of four holes so as to reduce the open area to 26% of the cross-sectional area, as shown in Table 3. One configuration was obtained by blocking alternate circular holes (Geometry A). Another variation involved blocking alternate pairs of holes (Geometry B), whereas the third geometry consisted of four neighboring holes arranged in a semicircle (Geometry C). These geometries were intended to yield information about the critical size of jet structure required for this type of initiation. As shown in the table, initiation was not achieved under any circumstances. Since multiple jet interactions (and hence small-scale turbulence) were present, as in the experiments with the original grid plate, the data likely suggest that the minimum volume requirement for the onset of detonation was not being satisfied.

Numerical Results

The results presented to this point have focussed on the combustion phenomena without regard to the detailed jet structure. This has been out of necessity in that cinematography was ineffective at yielding useful information about the jet. In order to further understanding of the phenomenon, two-dimensional, inviscid, explicit finite-difference

Table 3 Turbulent jet initiation of hydrogen-air mixtures (tests employing modified turbulence grid in exit plane).

Test No.	Amb. Pres. mbar	Amb. Temp. C	Chamber F_2 %	Chamber Pres. MPa	Chamber Grid Type[a]	Bag H_2 %	Bag Diam. m	Outcome
8767	931	17	24.0	1.96	A	31	1.94	**Deflagration:** Diaphragm opened in two halves. Two small ignition centers formed adjacent to ends of diagonal diaphragm slit but these were short lived. Mild exothermic reaction observed in bag. P_{max} = 4.1 bar.
8768	930	18	24.0	1.96	B	32	1.94	**Deflagration:** Diaphragm opened in two halves. Two arcs of ignition centers formed along periphery of diaphragm opening. These initially grew in size and luminosity but then attenuated. Mild exothermic reaction observed in bag. P_{max} = 0.8 bar.
8769	933	21	24.0	1.96	C	30	1.94	**Deflagration:** Diaphragm opened in three pieces. Explosion nucleus formed in vicinity of slit between two diaphragm quarters. Transmitted shock wave accelerated around periphery to opposite side of diaphragm opening but failed rather suddenly at that location. Mild exothermic reaction observed in bag. P_{max} = 1.8 bar.

[a] Grid geometries shown below; ● — denotes blocked grid plate hole.

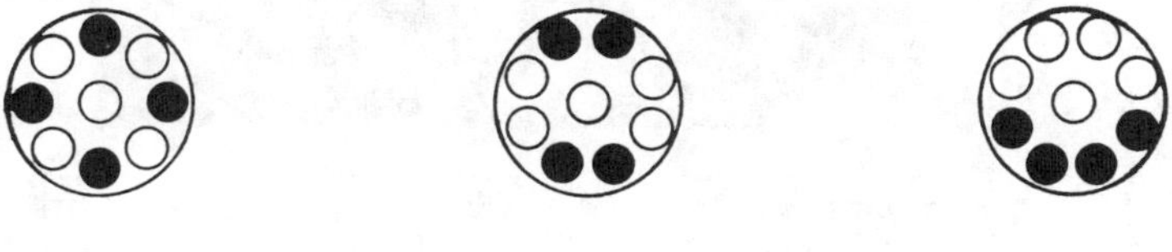

calculations were performed using the flux corrected transport (FCT) algorithm developed by Boris and Book.[17] Owing to the inviscid nature of the calculations, it was not our aim to quantitatively predict the turbulent and entrainment properties of the transient compressible jet, but it was rather to shed light on the inviscid gasdynamic features of these jets and on the breakup of vortices that play a central role in the overall mixing processes.

Two classes of problems were considered in the numerical investigation. The first class is the "starting jet," which is generated by impulsively applying constant choked conditions at the chamber exit. Such calculations should adequately simulate the early-time development of a jet generated by the sudden discharge from a large pressurized vessel. The second class is the "fully transient jet," which involves the

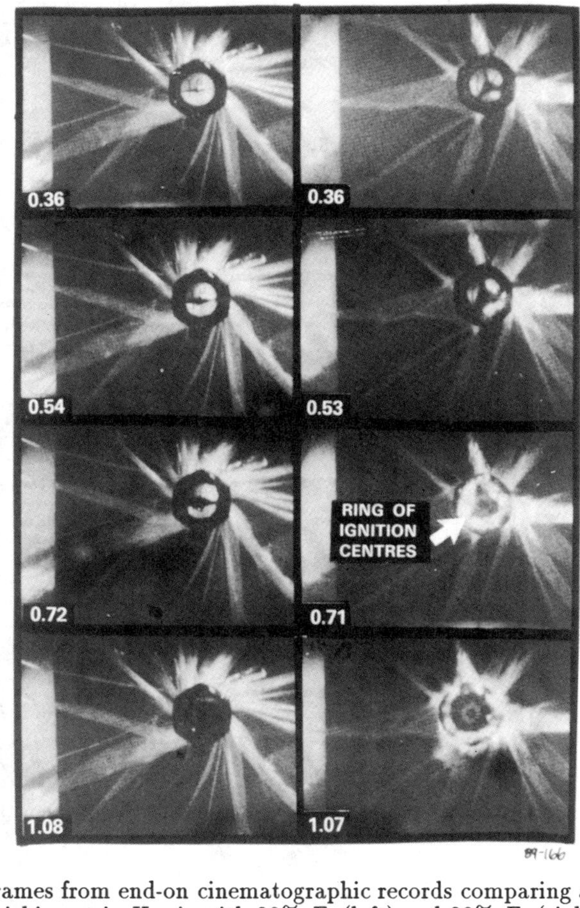

Fig. 8 Frames from end-on cinematographic records comparing attempts to initiate stoichiometric H_2-air with 20% F_2 (left) and 30% F_2 (right). Time is shown in milliseconds.

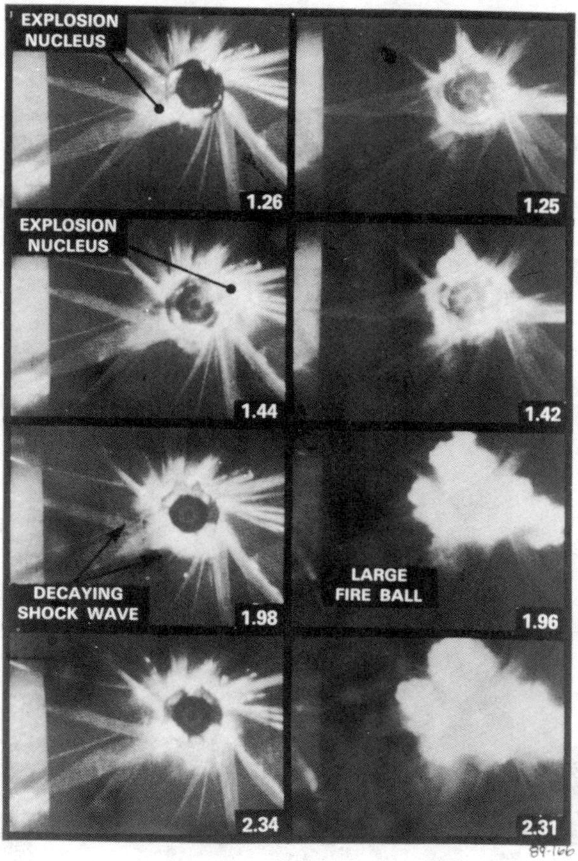

time-dependent venting of a pressurized vessel into the atmosphere. Calculations of the latter type describe the transient processes occurring inside and outside a vessel of finite size.

The calculations were performed using a fine grid with a mesh size of $D_o/30$. This grid, used to capture the jet, was itself surrounded by an expanding coarse grid that increased in size by a factor of 1.3 at each successive grid point. This was done to accommodate the leading shock without substantially increasing the total number of grid points required. In the fully transient case, the vessel dimensions were chosen to match those of the experimental apparatus; i.e., a cylinder of diameter D_o and length $2D_o$ that is vented at one end through an orifice plate of thickness $D_o/5$. Since the two-dimensional code cannot model the three-dimensional flow emerging from the series of circular holes in the grid plate, a single orifice hole of diameter $3D_o/4$ was used. This diameter provides the same open area as the actual grid plate.

Although it is recognized that this particular geometry (i.e., a single orifice hole) did not lead to transition in the experiments, it is nonetheless concluded that numerical calculations employing this geometry are largely valid for the purposes of describing the macroscopic features of the three-dimensional jet. Complete details of the calculations are provided by Thibault and Otvos.[18]

Figure 9 displays a sequence of pressure profiles for a starting jet corresponding to a chamber pressure ratio P/P_o of 20. Both injected and quiescent fluids are air. Results are shown for five values of nondimensional time $t^* = c_o t/D_o$. At the early time of $t^* = 2.28$, the nearly spherical shock wave generated by the jet is visible. The initial isentropic expansion of the jetting gases is followed by a well-defined Mach

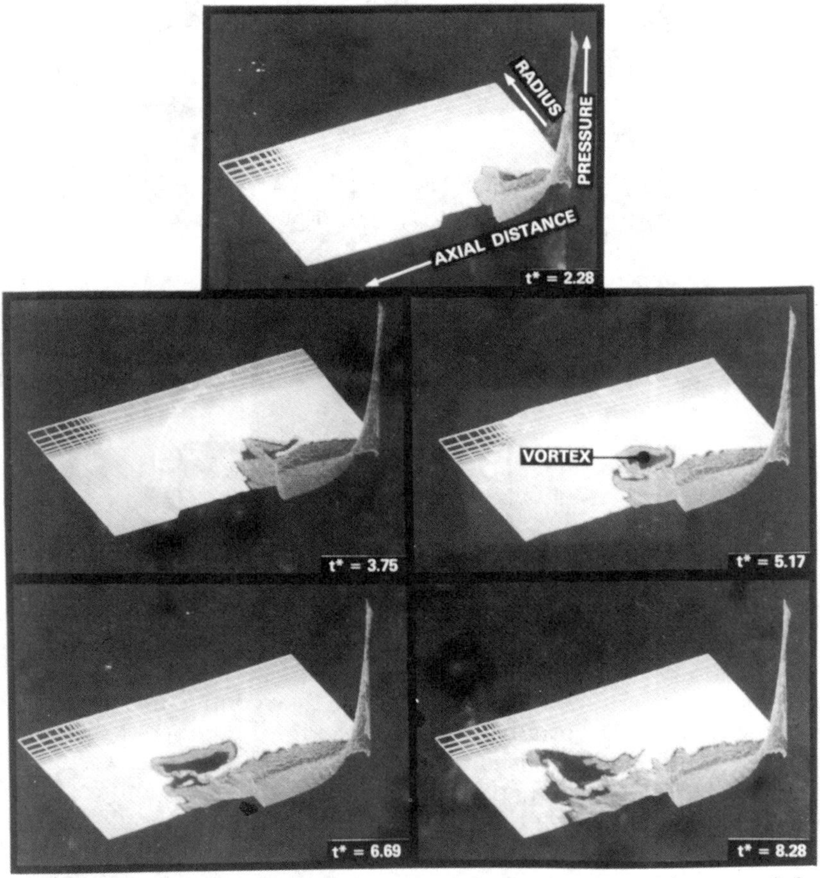

Fig. 9 Three-dimensional hidden-line pressure plots showing features of the "starting jet" for air issuing into air (P/P_o = 20). The grid resolution is 5 mm.

"disc" located approximately $3D_o$ downstream of the chamber exit and having a diameter of about $2D_o$. During the formation of the Mach disc, a vortex is formed within the triple shock structure between the recompression shock and the reflected wave. This vortex entrains a pocket of vented gas, which, at time $t^* = 5.17$, is very regular and almost circular in shape. The vortex resides in a relatively complex shock structure, part of which is associated with the reflection process responsible for the Mach disc.

A close-up view reveals the roll-up of cold and dense fluid into the vortex. Velocity vector plots show that most of this entrained fluid has been shocked by the recompression wave rather than by the Mach disc. The overpressure and temperature ratio in the vortex core are approximately -0.85 and 0.4, respectively, the latter of which corresponds to a temperature of about 120 K for an initial temperature of 300 K. This low temperature results from the combined effects of the initial expansion, the subsequent heating due to recompression, and, finally, the centrifugal expansion in the vortex core. Such a very low temperature is, of course, exaggerated due to the fact that the calculations are inviscid and do not take into account the turbulent mixing that would increase the temperature in the vortex core.

The region downstream of the Mach disc is also interesting due to the existence of two branches of vessel gas that emanate from the main front (clearly visible at $t^* = 5.17$). These protrusions may be the result of either multiple slip streams generated by the complicated shock reflection process or Rayleigh-Taylor instabilities occurring due to the positive pressure gradient interacting with the negative density gradient across the vessel/quiescent gas interface. The final profile in the figure ($t^* = 8.28$) shows that the pocket of vessel gas trapped in the vortex eventually loses its circular shape, taking on a more irregular outline.

The corresponding venting vessel calculations, which were carried out to extended nondimensional times, revealed early-time characteristics that were very similar to those of the starting jet. However, at later times, the original clockwise rotating vortex was seen to break up into a series of counterclockwise rotating secondary vortices. These shed from the main vortex at approximate times $t^* = 14$, 16, and 18, and were sequentially deposited along the core of the jet at later times.

In order to provide at least minimal validation of these calculations, a series of tests was conducted in which air in the high-pressure chamber was seeded with a small amount of fine white carbon tracer just prior to diaphragm rupture. A calibrated black and white checkered backdrop and front scattered illumination were used to enhance visualization of the jet/quiescent gas contact surface by high-speed cinematography. In general, the observed jet development was seen to be in good agreement with the numerical results, even revealing some of the finer details such as the shedding of secondary vortices. Owing to the finite opening time of the diaphragm, the head of the jet was seen to be somewhat more pointed than predicted by numerical calculations. In addition, the end-on camera revealed a jet cross section similar in shape to a four-leaf clover. This was a direct result of the diaphragm opening in four petals and was a characteristic that persisted, albeit to diminishing extents, over the interval of observation (some 20 chamber diameters).

Figure 10 shows a comparison between the numerical predictions and the experimentally observed contact surface trajectory after being corrected for the effect of camera viewing angle on the offset between the jet axis and calibrated backdrop. Despite the finite diaphragm opening time, agreement between the data and both starting jet and venting vessel calculations is remarkably good. Worsening agreement with the venting vessel results at later times could be due to the omission of energy dissipation mechanisms (turbulence) in the model and perhaps to the single-hole orifice approximation for the turbulence grid plate.

Although shock waves were not apparent in the cinematographic records from these tests, some data could be extracted from the F_2-air injection experiments that did not result in detonation initiation. In a number of those tests the decaying transmitted shock wave induced a travelling deflection point at the plastic bag boundary. Based on this observation, the axial position of the shock wave could be inferred by

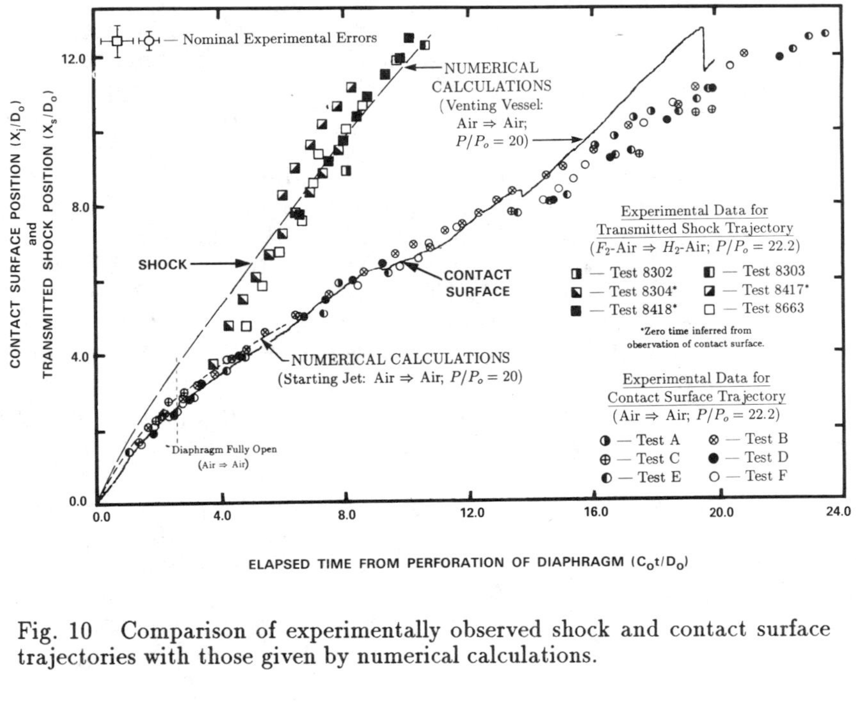

Fig. 10 Comparison of experimentally observed shock and contact surface trajectories with those given by numerical calculations.

assuming that it was spherical in shape. Figure 10 shows reasonable overall agreement between predicted and measured shock trajectories. Poor agreement at early times is likely the result of a smaller effective orifice diameter prior to the diaphragm achieving a fully open state.

A comparison between the chemical kinetic events (Fig. 7) and the gasdynamic development of the jet (Fig. 9) at similar nondimensonal times strongly suggests that the onset of detonation is tied to vortex development. The range of nondimensional times over which ignition and transition to detonation occur is approximately $3.1 \leq t^* \leq 6.7$. This is precisely the range over which a well-defined, sizeable, approximately toroidal vortex is observed. Outside this range, the vortex is either very small or large but fragmented. The numerical calculations show the presence of negative gradients in temperature and F_2 concentration as the vortex is traversed from its outer periphery towards its center. The combination of decreasing temperature and fluorine concentration implies the existence of increasingly longer induction times as the vortex center is approached. Thus, one possible explanation for the initiation phenomenon is that ignition takes place at the periphery of the vortex, generating a compression wave that amplifies as it propagates inward. If ignition takes place more or less simultaneously around the vortex periphery, the result would likely be an accelerating, toroidally-converging shock wave. Such a wave would generate very high pressures and temperatures inside the core and, additionally, a strong outwardly-propagating reflected shock wave subsequent to the incident wave focussing on the circular axis of the torus.

The high-speed cinematographic records provide some support for this hypothesis in that explosion nuclei were seen, almost without exception, to form off axis in a location corresponding approximately to that of the vortex core. In some tests, one or several isolated nuclei were observed while, in other experiments, a circular ring of nuclei were apparent. However, despite these observations, a toroidally-expanding detonation wave was never observed. Most often the detonation wave was spherical in shape and emerged from the head of the jet. Although somewhat ambiguous, the origin of such spherical waves appeared to correspond roughly to the point where the jet axis intersected a transverse plane through the torus. Thus, one might speculate that the exploding toroidal volume is responsible for a shock wave that converges on the jet axis. The subsequent reflection of this wave could generate a Mach disc that would propagate down axis through the gases in the jet head, eventually emerging as a spherical detonation wave. This postulated sequence of events is presented in schematic form in Fig. 11. The numerical calculations show the existence of negative temperature and F_2 concentration gradients near the head of the jet. It is possible that the SWACER mechanism might play a role in amplifying such a Mach disc.

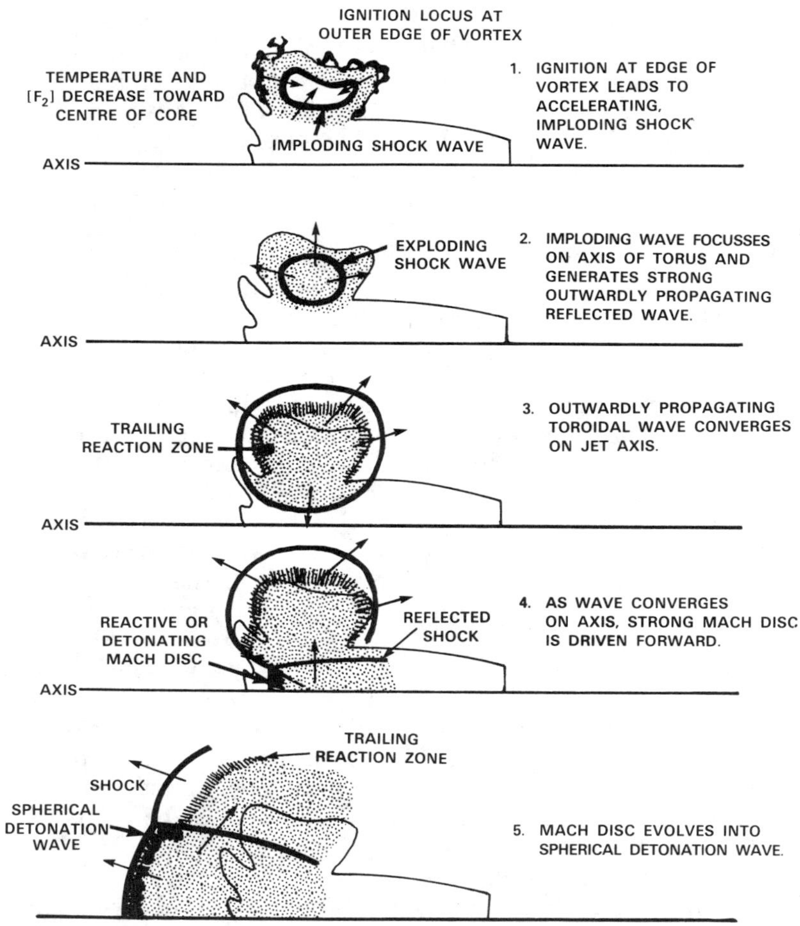

Fig. 11 Hypothesized sequence of events leading to chemical initiation of H_2-air detonation by a turbulent F_2-air jet.

Conclusions and Recommendations

Successful initiation of detonation in hydrogen-air mixtures near stoichiometric conditions has been achieved by a turbulent fluorine-air jet. High-speed cinematography and the results of numerical calculations to describe the turbulent jetting process suggest that transition to detonation could be the result of shock wave amplification inside a toroidal vortex generated by the jetting gases. Amplification would appear to be possible over a small time interval during which substantial gradients in both temperature and F_2 concentration extend over a sizeable volume. Photographic evidence suggests that the resulting explosion in the torus does not lead to detonation directly, but it instead generates a shock wave that converges on the jet axis, giving rise to a Mach disc that evolves into a spherical detonation wave.

The combination of a turbulence grid and a rupturing diaphragm that has a significant opening time with respect to the chemical/gas-dynamic phenomena of interest makes it difficult to quantify many of the important features of the process. Future investigations should either focus on "cleaner" geometries or employ slower fuel-fluorine kinetic systems so that combustion takes place well down axis from the chamber exit. Since the characteristics of the jet are very much a function of the ratio of molecular weights of the gases inside and outside the chamber, a combination of experiments and numerical calculations employing light and heavy chamber diluent gases (e.g., helium and sulphur hexafluoride) would likely provide further insight into the phenomenon.

Finally, it is recommended that one-dimensional numerical calculations be performed with full H_2–F_2 kinetics incorporating temperature and F_2 concentration gradients similar to those existing in the turbulent jet structure. Kinetic schemes for such calculations exist (e.g., Kapralova et al.,[19] and Sullivan et al.[20]), but they might not be applicable to the range of parameters in the present investigation.

Acknowledgments

The authors gratefully acknowledge the field assistance and valuable dialogue with Michael Gaug, Robert Dewar, Keith Gerrard, Gayle Hall, and Stephen Ward. The numerical modelling contributions by Tom Otvos and Mohammed Hassam are greatly appreciated. Thanks is also extended to members of the DRES Photo Group and Electronic Design and Instrumentation Group, with special thanks to Scott Trebble and Chuck Sutherland.

References

[1]Lee, J.H., Knystautas, R., and Yoshikawa, N., "Photochemical Initiation of Gaseous Detonations," Acta Astronautica, Vol. 5, No. 11/12, 1978, pp. 971–982.

[2]Zeldovich, Ya. B., Librovich, V.B., Makhviladze, G.M., and Sivashinsky, G.I., "On the Development of Detonation in Non-uniformly Preheated Gas," Astronautica Acta, Vol. 15, 1970, p. 313.

[3]Yoshikawa, N., "Coherent Shock Wave Amplification in Photo-chemical Initiation of Detonations," Ph.D. Dissertation, Dept. of Mechanical Engineering, McGill Univ., Montreal, Canada, 1980.

[4]Knystautas, R., Lee, J.H., Moen, I.O., and Wagner, H. Gg., "Direct Initiation of Spherical Detonation by a Hot Turbulent Gas Jet," Proceedings of the 17th Symposium (International) on Combustion, The Combustion Institute, Pittsburgh, PA, 1979, p. 1235.

[5]Moen, I.O., Bjerketvedt, D., Engebretson, T., Jenssen, A., Hjertager, B.H., and Bakke, J.R., "Transition to Detonation in a Flame Jet," Combustion and Flame, Vol. 75, No. 3-4, 1989, pp. 297–308.

[6]Mackay, D.J., Murray, S.B., Moen, I.O., and Thibault, P., "Flame-jet Ignition of Large Fuel-air Clouds," Proceedings of the 22nd Symposium (International) on Combustion, The Combustion Institute, Pittsburgh, PA, 1988, pp. 1339–1353.

[7]Ungut, A. and Shuff, P.J., "Deflagration to Detonation Transition from a Venting Pipe," Submitted for publication to Combustion Science and Technology, 1989.

[8]Lee, J.H., and Moen, I.O., "Fundamental Mechanisms of Unconfined Detonation," Abstracts from the 1978 AFOSR Contractors Meeting on Unconfined Detonations and Other Explosion Related Research, Atlantic Research Corp. Rept. AFOSR-TR-78-1426, 1979.

[9]von Elbe, G., "A Study of the Effect of Free Radical Generation by Fluorine and Fluorine Producers on Hydrocarbon Combustion," Atlantic Research Corp. Rept., Sept. 30, 1974.

[10]Urtiew, P.A., Lee, E.L., and Walker, F.E., "Chemical Initiation of Gaseous Detonation in a Small Spherical Volume," Lawrence Livermore Lab. Rept. UCRL-79271, June 15, 1977.

[11]Tulis, A.J., "Induced Heterogeneous Detonation in Hypergolic Fuel-oxidizer Dispersions," Eastern Section of the Combustion Institute Fall Technical Meeting, Miami Beach, FL, Nov. 1978.

[12]von Elbe, G., and McHale, E.T., "Chemical Initiation of FAE Clouds," Abstracts from the 1978 AFOSR Contractors Meeting on Unconfined Detonations and Other Explosion Related Research, Atlantic Research Corp. Rept. AFOSR-TR-78-1426, 1979.

[13]Sayles, D.C., "Method of Generating Single-event Unconfined Fuel-Air Detonation," U.S. Patent 4,463,680, Aug. 7, 1984.

[14]Funk, J.W., and Murray, S.B., "The DRES Large-scale Fuel-air Explosives Testing Facility," Defence Research Establishment Suffield Memorandum SM No. 1051, Ralston, Canada, 1982.

[15]Funk, J.W., Murray, S.B., Moen, I.O., and Ward, S.A., "A Brief Description of the DRES Fuel-air Explosives Testing Facility and Current Research Program," Proceedings of the International Conference on Fuel-Air Explosions, SM Study No. 16, Univ. of Waterloo Press, Waterloo, Canada, 1982, p. 565.

[16]Knystautas, R., and Lee, J.H., "Catalytic Initiation and Transmission of Detonation, Flame Acceleration and Transition to Detonation in Fuel-air Explosive Clouds (FAE III)," Defence Research Establishment Suffield Contractor Rept. 4/85 by McGill Univ., Montreal, Canada, 1984.

[17]Boris, J.P., and Book, D.L., "Flux Corrected Transport III: Minimal-error FCT Algorithms," Journal of Computational Physics, Vol. 20, 1976.

[18]Thibault, P.A., and Otvos, T., "Numerical Simulations of Transient Compressible Jets," Defence Research Establishment Suffield Contractor Rept. 37/86 by Combustion Dynamics Ltd., Medicine Hat, Canada, 1986.

[19]Kapralova, G.A., Margolina, E.M., and Chaikin, A.M., "The Effect of Fluorine, Hydrogen, or Inert Gases on the Upper Self-ignition Limit in the Reaction of Fluorine with Hydrogen," Kinetics and Catalysis, Vol. 10, No. 1, 1969, pp. 23-27.

[20]Sullivan, J.H., Feber, R.C., and Starner, J.W., "Mechanisms and Explosive Types of Behavior in Hydrogen-fluorine Systems," Journal of Chemical Physics, Vol. 62, No. 5, 1975, pp. 1714–1725.

Detonation of Unconfined and Semiconfined Charges of Gaseous Mixtures

A. A. Borisov and S. V. Khomik
USSR Academy of Sciences, Moscow, USSR
and
V. N. Mikhalkin
Moscow Institute of Physical Engineering, Moscow, USSR

Abstract

Critical diameters of detonation in cylindrical unconfined charges were measured for a number of fuel-oxygen and acetylene-air mixtures. It is found that when the charge diameter approaches the critical value the detonation velocity drops appreciably while the detonation wave propagates at the charge periphery in a galloping manner with alternating regions of locally overdriven detonations and those with completely disappeared cellular structure. An empirical relationship between the critical diameter of an unconfined cylindrical charge and the critical diameter for detonation transition from a tube to a large volume is derived which shows that the ratio of these critical diameters decreases and tends to a constant value (about 1.5) as the mixture reactivity lowers. This allowed realistic estimates of the critical diameters for unconfined cylindrical fuel-air charges to be made. The relationship accounts for the mixing process at the charge-air interface. The effect of slight confinement on propagation of detonation waves in cylindrical charges is investigated. A submerged jet of a detonable mixture issuing from a tube into surrounding air is shown to be able to support detonation only over a distance of no more than four tube diameters irrespective of mixture reactivity. Critical layer thicknesses for detonation propagation in acetylene- and hydrdogen-air mixtures over a smooth rigid wall are also measured.

Introduction

Investigations of detonation propagation in unconfined charges of various shapes are of practical importance since in real accidents burning and detonating fuel-oxidizer charges are not necessarily of a spherical or semispherical geometry. More often they have an irregular shape including slab- or cylinder-type parts. Ability of such irregular charges to support detonation depends primarily on the possibility of detonation propagation through "bottle-necks", that is through those portions of a cloud that are the narrowest at least in one direction. Unfortunately information on critical dimensions of cylindrical or planar unconfined charges is practically lacking in the literature.

The previous studies,[1–6] among which the most comprehensive ones performed by Vasiliev[3–5] and coworkers should be especially mentioned, were carried out using very reactive fuel-oxygen mixtures and small laboratory scale setups. The most interesting cases of large fuel-air charges were not investigated at all.

There is one more point that needs further elaboration. This is a contribution of the mixing process at the charge boundary to the limiting conditions of detonation propagation. Though the mixing effect was generally believed to be not very important, a definite reduction in the critical diameter of an unconfined acetylene-oxygen charge was observed by Vasiliev and Zak[5] when the charge formation procedure was changed from that with quick removal of a thin rubber tube to the one with parallel mixture and air flows (the ratio $d_{cr.c.}/\lambda$ changed from 60 to 42, with $d_{cr.c.}$ being the critical diameter and λ the detonation cell size) and special precautions were taken to minimize turbulization of the boundary layer at the mixture-inert gas interface.

The only numerical modeling work available in the literature[7] revealed that a very complicated flow pattern with various types of pulsations of a noncellular nature should be observed in detonation waves propagating through unconfined layers or cylinders.

The present investigation was undertaken to obtain more reliable and complete experimental evidence of stationary detonation in unconfined shaped charges, to find the critical conditions for its propagation, and to elucidate the propagation mechanism and structure of detonation waves. The main attention was paid to mixtures of low enough reactivity (diluted fuel-oxygen or fuel-air mixtures). Since in many practical conditions detonating charges are not completely unconfined we also investigated the effect of partial confinement on the detonation wave.

Experimental

Unconfined cylindrical charges were obtained using the technique described elsewhere.[6] A metal vertical detonation tube (Fig. 1) consisting of a plane-detonation wave generator (Sec. 1) and test Sec. (2) was purged with a mixture under investigation (about 4–5 tube volumes) then a wire that tied the test section and the detonation generator tube was fused electrically and the test tube was let to fall down. At the instant when the test section touched the floor it closed the initiator circuit that fired either a high explosive charge placed in the detonation generator or a pyrotechnical igniter placed in a narrow, long tube filled either with a mixture being tested

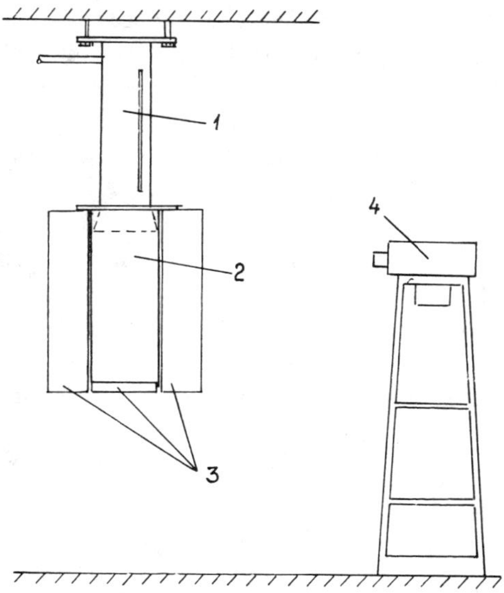

Fig. 1 Schematic of the setup for investigating detonation propagation in unconfined cylindrical charges: (1) plane detonation generator, (2) test section with a falling tube, (3) smoked foils, and (4) streak camera.

or a more reactive one and connected to the detonation wave generator. In both cases onset of detonation in the generator was checked using streak camera photographs or smoked foil records. The detonation generator tube was 70 mm in diameter and ended up with a conical section with a small divergence angle in order to smoothly expand the detonation wave front to the test-section diameter. In runs with low-reactivity mixtures a part of the detonation wave generator was filled with an oxygen-hydrogen or ethylene-oxygen mixture. The total length of the generator was varied between 800 and 900 mm.

The diameter and length of the gas column formed after the test section reached the floor varied in the range 37–200 mm and 500–900 mm, respectively. The free-charge life time before firing did not exceed 0.5 s. The wall thickness of the falling tube section ranged between 0.5 and 6.5 mm.

Methane, ethane, propane, ethylene, acetylene, and hydrogen were used as fuels and oxygen as an oxidizer, hydrogen-air and acetylene-air mixtures were also tested. Mixture composition was checked chromatographically.

In some experiments mixture was continuously fed into the tube during free fall of the test section (the flow velocity in the tube was 0.5–2m/s). No difference in the behavior of detonation waves in flowing and stagnant mixtures has been found which is an evidence of the absence of significant distortions of the cylindrical gas column. This was also confirmed by soot prints obtained with charges whose diameter was much larger than $d_{cr.c.}$

(they demonstrated that detonation cells reached the interface very closely), by consistency of the data on critical distances for gaseous slabs and cylinders whose preparation was completely different, and by runs in which a fixed piston was placed at the bottom end of the falling tube section that did not allow the gas to move with the tube.

A smoked plate (with a width equal to or less than the diameter of the mixture column and 1 mm thick) was placed along the charge axis. In some runs similar plates were also installed outside the falling tube section.

To imitate partial confinement we hung vertical metal strips or a wire spring at the charge periphery. Thin plastic film was also used as partial confinement in some experiments.

Critical planar-layer thickness was measured in a duct about 2 m long and of a maximal width of 320 mm formed by two vertical metal plates. The third wall of the duct was of a variable width and was placed at different distances from the edge of the plates, the fourth wall made of thin plastic film was removed quickly before firing the initiator. A mixture under investigation was purged through the duct and then detonation was initiated in it by a detonating cord occupying the whole duct width.

Results and Discussion

A typical streak camera record of a detonation wave propagating in a cylindrical unconfined charge is shown in Fig. 2. Fluctuations of the deto-

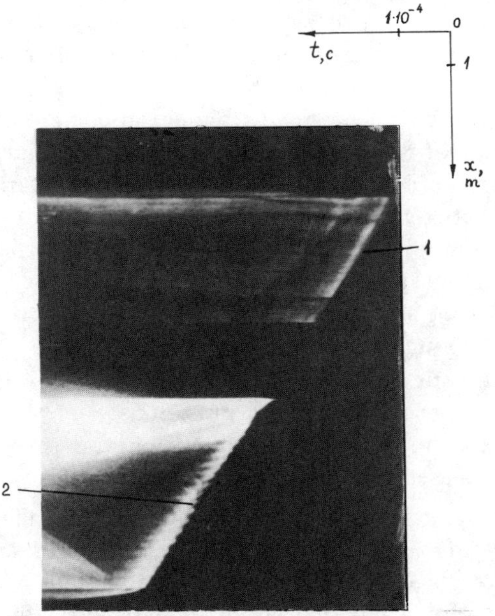

Fig. 2 Representative streak camera record of detonation in an unconfined charge: (1) detonation in the tube, (2) pulsating detonation in the unconfined charge. Stoichiometric ethylene-oxygen mixture, the charge diameter 60 mm.

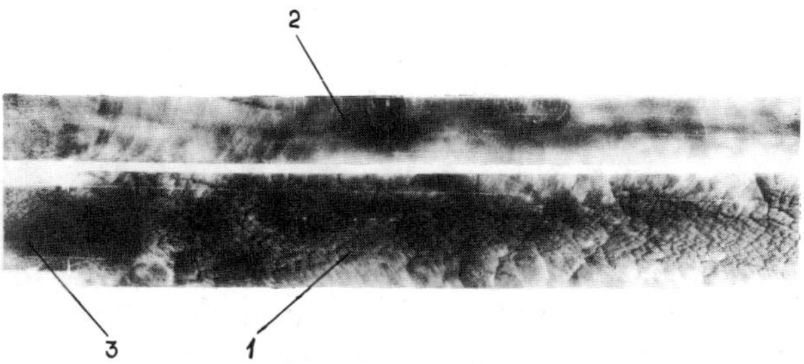

Fig. 3 Soot prints of detonation propagating in an unconfined column of a stoichiometric methane-oxygen mixture 106 mm in diameter: (1) in the gas column, (2) outside the gas column, and (3) in the tube.

nation front luminosity clearly seen in the bottom part of the picture, representing the unconfined charge, are of special interest. The scale of these fluctuations is much larger than the detonation cell size not discernible in the upper part of the photograph showing the confined detonation in the generator tube. The closer the charge diameter to the critical one the more pronounced are the luminosity fluctuations. Both the soot print (Fig. 3) and streak camera records exhibit three types of fluctuations in detonation waves propagating in unconfined charges. Apart from the normal detonation cells and fluctuations mentioned above that are due to periodical variations in the cell size at the charge periphery (or in many cases, especially in conditions close to the critical ones, to periodical disappearance of the cells near the charge boundary with the subsequent reinitiation of detonation) there exist smooth large-scale variations in the area of the region where the detonation cells are observed. The small-scale pulsations are obviously caused by a galloping mode of detonation propagation at the charge periphery; their longitudinal size is independent of the charge diameter; for different mixtures it varied between 23 and 45 mm. The wave velocity, as measured from the streak photographs, over the 0.15–0.2 small-scale pulsation length was 1.9–2.5 times the mean detonation velocity while at the end of a pulsation it dropped to 0.6–0.54 mean velocity. This type of detonation wave instability was predicted by numerical computations.[7]

The characteristic size of the large-scale pulsations was several tens of cm and usually no more than two such pulsations were present along the free-charge length. At the critical charge diameter large-scale pulsations and the regions with increased cell size penetrate practically to the charge axis.

The galloping mode of detonation propagation affects the detonation wave structure at the charge axis. Thus, the transverse cell size at the axis of an unconfined charge is larger than that in a tube (Fig. 4) and fluctuates with a frequency that approximately correlates with the frequency of the luminosity fluctuations. The lateral rarefaction waves reduce the mean detonation wave velocity to the greater extent the closer is the charge diameter to the critical one (Fig. 5). The maximal recorded relative velocity reduction due to energy and momentum losses is greater than in smooth tubes.[8] Since the intensity of the lateral rarefaction waves is higher at the periphery the mean cell size there is larger than that at the axis, which is

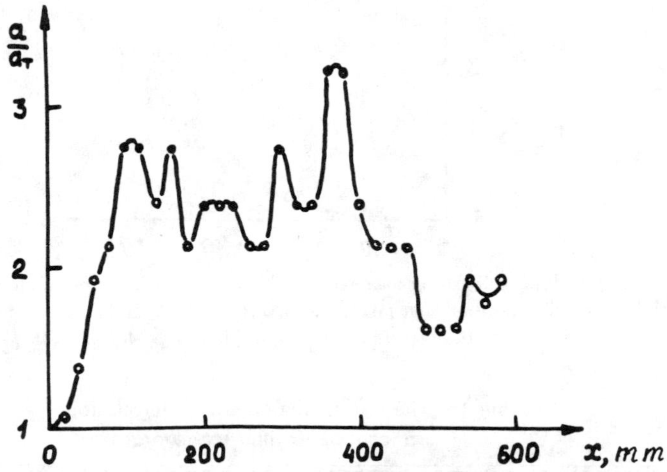

Fig. 4 Variation of the transverse cell size near the unconfined-charge axis along the charge. Each cell size was averaged over a distance of 70 mm. Stoichiometric methane-oxygen mixture, charge diameter 106 mm, a_t is the cell size in the tube.

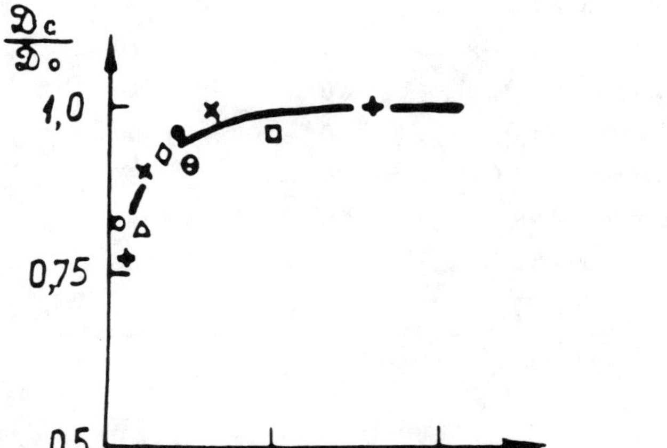

Fig. 5 Detonation velocity in unconfined cylindrical charges as a functiuon of charge diameter reduced by respective critical diameters of detonation transition into unconfined space: + propane-5$0_2$, O propane-7$0_2$, Δ propane-2.5$0_2$, θ propane-3.5$0_2$, × ethane-3.5$0_2$, ◇ ethane-4.4$0_2$, • ethylene-3$0_2$, □ ethylene-1.2$0_2$. D_o is the detonation velocity in the tube.

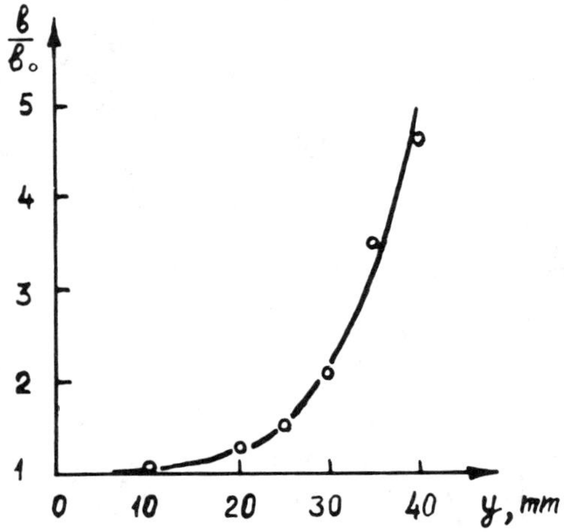

Fig. 6 Longitudinal cell size in an unconfined charge as a function of charge radius. Stoichiometric methane-oxygen mixture, charge diameter 106 mm, the cell size was averaged over a distance of several hundred mm along the charge axis, b_o is the cell size in the tube.

Table 1 The critical diameters of detonation unconfined cylindrical charges

Combustible Mixture	Unconfined Charge Diameter, mm		$d_{cr.c.}$ mm
	go	no go	
Methane + 2 O_2	106	82	894
Ethane + 3.5 O_2	70	60	94
Propane + 5 O_2	70	60	65
Ethylene + 3 O_2	60	50	55
Acetylene + 2.5 O_2	40	37	38
Acetylene + Air	200	155	178
Acetylene + 2.5 O_2^*			10–12
Acetylene + O_2^*			7

* Obtained by A.A. Vasiliev and D.V. Zak.[5]

seen in Fig. 6. The maximum longitudinal cell size at the periphery is close to the scale of the luminosity fluctuations.

In slightly subcritical conditions detonation usually decays during the second large-scale fluctuation. A typical behavior of the transverse cell size at the charge axis for a decaying detonation wave is presented in Fig. 7. There is some experimental evidence of a galloping detonation regime even at the charge axis near the critical charge diameter, however, transition from a stable detonation with a small layer at the periphery exhibiting fluctuations to a continuously decaying one occurs in a very narrow range of the experimental conditions.

The measured critical diameters are listed in Table 1 together with some data borrowed from the literature. It should be noted that a smoked foil placed along the charge axis slightly increases the critical diameters, which is apparently due to enhanced momentum and energy losses. The effect of mixture equivalence ratio on the critical diameters is demonstrated in Fig. 8 taking as an example propane-oxygen mixtures. As seen, the minimal critical diameter is observed in a slightly rich mixture. The critical diameters $d_{cr.c}$ reduced by the critical diameter for detonation transition from a tube to unconfined space, $d_{cr.}$, plotted against $d_{cr.}$ for all the mixtures studied form a unified curve (Fig. 9). The $d_{cr.c.}/d_{cr.}$ ratio tends to a constant value equal approximately to 1.5 at large $d_{cr.}$. There are two reasons for such a behavior of the aforesaid dependence. The first one is associated with the physico-chemical processes occurring behind diffracting detonation waves and the second reason is mixing at the charge-air boundary.

The temperature sensitivity of the reaction rate is the lower the higher the mixture reactivity and the more intense the detonation wave, since the dimensionless Arrhenius exponent E/RT behind the lead shock wave in this case is small (here E is the activation energy and T is the gas temperature). This implies that in highly reactive mixtures the reaction may persist behind the part of the detonation front diffracting at the tube edge, thus, contributing to the total energy release and supporting the detonation wave

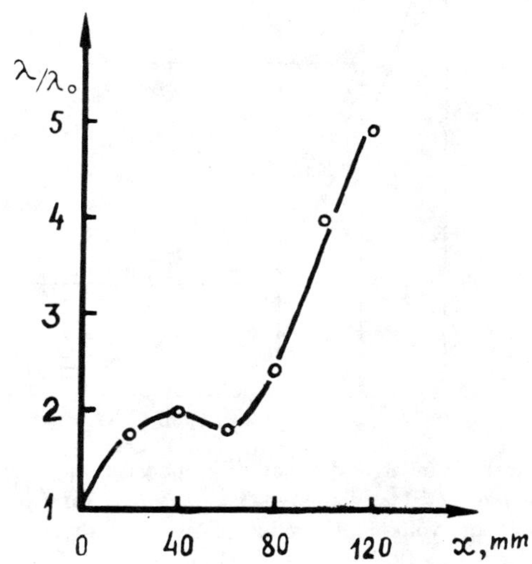

Fig. 7 Transverse cell size at the axis of an unconfined charge as a function of distance for decaying detonation. Stoichiometric methane-oxygen mixture, charge diameter 82 mm.

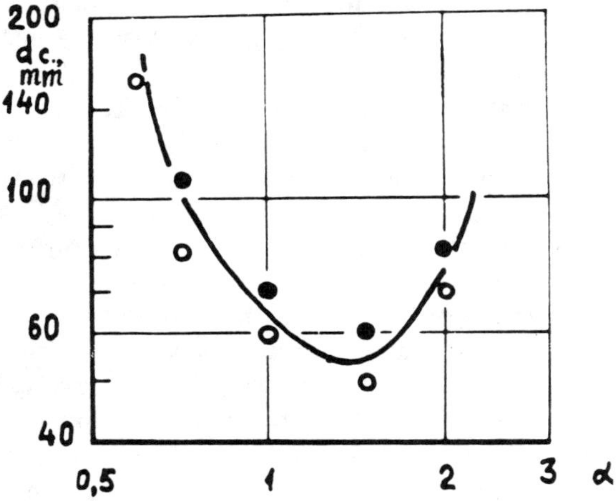

Fig. 8 The critical diameter of an unconfined charge versus equivalence ratio dependence for propane-oxygen mixtures: O no go, • go.

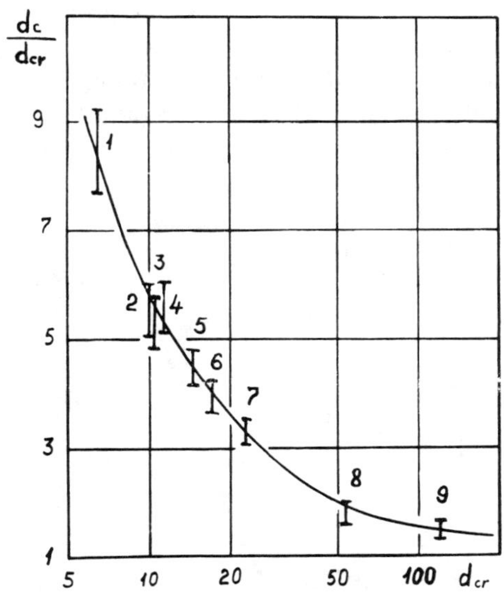

Fig. 9 The critical diameter of detonation propagation in an unconfined cylindrical charge versus the critical diameter of detonation transition from a tube into an unconfined mixture volume. The mixtures are: (1) ethylene+ 3 O_2; (2) propane + 3.4 O_2; (3) acetylene + 2.5(O_2 + N_2); (4) propane + 5 O_2; (5) ethane + 3.5 O_2; (6) acetylene + 2.5($O_2 + 1.33N_2$); (7) acetylene + 2.5($O_2 + 1.77N_2$); (8) methane + $2O_2$; (9) acetylene + air, stoichiometric.

that propagates from a tube to unconfined space filled with a mixture. There is no such an additional detonation support in cylindrical unconfined charges since no mixture is present outside the charge diameter (which is equal to the tube diameter), therefore, the difference between the two critical diameters should be large. With large values of E/RT this difference, on the contrary, must be very small since owing to a high temperature sensitivity of the reaction rate the temperature drop behind the curved portion of a diffracting wave will practically stop the reaction so that the wave will propagate as if no mixture is present in the unconfined space outside the gas column with a radius approximately equal to the tube diameter. It follows that the $d_{cr.c.}/d_{cr.}$ ratio should decrease as the detonation wave velocity diminishes or the effective activation energy grows.

Based on this idea we shall estimate the maximal possible value of these two critical diameters. Simple considerations and the experimental evidence[9] show that a detonation wave propagating in a tube survives upon entering an unconfined mixture volume only in the case when its reinitiation in the region influenced by the lateral rarefaction wave occurs before the rarefaction wave head spreading through the shocked mixture reaches the axis of symmetry. Hence the mixture volume, reaction in which is responsible for the distinction between the aforesaid critical diameters, is unconfined in a semisphere of a radius whose value is less than $1 + d/2$, with 1 and d being the distance travelled by the detonation front to the point where the lateral rarefaction waves merge at the tube axis and the tube diameter, respectively. To compensate for the lack of the additional energy available for supporting detonation the critical diameter of an unconfined cylindrical charge should be no greater than $21 + d_{cr.}$ (the inequality accounts for the fact that the diffracted wave spreads perpendicular to the tube axis at a velocity which is less than the detonation velocity). As estimates show, 1 is only insignificantly larger than $d_{cr.}/2c_s$, where c_s is the speed of sound in the shocked mixture, therefore, the value

$$d_{cr.c.} = 1 + D/C_s$$

can be taken as an upper bound of the critical diameter of an unconfined cylindrical mixture column (here D is the detonation velocity). The D/c_s ratio is 3 ± 0.2 for hydrocarbon-oxygen mixtures and about 2.4 for hydrocarbon-air mixtures; in hydrogen-air and hydrogen-oxygen mixtures it is about 2.2. Thus, the ratio of the critical diameters cannot exceed 3.4–4. Its real value is expected to be even lower since, first, not all the energy confined in the semisphere of the radius $1 + d/2$ reacts before detonation reinitiation and, second, from experiment and simple considerations it follows that reinitiation starts much earlier than the rarefaction wave reaches the tube axis even at the limit of detonation propagation. The authors of Ref. 9 have demonstrated that the transverse diffracted shock wave travels only a half of the tube diameter from the tube rim by the instant of detonation reinitiation, which lowers the maximal ratio of the critical diameters to about 2.5. In accord with this finding we have observed that the distance over which detonation decays in the limiting conditions is about 1.5 $d_{cr.}$.

Hence, there is no doubt that the large values of $d_{cr.c.}/d_{cr.}$ ratio obtained in the previous experiments are associated with mixing at the cylindrical charge boundary. Treatment of all our experimental data yielded the following empirical equation

$$d_{cr.c.}/d_{cr.} = A + B/d_{cr.} - C/d_{cr.}{}^2 \tag{1}$$

Here A, B, and C are equal to 1.6, 40 mm, and 7 mm^2, respectively. The above value of A is probably close to the limiting $d_{cr.c.}/d_{cr.}$ ratio with no mixing layer (the latter ratio must be strictly greater than unity since when the mixture is present outside the gas column with a radius equal to that of the tube it always contributes to the energy release behind a diffracting detonation wave prior to detonation reinitiation at the periphery, at least due to burning of the mixture near the contact surface separating the shocked unburnt gas and the detonation products). The second coefficient, B, might be interpreted as a doubled thickness of the unreacted layer at the charge periphery, but it should be expected to be somewhat greater than the mixing layer thickness because of the effect of the reaction behind the diffracting wave discussed above. The term $C/d_{cr.}{}^2$ may be associated with heat release in the mixing layer.

The soot prints show that the thickness of the layer in which no detonation cells are observed is 10–15 mm. This is consistent with the $B/2$ value and the estimated mixing layer thickness. Even in the case when an unconfined charge is formed by coflowing mixture and air jets, which is believed to be the best way of cylindrical-charge preparation,[6] the mixing layer thickness at the charge end estimated from the simple formula $x = (2D't)^{1/2}$ amounts to 2.8 mm. Here D' is the diffusion coefficient assumed to be 0.2 cm^2/s and $t = 0.2$ s. The value of $B/2$ calculated by Eq. (1) using the measured critical diameters and the above magnitudes of coefficients A and C is 2–2.5 mm. This also confirms the conclusion that the main reason why the $d_{cr.c.}/d_{cr.}$ ratios are abnormally large in reactive mixtures is a change in mixture composition at the charge boundary. Equation (1) allows estimation of $d_{cr.c.}$ based on measured $d_{cr.}$. The estimated values for fuel-air mixtures are presented in Table 2.

The effect of mixture dilution is especially prominent when a detonable mixture issues from a tube into air. Investigations of detonation in such submerged jets are of practical importance since they will allow appraising possible hazard of a detonable mixture discharged from a duct to the ambient atmosphere. It is obvious that detonation may propagate only in that part of the submerged jet, wherein, the fuel and oxidizer concentrations are within the detonability limits. According to Reference 10, the mixing layer in a submerged jet reaches the jet axis at a distance of about $4d$ from the tube edge. Since the concentration detonation limits are rather narrow this distance can be assumed to be an upper bound of a distance over which detonation propagates in a submerged jet, L. For tubes with diameters greater than $d_{cr.c.}$ L can be estimated from the equation

$$L \simeq d(4 - 3d_{cr.}.$$

Table 3 presents the values L measured from the soot prints and calculated by the above simple equation. The results are seen to agree fairly well.

It is of interest to compare the critical diameters of unconfined cylindrical charges with the heights of unconfined planar mixture layers critical for detonation propagation. When simulating a planar unconfined mixture slab in a duct, one must be sure that the duct is sufficiently wide and that its walls exert no effect on detonation propagation. For this purpose we measured critical layer heights in ducts of various width. As seen from Fig. 10, the critical layer height first diminishes as the duct width grows and then levels off when more than three detonation cells are present along the duct

Table 2 Estimated critical diameters of detonation in unconfined cylindrical charges of fuel-air mixtures

Mixture	$d_{cr.}$, mm	$d_{cr.c}$,*	$d_{cr.c.}$, mm
Ethylene – Air	320	510	550
Ethane – Air	670	1070	1110
Propylene – Air	700	1120	1160
Propane – Air	700	1120	1160
Hydrogen – Air	200	320	360

Note: The mixtures are stoichiometric, $d_{cr.c.^*}$ is the critical diameter of detonation propagation in a charge with no smearing of the charge boundary.

Table 3 The detonation propagation distance in a submerged mixture jet

Mixture	$d/d_{cr.}$	$L_{exp.}/d$	$L_{calc.}/d$
Propane + 4.5 O_2	1.5	1.14	1.5
Propane + 5 O_2	2.0	1.7	2.5
Propane + 4.5 O_2	4.3	2.43	3.3

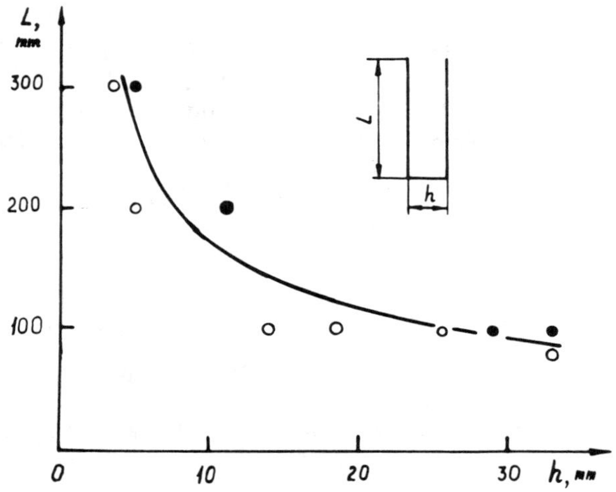

Fig. 10 The critical duct height as a function of duct width. Stoichiometric acetylene-air mixture.

width. This limiting value was assumed to be the critical height of an infinite unconfined slab. The critical layer size for a stoichiometric acetylene-air mixture turned out to be close to $d_{cr.c.}/2$ (detonation propagated in a layer 100 mm high and failed to propagate in a layer 80 mm high). In a hydrogen-air mixture detonation propagated in a layer 200 mm high and decayed in a layer 160 mm high, which is probably also close to the critical diameter of a cylindrical charge (no stable detonation was observed in an unconfined charge 300 mm in diameter). Thus, for fuel-air mixtures one cannot expect a great difference between the critrical layer heights and cylindrical-charge diameters because of a large radius of curvature of the charge. This certainly simplifies the procedure of critical diameter determination for fuel-air mixtures.

Critical diameters of unconfined cylindrical charges are affected even by very light confinement. Thus, placement along the tube axis of four metal strips, 25 mm wide, at the diameter of 80 mm within an unconfined methane-oxygen mixture charge 82 mm in diameter made possible detonation propagation through the charge (no detonation was observed in the same charge without strips). Practically no visible changes in detonation wave propagation pattern were noticed (neither in the smoked foil records nor in streak camera photographs) as compared to that in a rigid-wall tube for fuel-oxygen charges confined in a tube made of thin plastic film. Solid objects placed in an unconfined charge exert two opposite effects on detonation waves: on the one hand, they reduce the intensity of the lateral rarefaction waves and thereby facilitate detonation propagation and, on the other, they increase momentum and heat losses, thus decreasing the detonation velocity and destroying the detonation wave. For example, a spiral 70 mm in diameter with a pitch of 17.6 mm made of a 3 mm wire and placed in a stoichiometric methane-oxygen charge 106 mm in diameter reduced the detonation velocity by 2% as compared to that in a completely unconfined charge; a thin plate introduced along the unconfined-charge axis increased the critical diameter. However, in the case of detonation failure the spiral supported the velocity of the decoupled shock wave-combustion zone complex at a higher level, at least at early stages of the process.

Conclusion

From the investigation performed it can be inferred that:

1) Detonation velocity drops significantly and pulsations with a scale much larger than the detonation cell size become more prominent as a diameter of an unconfined cylindrical gaseous charge approaches its critical value.

2) The critical diameter of detonation in unconfined cylindrical charges cannot exceed about three critical diameters of detonation transition from a tube to unconfined space filled with a mixture, for fuel-air mixtures and diluted fuel-oxygen mixtures the ratio of these two critical diameters tends to 1.5.

3) The cell size in unconfined charges is greater than in a tube, which implies that influence of the lateral rarefaction waves penetrates to the charge axis, at least for charges whose diameter is not much greater than the critical one.

4) The critical heights of planar gaseous explosive layers are close to the critical detonation diameters of respective unconfined cylindrical charges.

5) Detonable mixtures issued from a tube into ambient air can support detonation only over a distance not exceeding four tube diameters.

References

[1]Dabora, E.K., Nicholls, S.A., and Morrison, "The Influence of a Compressible Boundary on the Propagation of Gaseous Detonations", Tenth Internat. Symposium on Combustion, The Combustion Institute, Pittsburgh, PA, 1965, pp. 817–830.

[2]Adams, T.G., "Do Weak Detonation Waves Exist?" AIAA Journal, Vol. 16, No. 10, 1978, pp. 1035–1040.

[3]Vasiliev, A.A., "On the Critical Diameter of Detonation in Gaseous Mixtures", Fizika Gorenia i Vzryva., Vol. 18, No. 3, 1982 pp. 98–104, (in Russian).

[4]Vasiliev, A.A. and Subbotin, V.A., "An Experimental Study into Propagation of Detonations in a Cylindrical Gaseous Charge," Dinamika Sploshnoi Sredy, issue 62, Novosibirsk, 1983, pp. 32–38, (in Russian).

[5]Vasiliev, A.A. and Zak, D.V., "Detonations of Gaseous Jets," Fizika Gorenia i Vzryva., Vol. 22, No. 4, 1986, pp. 82–88, (in Russian).

[6]Borisov, A.A., Mikhalkin, V.N., and Khomik, S.V., "On the Critical Diameters of Detonation in an Unconfined Cylindrical Charge," Dokl, Akad. Nauk SSSR, Vol. 291, No. 1, (1987), p. 88, (in Russian).

[7]Ivanov, M.F., Fortov, V.E., and Borisov, A.A., "Numerical Modeling of Detonation Development in Gaseous Volumes of Finite Thickness," Fizika Gorenia i Vzryva. Vol. 17, No.3, 1981, pp. 108–116, (in Russian).

[8]Zeldovich, Ya.B. and Kompaneets, A.S., Theory of Detonation, Moscow, Gostekhizdat, (in Russian).

[9]Edwards, D.H., Thomas, G.O., and Nettleton, M.A. , "The Diffraction of a Planar Detonation Wave at an Abrupt Area Change," J. Fluid Mech., Vol. 95, No. 1, 1979, pp. 79–96.

[10]Abramovich, G.N., Theory of Turbulent Jets, Moscow, Fizmatgiz, (in Russian).

Initiation of a Detonation Wave Due to Multistage Self-Ignition

S. M. Frolov,* B. E. Gelfand,† and S. A. Tsyganov‡
USSR Academy of Sciences, Moscow, USSR

Abstract

This paper deals with the theory of the "knock" phenomenon in internal combustion engines. The present theoretical approach is based on two ideas: 1) Sokolik's suggestion that the low-temperature, multistage, self-ignition of hydrocarbons exerts considerable effect on shock-wave formation in engines, and 2) Zel'dovich's fundamental idea of the possibility of shockless detonation initiation when conditions in the reactive mixture are close to self-ignition. We consider a finite volume of fuel/air mixture placed in a medium of lower chemical activity, and simulate a two-stage chemical release in it by two distinct kinetic equations of the Arrhenius type describing "cool-flame" and "hot" explosion stages of the process. The Euler equations for an unsteady flow of a combustible gas are integrated numerically in order to reveal the history of events in an initially quiescent and homobaric system. Three principal parametric domains are identified, namely, "cool," "hot," and "shock." The first contains only a "cool-flame" explosion because of a rarefaction-wave quenching effect accompanying the thermal explosion of the volume. In the second domain, chemical energy is released as in a single-stage thermal explosion without any noticeable interaction with the gasdynamic flowfield. Within the third "shock" domain exists a vigorous chemical reaction—gasdynamics coupling resulting in shock or even detonation formation in the volume. In this instance a rarefaction wave propagates through the volume after the first stage of reaction, thus producing conditions for rapid amplification of a pressure wave generated by a subsequent "hot" explosion stage. Thus, the two-stage thermal explosion of an unconfined volume may produce very strong shock waves in the ambient atmosphere. A simple coupling criterion is proposed and a good quantita-

*Researcher, Institute of Chemical Physics.
†Chief Researcher, Institute of Chemical Physics.
‡Head of Laboratory, Institute of Chemical Physics.

tive agreement is shown to exist between the criterion and the numerical calculations. Qualitatively, the criterion appears to account for numerous features of the "knock" phenomenon.

Introduction

In our recent publication on a possible role of concentration and temperature nonuniformities in the process of strong pressure wave generation under conditions close to mixture self-ignition,[1] we suggested an idea that multistage self-ignition of an unconfined gaseous explosive charge may be accompanied by generation of much more severe shock waves, compared to those generated under single-stage self-ignition of the charge. The present paper deals with further development of the idea. The following analysis may give useful information to interpret the "knock" phenomenon in internal combustion engines, in addition to suggestions given by Sokolik[2] and Zel'dovich, et al.[3]

Multistage Self-Ignition

For a number of detonable hydrocarbon/oxidizer mixtures one may distinguish two kinetic self-ignition types: high-temperature and low-temperature.[2] The former is typically characterized by high values of the activation energy corresponding to rupture of strong intramolecular links to form active reaction centers. As a result of subsequent self-acceleration of a chain reaction started by these centers, the heat addition rate may exceed the rate of mixture cooling because of heat transfer. Hence, the temperature will start to rise progressively, resulting in thermal acceleration of the reaction and eventually in explosion. The high-temperature self-ignition process is single-staged and usually arises as a "hot spot" event.

The low-temperature hydrocarbon self-ignition is multistaged. At the beginning of preflame processes, auto-oxidation of hydrocarbon molecules takes place that leads to the formation of peroxides. Peroxide accumulation continues until a certain critical concentration of them is achieved and then rapid decomposition of the intermediates occurs, giving rise to "cool flame" formation.

Heat release at the cool-flame stage of explosion is only 5 - 10% of the whole heat of reaction. Therefore, the corresponding increase in temperature is relatively small. Further oxidation of the mixture proceeds in the presence of active decomposition products until thermal chain-branching, or "hot" explosion, occurs. Consequently the hot explosion differs from a one-step high-temperature self-ignition by the fact that it involves self-ignition of a preconditioned, partially decomposed initial mixture, rather than of a pure initial mixture. Globally, this process is interpreted as bulk ignition of a part of an explosive charge.

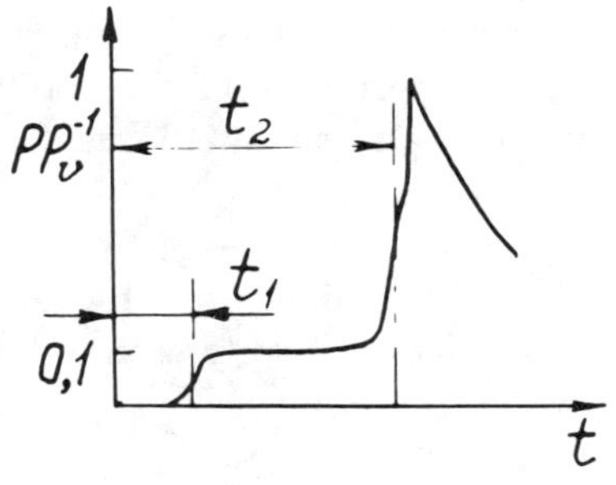

Fig. 1a Sketch of multistage explosion.

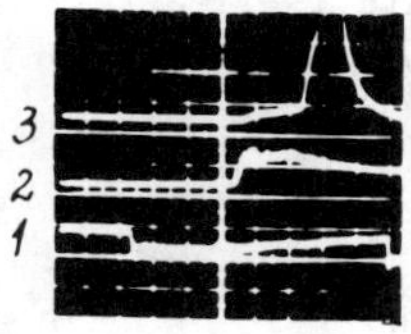

Fig. 1b Two-stage ignition of kerosene droplet with a promoting additive.

Shown in Fig. 1a is a sketch of a multistage explosion. Induction periods t_1 and t_2 represent, respectively, cool flame and hot explosion delays, and the quantity P_v denotes the final pressure of the explosion at a constant volume. Similar features of the self-ignition process have been observed in Ref. 4 for some liquid hydrocarbon fuels with additives promoting oxidation. Figure 1b shows the records of luminescence (1) and pressure (2,3) obtained during droplet ignition of kerosene plus 10% $(CH_3)_3$-C-O-OH in a shock tube. One may clearly distinguish the two explosion stages in the upper record.

An Approximate Criterion for Detonative Explosion

As we show below, low-temperature multistage self-ignition of a finite volume of a reactive mixture may give rise to detonation. Depending on the values of t_1 and t_2 (see Fig. 1a), evolution of a multistage explosion may be different. At the "cool-flame" stage (stage 1) the volume expands thermally, accompanied by a temperature drop and a respective variation of the intermediate species. Along with this process, active products accumulate, leading to self-acceleration of the reactions and triggering the "hot" explosion (stage 2). Following Ref. 1, one may distinguish between two possible modes of mixture burnout in the volume.

If stage 2 of the explosion takes place after thermal expansion and the pressure is equalized all over the volume (that is, $t_2 \gg t_1$) the thermal energy will be released in almost uniform spatial conditions. The subsequent self-ignition process will spread rapidly through the volume and the pressure wave generated in the outer region will be relatively weak. The same effect is to be expected if stage 2 develops almost immediately after the onset of the cool-flame stage (that is, $t_2 \approx t_1$).

Under certain conditions discussed later, a detonation-like wave may arise in accordance with the coupling mechanism of Ref. 1. The main role in amplification of the pressure wave generated after self-ignition of the charge core is played by the rarefaction wave, creating the conditions for nonuniform development of stage 2 of the explosion. Interaction of the pressure wave with the reaction front propagating towards the periphery of the volume leads to rapid amplification of the wave up to the intensity sufficient for triggering detonation.

As to the "knock" phenomenon, Sokolik[2] was the first to suggest the idea of the prevailing role of multistage self-ignition in detonation onset. However, Sokolik did not consider the outlined process of reaction-front/pressure-wave interaction and restricted the analysis only to "instantaneous" detonation (that is, constant volume explosion).

Mathematically, an approximate condition (coupling criterion) of rapid pressure wave acceleration due to coherent chemical energy release is given in Reference 1,

$$U_c \approx U_g \tag{1}$$

where U_c is the typical velocity of the chemical reaction front, and U_g is the typical gasdynamic velocity. For the problem of multistage self-ignition of a spherical volume of initial radius R_o, one may rewrite Eq. (1) in the following form:[1]

$$(dt_i/dR)^{-1} \simeq (\gamma R_* T_1)^{1/2}$$

where $t_i = t_2 - t_1$ is the local induction time of the "hot" explosion, R is a distance from the center of symmetry, γ is the ratio of specific heats, R_* is the gas constant, T is temperature, and index 1 denotes quantities after the "cool-flame" stage. The term dt_i/dR can be evaluated from an assumption that $t_i \simeq t_{ad}$, where t_{ad} is the adiabatic ignition delay, and that the typical temperature gradient formed in the system prior to the "hot" explosions is $\chi = \Delta T/R$. Here ΔT is a temperature drop at the volume boundary given approximately by the expression[1]

$$\Delta T \simeq \frac{\gamma - 1}{\gamma} T_1 \left(\sqrt{T_1/T_o} - 1 \right)$$

where T_o is the initial temperature distributed uniformly throughout the system. It has been shown in Ref. 5 that the temperature reduction ΔT after stage 1, required for subsequent pressure wave amplification, is of the order of the characteristic pre-ignition temperature rise $R_* T_1^2/E_2$, where E_2 is the activation energy of the "hot" explosion.

The final form of the approximate coupling criterion for the problem discussed is

$$\alpha_1 = (\delta - 1)/(\gamma - 1) \tag{2}$$

where

$$\delta = \beta_2 / ln(\alpha_2 k_2 R_o / (\gamma R_* T_o)^{1/2})$$

and $\alpha_1 = Q_1/R_*T_o$, $\beta_2 = E_2/R_*T_o$, $\alpha_2 = Q_2/R_*T_o$, and Q_1 and Q_2 are, respectively, the values of heat release typical for stages 1 and 2.

Equation (2) allows for evaluation of the minimum radius $R_o = R_{o\min}$ required for generation of a strong shock wave at the volume boundary after two-stage self-ignition. The phenomenon will also take place for $R_o > R_{o\min}$, other conditions being equal. Note that we consider here an unconfined explosion. Confinement may exert a pronounced effect on the process.[1] For $R = D$, where D is the typical confinement dimension, the constant-volume explosion exhibits no unusual phenomena. In Fig. 2, reproduced here from Ref. 1, Eq. (2) is represented by a dashed line extending from the origin of coordinates.

Results of Numerical Calculations

It is natural to verify the criterion given by Eq. (2) by means of numerical integration of the gasdynamic equations, together with the equations of chemical kinetics. The mathematical details are the same as in Ref. 1. Note only that stages 1 and 2 of the explosion are modeled by Arrhenius type reactions of the first order, without regard for the effects associated with viscosity, heat conductivity, and mass diffusion.

A numerical solution of the problem indicates that within the time interval for the rarefaction wave to travel from the contact surface to the center of the volume, the pressure wave is amplified only within the shaded segment in Fig. 2. Near the lower boundary line of the segment, the pressure in the shock wave increases to a maximum that may approach the peak pressure in the GJ detonation wave. The upper boundary line of the segment is somewhat conditional, since it is drawn to meet the requirement that the peak static pressure in the chosen volume be no more than 10% larger than

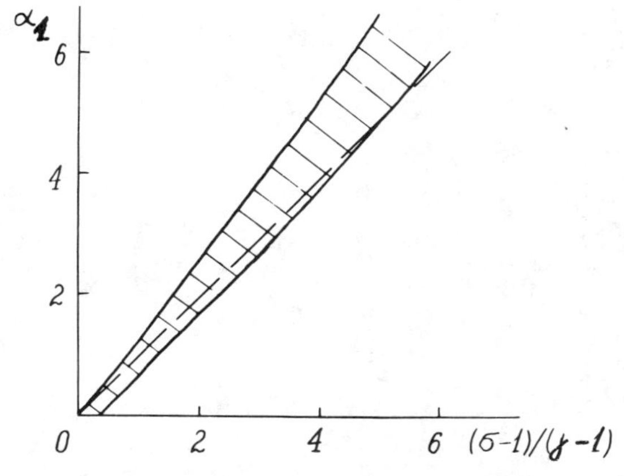

Fig. 2 The parametric region of pressure wave amplification for multistage self-ignition.

the pressure of a constant-volume adiabatic explosion. Thus, self-ignition of an unconfined explosive charge represented parametrically by a point in the shaded segment in Fig. 2 results in generation of a shock wave that is much stronger than those produced by a constant-volume explosion.

Location of segment boundaries is almost independent of the symmetry of the chosen volume (planar, cylindrical, or spherical). It should be emphasized that this holds only for time intervals that are less than the time taken for the rarefaction wave to travel from the contact surface to the center of symmetry. If one takes into account later expansion phases of a volume of spherical (or cylindrical) symmetry, an interesting feature comes into effect, consisting of a possible explosion initiation at the center of symmetry after arrival and reflection of a secondary shock wave. The converging secondary shock wave is formed because the gas passing through a spherical lead shock front expands to a greater degree than that passing through a planar shock. As a result of the explosion induced by the secondary shock, an accelerating complex "pressure wave/ignition front" may propagate through the volume and eventually give rise to detonation. This widens the shaded segment in Fig. 1 for spherical (and cylindrical) symmetry of the volume. For example, in a spherical volume characterized by the value of $(\delta - 1)/(\gamma - 1) = 2.8$, the lower boundary line of the shaded segment drops to the value $\alpha_1 \simeq 2$.

It is clear from Fig. 2 that there is a good agreement between the approximate results given by Eq. (2) (dashed line) and numerical calculations. Existence of a parametrical region instead of a single line is associated with two factors that are not taken into account when Eq. (2) has been derived. The first is a wide spectrum of ignition-delay gradients in rarefaction waves propagating to the center of the volume at $t > 0$. The second is associated with the ability of a pressure wave to modify and adjust the phase of local explosions that brings about its further amplification. The lower boundary line of the

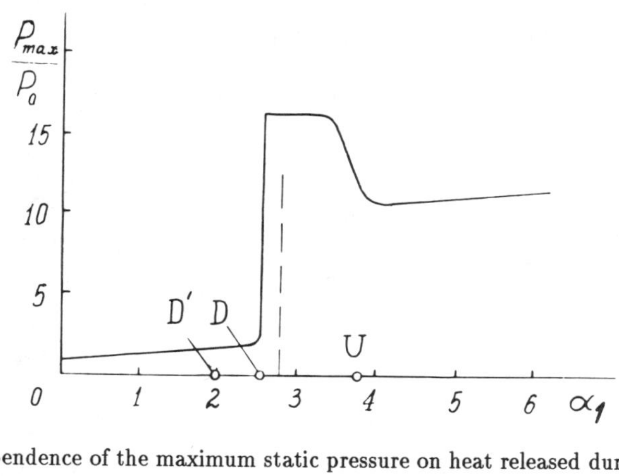

Fig. 3 Dependence of the maximum static pressure on heat released during stage 1 of explosion.

dashed segment in Fig. 2 does not emanate from the origin of coordinates. This means that for $T_1 = T_o$ and $P_1 = P_o$ (that is, a single-stage explosion) there exists an interval of initial temperature T_o such that the pressure wave entering the ambient atmosphere will be more intense than the wave generated by a constant-volume explosion. For example, for $T_1 = T_o = 550$ K, $Q_1 = 0$, $Q_2/R_*T_o = 10.7$, $E_2/R_*T_o = 18.7$, $k_2 = 10^{10}s^{-1}$, $R_o = 0.1$ m, $\gamma = 1.4$ (that is, $\delta = 1.1$), the final pressure of explosion at a constant volume is about $5.3P_o$, which results in the generation of a shock wave with overpressure of about $2P_o$. When an unconfined gaseous charge of the same size self-ignites, the intensity of the shock wave entering the ambient atmosphere is as large as $4P_o$. This becomes possible at the expense of a pressure wave formed inside the charge and its further acceleration due to staged self-ignition of the mixture during wave propagation towards the contact surface.

Clearly, the phenomenon is more prominent for multistage self-ignition. Figure 3 shows the dependence of the maximum static pressure $P_{\max}/P_o$ at the contact surface of an unconfined explosive charge ($\delta = 2.12$, $\gamma = 1.4$, $T_o = 300$ K, $Q_2/R_*T_o = 20$) on the value of chemical energy release α_1 at stage 1 of the explosion. Points D and U on the X-axis correspond to the lower and upper boundaries of the shaded segment in Fig. 2. The vertical dashed line in Fig. 3 represents α_1 given by Eq. (2). Figure 3 accounts for only the first passage of the rarefaction wave from the contact surface to the center of the symmetry. At points lying to the left of point D, stage 2 of the explosion coming into effect later is not taken into account. The two-stage self-ignition proceeds like a constant-volume explosion to the right of point U. The region between points D and U is noteworthy since a shock or even detonation wave may arise in the volume in this range and the static pressure may achieve the value in the CJ detonation wave. Because of the instability of the unidimensional "shock wave reaction front" complex, strong pressure oscillations were observed in numerical calculations; therefore, Fig. 3 shows the average values of $P_{\max}/P_o$.

Note that a detonation wave is formed in and propagates through a medium characterized by an elevated temperature $T_1 > T_o$. This is an explanation of relatively low static pressures at the leading shock front. Nevertheless, the maximum total pressure in the region is much higher than typical of a constant-volume explosion. It is worth noting that the abnormal pressure rise may be achieved also in the parametric region lying to the left of point D in Fig. 3. For example, taking into account the secondary shock wave in a volume of spherical symmetry results in a shift of point D to position D'; the average maximum pressure in region DD' is $P_{\max}/P_o \simeq 16$. Furthermore, after several journeys of the rarefaction wave, the "hot" explosion may be interpreted as a single-stage explosion occurring in nearly uniform conditions. In this case, appearance of shock waves of an elevated intensity is possible in accordance with the above-mentioned mechanism, as

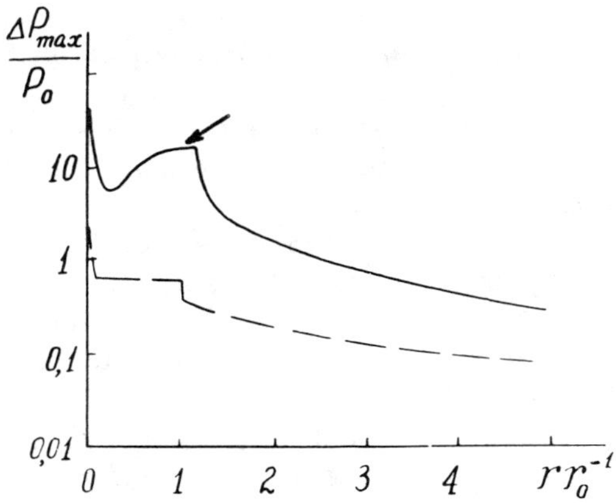

Fig. 4 Dependence of the maximum overpressure on distance from the explosive charge center. Arrow indicates a detonation wave.

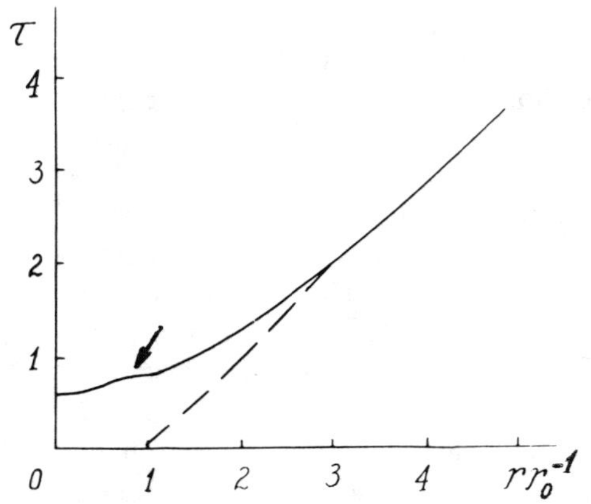

Fig. 5 Time/distance diagram for shock waves under two-stage explosion. Arrow indicates a detonation wave.

well as with the mechanism studied by Zel'dovich et al.,[3] because of residual spatial nonuniformities of the "hot" explosion induction time.

Figure 4 is a plot of the maximum, overpressure $\Delta P_{\max}/P_o$ versus distance from the center of a spherical, gaseous, explosive charge characterized by the following set of the parameters: $\delta = 1.6$, $\gamma = 1.2, T_o = 300$ K, $Q_1/R_*T_o = 3$, $Q_2/R_*T_o = 30$, $\gamma_o = 1.4$ (γ_o is the ratio of specific heats in the ambient atmosphere). It is seen in Fig. 4 that during thermal expansion of the charge after stage 1 of the explosion (dashed line), the detonation wave is formed inside the volume. The averaged overpressure in the wave is $\Delta P_{\max}/P_o \simeq 15$ (solid line). The wave reaches the contact surface, displaced by $0.15R_o$ at the time elapsed from the beginning of the process, and attenuates rapidly. Figure 5 is the time vs distance diagram for the corresponding shock fronts. It follows that in the close vicinity of the charge ($R/R_o < 3$) the shock wave has a two-front configuration. Note for comparison that a constant volume explosion would result in generation of a shock wave with the overpressure $\Delta P_{max}/P_o \simeq 3.5$ in the ambient atmosphere, as against the $\Delta P_{max}/P_o \simeq 15$ typical of the unconfined multistage explosion.

Conclusion

Our study shows that multistage self-ignition of a reactive mixture may result under certain conditions in spontaneous onset of strong shock and detonation waves. The mechanism of wave generation is not associated with flame acceleration, as is usually the case under deflagration-to-detonation transition in tubes. It consists in self-correlated propagation of pressure waves and chemical reaction fronts.

Realization of such a mechanism in engines seems to be rather probable. The high-speed camera technique used in Ref. 6 for visualization of knock combustion in engines does reveal such features of the process as spontaneous ignition fronts and registers onset of shock waves in local portions of unburnt pockets of the charge.

References

[1]Zel'dovich, Ya.B., Gelfand, B.E., Tsyganov, S.A., Frolov, S.M., and Polenov, A.N., "Concentration and Temperature Nonuniformities of Combustible Mixture as a Reason for Pressure Wave Generation," Progress in Astronautics and Aeronautics, Vol. 114, Dynamics of Explosions, edited by A.L. Kuhl et al., AIAA, New York, 1988, pp. 99–123.

[2]Sokolik, A.S., Combustion in Transport Compression Engines, Moscow, Izd. Akad. Nauk USSR, 1951, pp. 185–211.

[3]Zel'dovich, Ya.B., Librovich, V.B., Makhviladze, G.M., and Sivashinsky, G.I., "On the Development of Detonation in a Nonuniformly Preheated Gas," Astronautica Acta, Vol. 15., 1970, pp. 313–318.

[4]Gelfand, B.E., Kalinin, V.N., Petrunin, A.B., Stepanov, V.V., Timofeev, E.I., Tsyganov, S.A., Zacharkin, A.I., and Zhigtch, A.F., "Ignition of Droplets of Boron-organic Compounds in Air Behind Shock Waves," Dokl. Akad. Nauk USSR, Vol. 240, No. 3, 1978, pp. 627–629.

[5]Gelfand, B.E., Frolov, S.M., and Tsyganov, S.A. "A Criterion for Shock and Detonation Wave Generation in a Reactive Medium," Khimicheskaya Fizika, Vol. 8, No. 5, 1989, pp. 118–128.

[6]Livengood, J., and Leary, W. "Self-Ignition under a Rapid Compression," Ind. Eng. Chem., Vol. 43, 1951, pp. 2797–2805.

Critical Energy of Direct Detonation Initiation in Gaseous Mixtures

A. A. Borisov* and S. V. Khomik*
USSR Academy of Sciences, Moscow, USSR
and
V. R. Mikhalkin† and E. V. Saneev†
Moscow Institute of Physical Engineering, Moscow, USSR

Abstract

Minimal energies of direct plane detonation initiation in various fuel/air mixtures were measured in tubes. Plane-detonation waves were initiated by explosion of either condensed high explosives or an acetylene/oxygen mixture. A correlation is found between the initiation energies for plane and spherical detonation waves that corresponds to the well-known dependence of minimal initiation energy on the reaction time, $E_i \alpha \tau^i$, where i is the index of symmetry. The energy of detonation initiation by gaseous explosive is higher than that by condensed explosive. A semi-empiric formula is suggested for estimating the critical detonation initiation energy of charges of any geometry on the basis of measured critical diameters for detonation transition from a tube to an unconfined volume. The formula agrees fairly well with the available experimental data.

Introduction

Knowledge of parameters of direct detonation initiation in gaseous mixtures is of great practical importance since they are extensively used when appraising the probability of occurrence of the most detrimental explosions in unconfined and confined gaseous mixture volumes under various conditions. Unfortunately, the most important parameter, the minimal energy of unconfined detonation initiation, can only be easily measured in laboratory-scale esperiments for very reactive combustible mixtures, such as hydrocarbon/oxygen or acetylene/air. This is due to the fact that so-called critical radii for detonation initiation in most of the mixtures encountered in

---

*Institute of Chemical Physics
†Institute of Physical Engineering

practice are so large that assessment of their detonability requires very expensive, large-scale field tests.

For this reason, the methods of estimating the minimal energy of direct detonation initiation in fuel/air mixtures, based on either easily measured detonation parameters or numerical modeling, have received special attention in the literature. All of the methods essentially employ the idea first put forward by Ya. B. Zeldovich et al.,[1] which relates the chemical reaction time in a detonation wave and the critical energy of detonation initiation. According to this idea, the critical energy is proportional to the chemical reaction time in power 1, 2, or 3 for plane, cylindrical, or spherical geometry respectively .

The semi-empirical methods (see, for example, Refs. 2 - 3) employ either detonation cell size or measured ignition delays (or other related kinetic parameters) as basic data for calculating the minimal initiation energy. The evaluated energies are usually scattered in a wide range because of both the appropriate nature of the correlation equations and the spread in the original experimental data. Moreover, numerical modeling shows that the initiation energy is not a unique function of the mixture properties; it depends on the initiator characteristics as well. This is also observed in experiments.[4-6]

Purely numerical methods (see, for example, Ref. 7) also have a disadvantage since they require knowledge of the chemical kinetic parameters that are frequently measured to within a factor of two or even with a much worse accuracy. Therefore, there still exists a need for simple but accurate enough methods of initiation energy assessment.

In the present study we consider two such methods applicable to both the high- and the low-reactive mixtures. The first is based on measurements of minimal energies of direct-plane detonation initiation in tubes where the critical detonation-initiation distances of several meters are not a problem. The second method employs the experimental fact that the critical diameter for detonation transition from a tube into a large volume through a cone ceases to depend on the divergence angle when its value exceeds about 60 deg. This allows for evaluation of the energy necessary for initiation of the reaction behind the lead shock wave.

Direct Wave Detonation Initiation by High Explosives

The schematic of the setup is shown in Fig. 1. The mixture under investigation was purged through a vertical tube 0.145 m in diameter and 2.2 m long, equipped with pressure gauges and windows for taking streak camera photographs. About 4 - 5 tube volumes of the mixture were purged to ensure its uniform distribution in the tube. The mixture was also analyzed chromatographically. To eliminate ringing of the transducers, the initiating charge was placed either inside the tube at about 10 cm from the end plate (not touching the tube walls) or in a special section separated by an air gap from the main tube. Detonating cord spirals, sheets of plastic high explosive,

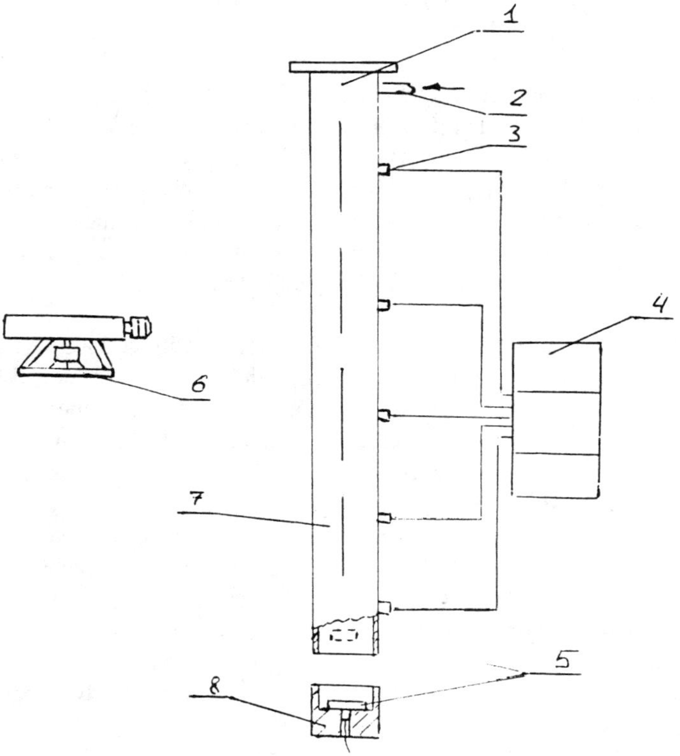

Fig. 1 Schematic of the experimental setup with condensed explosives: 1) detonation tube, 2) mixture inlet, 3) pressure pickups, 4) electronic chronometer, 5) explosive charge, 6) streak camera, 7) windows, and 8) initiator sector.

or sensitized nitromethane spilled onto the end plate were used as initiating charges in order to generate a more or less planar shock wave from the very beginning of its propagation. The differing geometry of the charges and various explosives used necessitates unification of the charge energy assessment. For that purpose we utilized, as did the authors of Ref. 8, the relations derived for a point explosion and calculated the effective energy from the measured shock wave velocities by the formula

$$E_{eff} = \rho S u^2 [\sqrt{L_1} + \sqrt{L_2} + \frac{\sqrt{L_1 L_2}}{\sqrt{L_1} + \sqrt{L_2}}]^2$$

where ρ is the initial gas density, S is the tube cross-sectional area, and u is the wave velocity measured at the base between transducers positioned at distances L_1 and L_2 from the charge. The effective initiation energy was

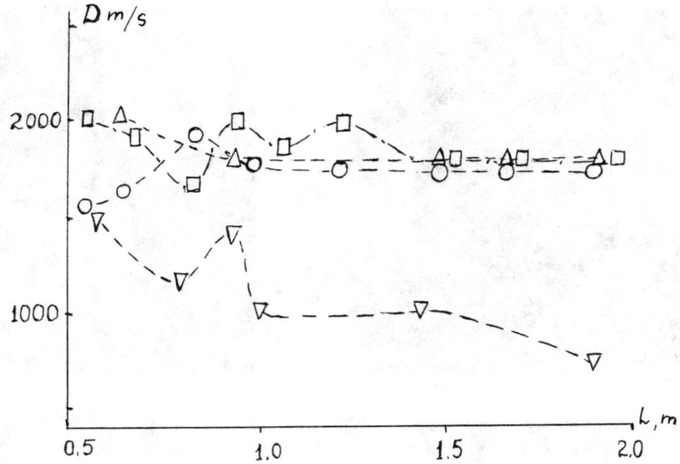

Fig. 2 Detonation velocity as a function of distance for different energies of initiating charges:

▽ 2.3, ○ 3.1, □ 4.6, and △ 6.6 MJ/m.^{2}

averaged over a distance required for onset of a steady Chapman-Jouguet (C-J) detonation. The energies thus calculated were slightly lower (by about 20% - 30%) than the chemical energy of the charge.

Figure 2 shows the detonation velocities measured from the *x-t* diagrams as functions of distance for a stoichiometric propane/air mixture and for initiation energies in the range of 2.3 - 6.6 MJ/m^2. As seen from the graph, the minimal detonation initiation energy is close to 3.1 MJ/m^2. The nonsteady wave velocity approaches the C-J level from both above and below. The streak camera photographs in Fig. 3 demonstrate that with charges close to the minimal ones, detonation propagates in a galloping manner (Fig. 3a). Thus, the loss of stability of detonation waves and their galloping propagation is a common feature of the process near the limits, irrespective of the nature of the limiting phenomenon, concentration limit,[9] or minimal induction energy.

The minimal initiation energies measured using this technique for various propane/air mixtures are presented in Fig. 4, curve 2. The minimum of the bell-shaped curve is shifted towards rich mixtures. Practically speaking, the measured energies do not depend on the type of initiating charge (see Table 1), which indicates that the method employed ensures generation of nearly plane initiating waves.

The minimal energies of detonation initiation by plane shock waves measured for different mixtures were used to check the well-known theoretical correlation between the energies of detonation initiation in charges of differing geometry. The energy for initiation of spherical detonations in Fig. 5 is borrowed from literature sources (see Table 2). It is obvious that the expected correlation does exist in a wide range of initiation energies since the slope of

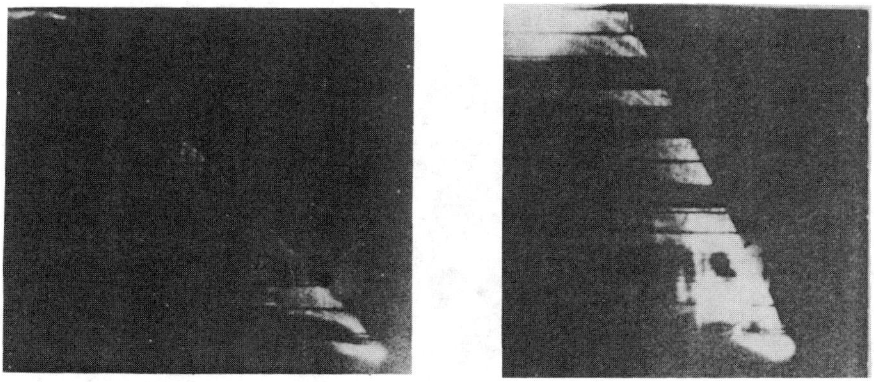

Fig. 3 Streak camera photographs of detonation initiation in propane/air mixtures by charges of different energy: a) a galloping regime (charge energy is close to its critical value), b) steady detonation (large initiator energy).

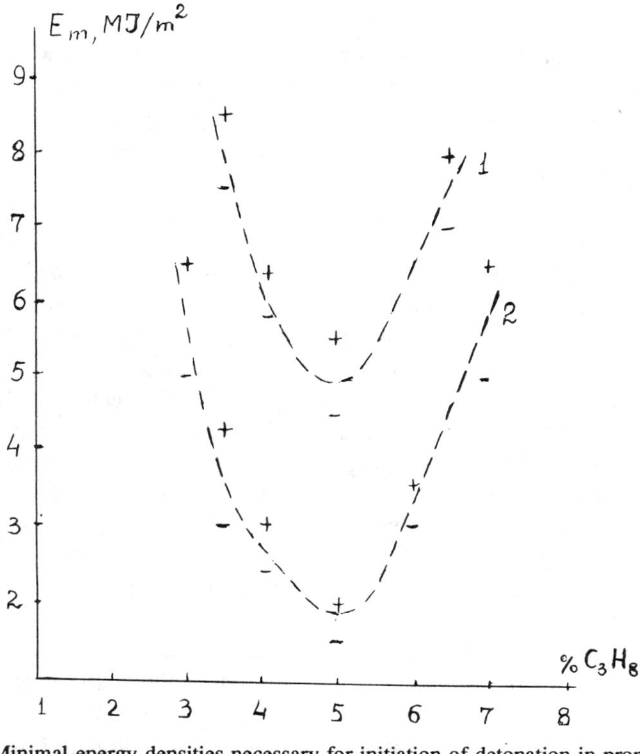

Fig. 4 Minimal energy densities necessary for initiation of detonation in propane/air mixtures of various compositions: 1) initiation by gaseous charges, and 2) initiation by condensed explosives; +, detonation, – no detonation.

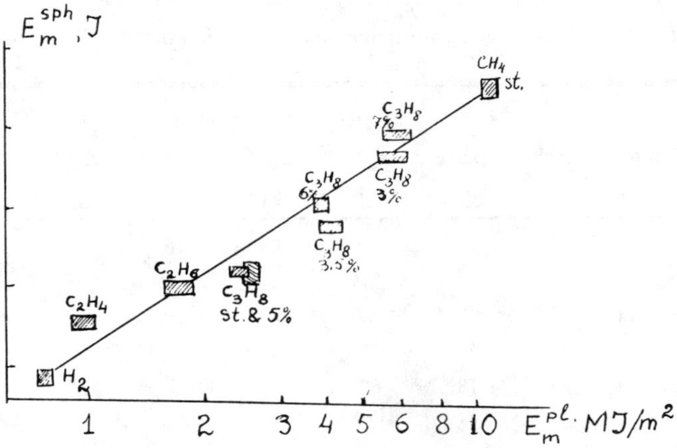

Fig. 5 Correlation between the critical energy density for plane detonation initiation and energy of spherical detonation initiation in fuel/air mixtures.

Table 1. Initiation Energies Obtained with Various Charges

Charge	Location	E_{eff}, MJ/m^2	
		no go	go
Detonating cord, spiral	Inside the tube	2.6	3.2
Nitromethane	Outside	1.25	2.7
Explosive sheet	Outside the tube	2.3	3.1

the straight line that fits the experimental data is close to three, as predicted by the Zeldovich hypothesis. Confirmation of the correlation allows for a rather reliable prediction of the energy of direct initiation of spherical detonation in methane/air mixtures. The minimal energy for plane detonation initiation in this mixture is about 10 MJ/m, which gives for the spherical case an energy of about 40 MJ (about 10 kg of TNT), that is, of the same order as the estimate reported in Ref. 10. The measured critical energy densities for plane detonation initiation are somewhat higher than those reported in Refs. 8 - 11.

Initiation by Detonation of a Gaseous Mixture

Since the initiation energy is anticipated to depend on the dynamics of energy deposition and the amplitude of a shock wave at early stages of its propagation, it would be interesting to measure minimal initiation energies for energy sources other than condensed high explosives (that are close to a point strong source). For that purpose we conducted experiments with gaseous

Table 2. Critical energy for initiation of detonation in the fuel-air mixture

Mixture	d_{cr}, cm	Source	Spherical detonation E_m^{cal}, kJ	E_m^{exp}, kJ	Source	Plane detonation E_m^{cal}, MJ/m^2	E_m^{exp}, MJ/m^2
H_2+air	13[a]	15	2.6	3.4	16	0.3	0.7
	20	17	9.6	4.7		0.5	
C_2H_2+air	11.5[a]	18	3.2	6.3	19	0.3	
	12	17	3.6	5.8			
C_2H_4+air	31	18	64	43	6	1.2	1.0
	38	17	116	64		1.5	
C_2H_6+air	59[a]	15	427	130	20	2.2	1.8
	67	17	642	170		2.5	
C_3H_8+air	60[a]	15	483	210	21	2.5	2.7
	70	17	760	340		3.0	

[a] estimated based on the formula $d_{cr}=13\lambda$.

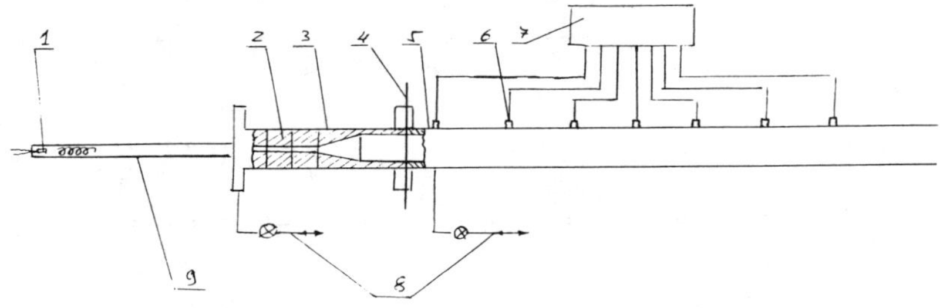

Fig. 6 Schematic of the experimental setup for initiation of gaseous charges: 1) pyrotechnical igniter, 2) inserts of a variable length, 3) initiator section, 4) diaphragm removed before firing, 5) test section, 6) pressure pickups, 7) electronic chronometers, 8) mixture inlets, and 9) narrow tube with a wire spiral.

detonation as an energy source. The setup comprised a test tube 8.5 m long and 70 mm in diameter, equipped with pressure or ionization pickups used for detonation velocity measurements, and an initiation section. The latter section consisted of a long narrow tube (17 mm in diameter and 1.65 m long) in which a detonation wave formed in the initiating gaseous mixture ignited by a weak source, and of a 70-mm-diam tube with a conical section and a variable number of inserts of a smaller internal diameter (see Fig. 6). The initiation section was filled with a stoichiometric acetylene/oxygen mixture. The diaphragm separating the two sections was removed before firing the mixture.

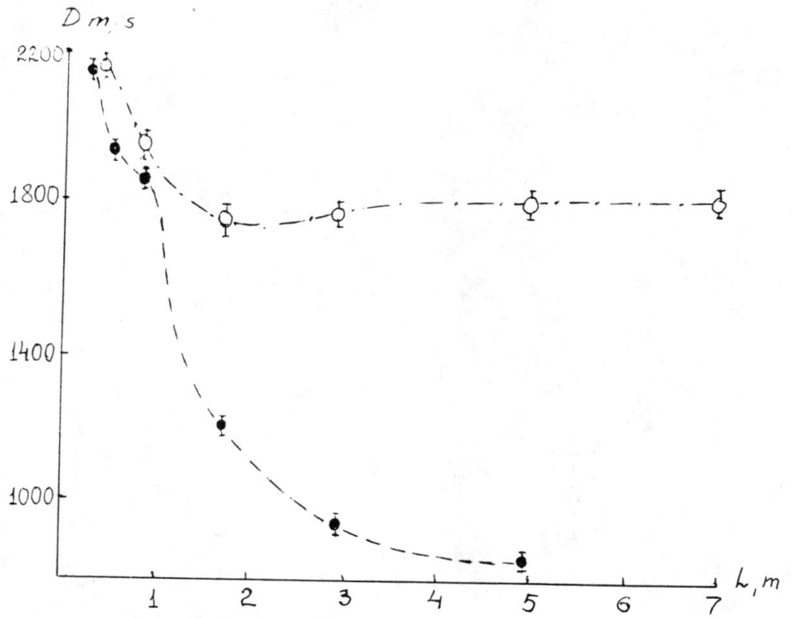

Fig. 7 Detonation velocity in a stoichiometric propane/air mixture vs distance. Energy density of the gaseous initiator: ○ 6.6 and ● 5.6 MJ/m^2.

The total energy contained in the initiating mixture varied between 2 - 15 MJ/m^2.

Propane/air mixtures of various compositions and a stoichiometric ethylene/air mixture were used as test mixtures. Propagation of shock and detonation waves was monitored by measuring the wave velocity as a function of distance at seven bases and in some experiments by analyzing the smoked foil prints and measuring the detonation cell size at different tube sections.

Typical wave velocity vs distance dependences for supercritical and subcritical initiation charges are presented in Fig. 7. It is seen in the figure that the steady C-J detonation sets in at a distance greater than 1.5 m (see also Figs. 8 and 9); that is, the characteristic time of nonsteady processes preceding the detonation onset (equal approximately to 1 ms) is commensurate with the energy deposition time, calculated as the time taken for the rarefaction wave to travel back and forth in the detonation products of the initiating charge.

It is worth noting that the cell size changes very little in Fig. 9 when the wave velocity increases from the minimum to the C-J level. This is probably due to the fact that the velocity minimum is very shallow. The time (and distance) to steady C-J detonation decreases when the energy deposition time is shorter and the density of the initiator energy is greater (see Fig. 2).

A comparison of the two types of initiators in Fig. 4 demonstrates that E_m for the gaseous initiator are about two times that for high explosives. The

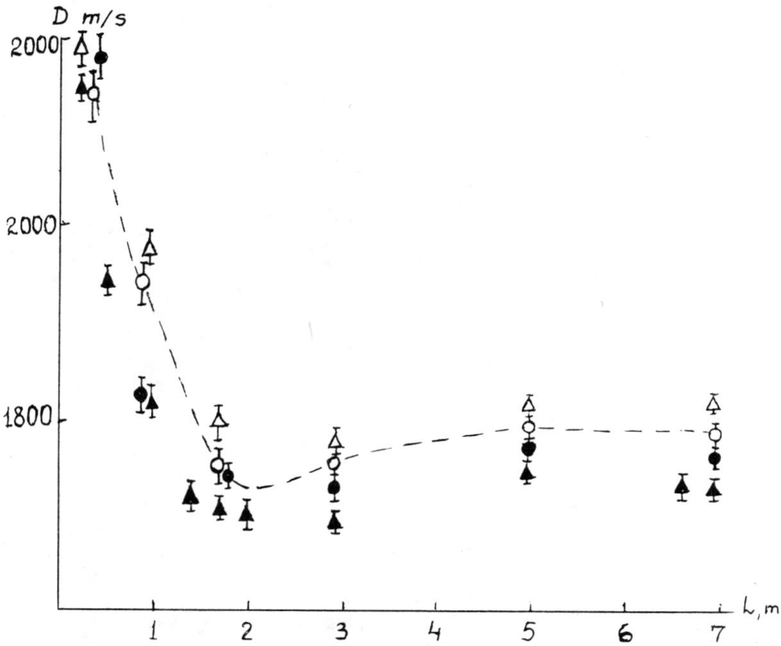

Fig. 8 Detonation velocity as a function of distance for propane/air mixtures of various composition: ○ 4%, ● 6.6%, △ 5%, and ▲ 3.5% C_3H_8.

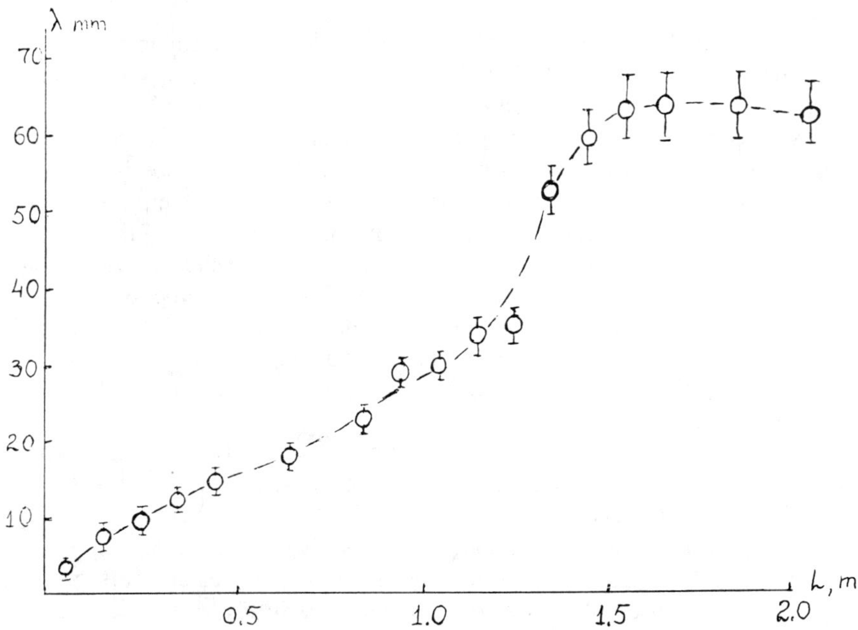

Fig. 9 Transverse cell size in detonation waves in stoichiometric propane/air mixtures initiated by a gaseous charge as a function of distance from the diaphragm.

same ratio of the minimal initiation energies was obtained for a stoichiometric ethylene/air mixture (E_m for the gaseous initiator is 2.4 MJ/m^2). This discrepancy may be accounted for by, first, a relatively low initial intensity of the shock wave and, second, a longer time of energy deposition in the case of detonating gas. The importance of the first reason is confirmed by the fact that detonation in a hydrogen/air mixture is initiated by larger gaseous charges than detonation in an ethylene/air mixture, whereas the situation with solid explosive initiating charges is the reverse. It seems reasonable to assume that when the initial shock wave amplitude is high enough to initiate the reaction in the acceptor mixture, the minimal initiation energy is determined by the work done by the detonation products. This conclusion follows from a comparison of the fractions of chemical energy that transform into work done by the detonation products of high explosives and gaseous mixtures. The ratio of these fractions is close to 1:0.5; that is, to the inverse ratio of the energies of direct detonation initiation by these charges.

Estimation of the Minimal Energy to Initiate Direct Detonation

To estimate the minimal energy required to initiate direct detonation in charges of various geometries, we used the relationship between the minimal initiation energy of spherical detonation and the critical diameter for detonation transition from a tube into a large mixture volume (d_{cr}) through a cone. The critical diameters were measured in a detonation tube described in Ref. 12. The results for various mixtures are presented in Fig. 10, together with the data of Ref. 13. As seen from the figure, the critical diameter ceases to depend on the cone angle when its value exceeds 60 deg, which simply means that with cone angles greater than 60 deg a part of the mixture at the

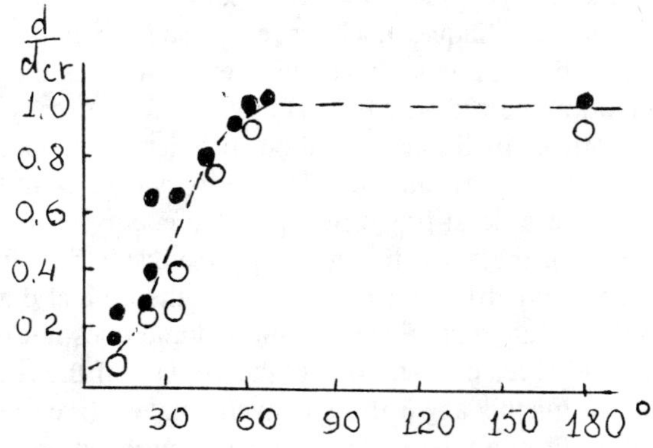

Fig. 10 The critical diameter for detonation transition from a tube to a large mixture volume as a function of the cone angle (mixtures: hydrogen/oxygen and methane/oxygen); open and closed circles pertain to "no-go" and "go" runs, respectively.

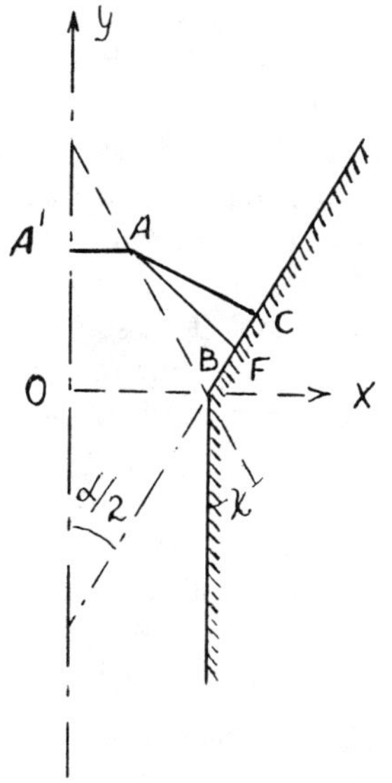

Fig. 11 Schematic of detonation wave diffraction in a cone.

periphery does not contribute to re-initiation of detonation, because the amplitude of the diffracting wave is too low and the ignition delays are too long in this region.

Propagation of a diffracting detonation wave is shown schematically in Fig. 11. Near the tube axis, a nondiffracted plane detonation front AA′ propagates; its area is decreasing continuously due to the rarefaction wave spreading inward toward the gas column contained in the cylindrical volume confined within the tube diameter. The rarefaction wave trajectory is shown schematically by the line AB. The line AF separates the unreacted mixture compressed in the lateral oblique shock wave AC and the detonation products. At the critical and subcritical conditions the line AB reaches the axis prior to reaction re-initiation in the region AFC. Hence, the total energy in this region of the gas at the critical conditions should be sufficient to support detonation; that is, one may reasonably assume that the energy contained in the region AFC is either equal, or at least proportional, to the energy of direct detonation initiation in volumes of the respective geometry (semispherical for a cylindrical tube; semicylindrical for a slot). Since the critical diameter for detonation transition to an unconfined mixture volume does not depend on the cone angle at its value greater than 60 deg, the volume of the region AFC should be calculated for this angle if the conventional d_{cr} (measured for detonation transition from a tube without a cone) is utilized.

Thus the energy of direct detonation initiation can be written as

$$E_{cr} = K(\Delta E_{int} + \Delta E_{kin})$$

Here K is a constant coefficient, and ΔE_{int} and ΔE_{kin} are the specific internal and kinetic energies of the shocked gas. To find the shock wave parameters and the gas volume, we assumed that the line AC is straight and perpendicular to the cone wall, the reaction zone length behind the plane detonation wave is much less than the critical diameter, the line AF is also straight, BF is the distance traveled by a gas particle positioned initially at point B behind the shock front, and no reaction takes place in the region AFC.

Thus, the oblique shock wave velocity is

$$D_\alpha = D_0 \cos(\chi + \alpha/2)/\cos\chi$$

The shocked gas volume and its specific energy are

$$V = \pi d_{cr}^3 (D_0/U + \text{cotan}\ \alpha/2) \sin \alpha/2 < (D_\alpha/U - u_\alpha/U)[2 + \sin \alpha/2 (D_\alpha/U + u_\alpha/U)]/24$$

and

$$\Delta E_{int} + E_{kin} = P[1 - (P_0/P)(\gamma+1)/(\gamma_0-1)]/(\gamma-1) + \rho u_\alpha^2/2$$

Here U is the component of the velocity of the lateral rarefaction wave perpendicular to the direction of detonation propagation; P, ρ, u_α, and γ are the pressure, density, gas velocity, and the ratio of the specific heats behind the shock front. The zero index labels the quantities in the initial gas mixture. The parameters of the detonation and shock waves were calculated according to the standard program.[14] The calculated critical energies practically coincide with their experimental values when K is taken to be 1.03 (the dispersion σ=1). The greatest error in calculations is introduced by the uncertainty in the critical diameter. The data are presented in Table 2.

One may hope that the minimal initiation energy density for the plane case may be found by dividing the total energy in the shocked gas by the area of the surface described by the line AC. The energies thus estimated are also presented in Table 2 and compared with the measurements.

Conclusion

Thus, the results of the investigation performed demonstrate that:

1. The critical energy density necessary for initiation of direct detonation of gaseous mixtures by plain explosive charges depends to a very minor degree on the type of charge and the tube diameter. (Experiments were carried out in tubes of two different diameters.)

2. A correlation exists between the critical energy density for plane charges and the energy required for initiation of spherical detonation, which is consistent with the theoretical predictions by Zeldovich (the cubic root dependence).

3. The critical energy for initiation of spherical detonation in a stoichiometric methane/air mixture estimated from the measured critical energy density for plane-wave initiation is about 40 MJ.

4. In accordance with the work that can be done by the detonation products on the mixture to be initiated, the gaseous charges are about two times less efficient than condensed explosive charges in the initiation of plane detonation waves in fuel/air mixtures. The gas charge efficiency should naturally be expected to be greater at higher initial pressures of the initiating mixture.

5. Both the critical energy density for plane detonation initiation and the critical energy of spherical detonation initiation may be estimated with reasonable accuracy, based on measured critical diameters for detonation transition from a tube to unconfined volume and on the relation derived in the present study.

References

[1]Zeldovich, Ya. B., Kogarko, S. M., and Simonov, N. N., "Experimental Investigation of Spherical Gaseous Detonation," Zh. T.F., Vol. 26, No. 8, Aug. 1956, pp. 1744-1768.

[2]Matsui, H. and Lee, J. H., "On the Measure of Relative Detonation Hazards of Gaseous Fuel/Oxygen and Air Mixtures," Proceedings of the 17th Symposium International on Combustion, The Combustion Institute, Pittsburg, PA, 1979, p. 1269.

[3]Shulenin, Yu. V. and Bokhon, Yu. A., "The Minimal Energy of Initiation of Unconfined Gaseous Detonation," Dokl. Akad. Nauk SSSR, Vol. 257, No. 3, Mar. 1981, pp. 680-683.

[4]Lee, J. H., "Initiation of Gaseous Detonation," Ann. Rev. Phys. Chem., Vol. 28, 1977, pp. 75-104.

[5]Sichel, M., "A Simple Analysis of the Blast Initiation of Detonations," Acta Astronautica, Vol. 4, 1977, p. 409.

[6]Bull, D. C., Elsworth, J. E., and Hooper, G., "Initiation of Spherical Detonation in Hydrocarbon-Air Mixtures," Acta Astronautica, Vol. 5, Nos. 11-12, Nov.-Dec. 1978, pp. 997-1118.

[7]Westbrook, C. K., Pitz, W. J., and Urtiew, P. A., "Chemical Kinetics of Propane Oxidation in Gaseous Detonations," Progress in Astronautics and Aeronautics, New York, Vol. 94, 1984, pp. 151-174.

[8]Ohyagi, S., Yoshihashi, T., and Harigaya, "Direct Initiation of Planar Detonation Waves in Methane/Oxygen Mixtures," Progress in Aeronautics and Astronautics, New York, Vol. 94, 1984, pp. 3-32.

[9]Borisov, A. A., Gelfand, B. E., Loban, S. A., Mailkov, A. E., and Khomik, S. V., "Investigation of Concentration Limits to Detonations of Fuel/Air Mixtures in Smooth and Rough Tubes," Khimicheskaya Fizika, Vol. 1, June 1982, pp. 848-853.

[10]Bull, D. C., Elsworth, J. E., Hooper, G., and Quinn, C. P., "A Study of Spherical Detonation in Mixtures of Methane and Oxygen Diluted by Nitrogen," Journal of Physics D: Applied Physics, Vol. 9, 1976, pp. 1991-2000.

[11]Wolański, P., Kauffman, C. W., Sichel, M., and Nicholls, J. A., "Detonation of Methane/Air Mixtures," Proceedings of the 18th Symposium International on Combustion, The Combustion Institute, Pittsburg, PA, 1981, pp. 1651-1660.

[12]Gubin, S. A., Kogarko, S. M., and Mikhalkin, V. N., "Experimental Studies into Gaseous Detonations in Conical Tubes," Fizika Gorenia i Vzryva, Vol. 18, No. 5, 1982, pp. 111-117.

[13]Kogarko, S. M., "On the Possibility of Detonation of Gaseous Mixtures in Conical Tubes," Izvestia Akad. Nauk SSSR. Otdelenie Khimicheskikh Nauk, No. 4, 1956, pp. 419-426.

[14]Gubin, S. A., Mikhalkin, V. N., Odintsov, V. V., and Shargatov, V. A., "Calculation of Parameters and Composition of Detonation Products in Low-Density Mixtures of Various Aggregate States," Khimicheskaya Fizika, Vol. 2, March 1983, pp. 420-427.

Limit Criterion of Detonation in Circular Tubes

Gabrielle Dupré*
National Centre of Scientific Research and University, Orléans, France
and
Olivier Péraldi,† Josiane Joannon,‡ John H. S. Lee,§ and Rom Knystautas§
McGill University, Montreal, Quebec, Canada

Abstract

Detonations in lean near-limit mixtures of H_2/air at atmospheric initial pressures are studied in a detonation loop comprised of five tubes of internal diameter of 152, 97, 74, 49, and 38 mm. The length of the tubes, initially equal to 9.4 m, has been increased to 11 m. The mixtures are prepared by the method of partial pressures in the evacuated loop, and mixing is achieved by a bellows-type recirculating pump. Ignition at the end of the largest 152-mm-diam tube is by a solid explosive charge. Diagnostics consist of velocity measurements by a series of ionization probes spaced at 0.5-m intervals at the beginning and at the end of each of the five tubes of the loop. The tubes are connected via a 180-deg semicircular bend. To eliminate the strong perturbations associated by the sudden area change from one tube to the other and by the curved bend itself, a tube section with a length over diameter ratio of 6 and consisting of an acoustic attenuating wall liner is placed at the beginning of each tube. To induce transition, a short section of rough-walled liner is inserted into the tube just after the damping section. The wall roughness is generated by a 2.5-mm-deep spiral in the liner or by the length of ring obstacles.

The present results indicate that contrary to the previous postulate that a detonation cannot propagate for a ratio of cell size over tube diameter λ/d larger than π,

*Chargée de Recherche, Research Center on the Chemistry of Combustion and High Temperatures (C.R.C.C.H.T., C.N.R.S.).

†Research Associate, Department of Mechanical Engineering (Engineer, currently with Matra Défense, Vélizy, France).

‡Graduate Student, Department of Mechanical Engineering.

§Professor, Department of Mechanical Engineering.

transition to detonation has been observed for $\lambda/d > \pi$. However, results obtained to date indicate that no transition occurs for $\lambda/d \geq 2\pi$. The relationship between cell size and tube diameter for a single-headed spin in a circular tube is not clear, since single-headed spin structure is unique to the circular tube geometry. Hence, it is difficult to define the structure in the limit $\pi \leq \lambda/d \leq 2\pi$. Nevertheless, it appears reasonable to choose the ratio of the chemical length scale λ to the geometrical length scale of the tube d to define a limit criterion. The present study also argues that although a nonattenuating "detonation-like" phenomena can be generated in a tube if a sufficiently strong igniter is used, true detonation should correspond to conditions in which transition can occur and, hence, where transition and detonation limits are similar. The present results also imply the necessity of transverse waves and that their generation and amplification are essential to the formation and self-sustenance of a detonation.

Introduction

Mixtures of combustible gases and vapors are often present in the piping system of industrial processing equipment. The possibility of detonations occurring is of a great concern. The propagation limits of detonation in tubes have not been established. Theoretical models based on a one-dimensional structure with heat and momentum losses fail to predict the limits. It is clear that the three-dimensional cellular structure of the detonation must be considered to describe adequately the propagation mechanism of the detonation realistically. Thus far, numerical simulation of cellular detonation has been limited to two-dimensional geometry. Furthermore, the recent work of Sugimura et al.[1] indicates that the resolution required to describe the wave instability (hence the limit mechanism) may be beyond current computing capacities even for the two-dimensional case.

Recognizing the cellular nature of detonation, various postulates have been advanced to link the cell size λ to the tube diameter d for a limit criterion. Shchelkin[2] suggested that $\lambda = d$, whereas Dove and Wagner[3] proposed that the single-headed spin mode is the lowest stable mode and, hence, represents the detonation limit. In a two-dimensional rectangular channel of height H, the single-headed spin mode then gives a limit criterion of $\lambda = 2H$. However, there are strong indications that the detonation process is three-dimensional and that the limit criterion $\lambda = 2H$ must apply to the minimum dimension of the cross-sectional area. For round tubes, Moen et al.[4] arrived at the single-headed spin limit of $\lambda \simeq 1.7d$ on the assumption that the spin pitch corresponds to the cell length. Lee[5] and Dupré et al.[6] suggested that $\lambda = \pi d$ as a limit criterion on the basis that the circumference πd represents the largest physical length scale that should be linked to the chemical length scale λ for a resonant coupling between the chemical and gasdynamics processes. It should be noted that $\lambda = \pi d$ does not represent a single-headed spin in a circular tube. The wave configuration corresponding to $\lambda = \pi d$ is illustrated in an earlier paper by Denisov and Troshin[7] to consist of two transverse waves sweeping alternately in the x and y directions; z is the direction of the tube axis. A single-headed spin configuration would then correspond to $\lambda = 2\pi d$. However, the relationship between the cell size and tube diameter for single-headed spin is unclear, since the spin structure is unique to the circular tube geometry. On the basis of some experimental data of Knystautas et al.[8], Lee further suggested that the abrupt transition to detonation is governed by the $\lambda \simeq d$ criterion. Thus, transition requires a more sensitive mixture (that is, a mixture with a lower cell size) than for propagation. All of the above postulates have not been rigorously confirmed by experiments.

The determination of limits in tubes is an extremely difficult task. Near-limit mixtures are insensitive. Thus, if a weak ignition source is used, transition from flame to detonation may not be possible even in very long tubes. If a very strong initiator is used (e.g., a solid explosive charge or a high-pressure, detonation-driven section), the strong transverse perturbations generated may never decay (at least over the length of the tube used). As pointed out by Manson[9], "detonations have a very long memory of their initial conditions." It is well established now that transverse waves are necessary to sustain detonation propagation.[10] Thus, if the igniter generates strong transverse perturbations on the front and the perturbations do not attenuate due to energy supply by the chemical reactions to compensate for any losses, a detonation-like complex of three-dimensional interacting waves with chemical reactions supporting them can propagate without attenuation. This could account for the ability of Wolanski et al.[11] to detonate methane/air mixtures in a 5x5-cm channel even though the cell size for stoichiometric CH_4/air mixture is of the order of 70 cm. Further support of the above argument can be found in the

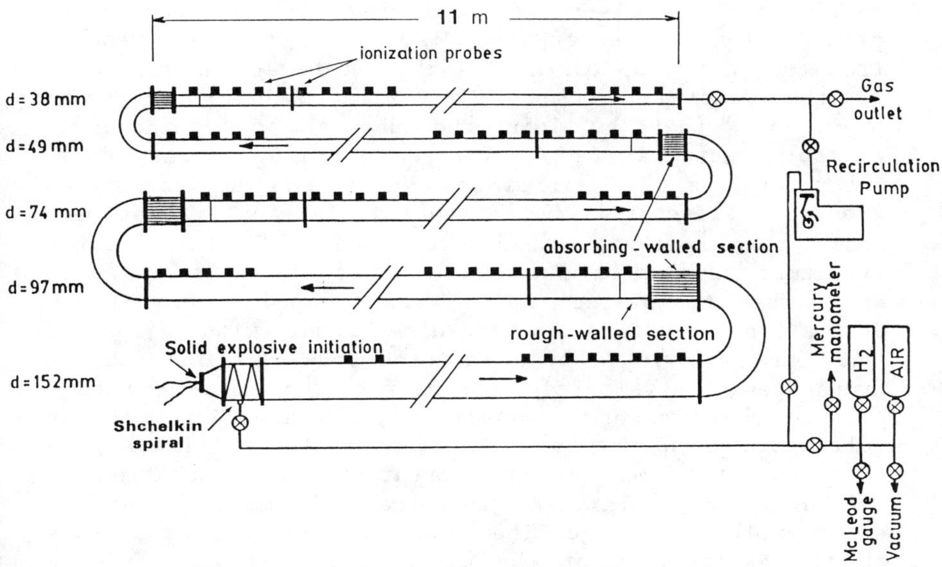

Fig. 1 Schematic of the detonation loop and associated diagnostics.

experiment of Saint-Cloud,[12] who demonstrated that if the three-dimensional wave system of a limit detonation were to be perturbed by an obstruction in its path, the wave would fail. Thus, it appears that it is important to study the near-limit propagation in tubes to elucidate further on the near-limit phenomena and clarify some of these proposed criteria. The present study describes some recent results of such an investigation.

Experimental Details

Five circular tubes [152, 97, 74, 49, and 38-mm internal diam and 9.4-m long (in most recent experiments 11-m long)] are connected together by semicircular bends to form a continuous loop of decreasing tube diameter (Fig. 1). Since the bends are of a constant cross section matching the larger tube, there is a sudden area change as the detonation propagates past the bend from a larger to the next smaller diameter tube. It has been found that the strong perturbations produced by the abrupt area change (as well as the complex perturbations generated by the bend itself) remain coupled to the detonation, resulting in large fluctuations.[6] Thus, a damping section with a length over diameter ratio of 6 is added to each tube following the bend to eliminate the transverse perturbations. Moreover, the diameter is smoothly reduced from the larger to the next smaller tube at the beginning of the damping section (Fig. 2). It has been found previously[10] that, by lining the tube wall with an acoustic absorbing material (e.g., a layer of wire screens, foam, fiberglass, etc.), transverse waves on the detonation front can effectively be eliminated. In fact, it has been shown that even the transverse waves of the cellular structure can be eliminated to cause the wave to fail. When the transverse waves are damped, the failed detonation structure is one-dimensional with a planar shock front followed by a reaction zone, and the complex propagates at a velocity about half the normal C-J value.

In the present experiment a 1.25-cm-thick layer of rolled stainless steel wire screens protected with a grid is placed to damp out the perturbations. Detonation is initiated in the largest tube (i.e., 152-mm i.d.) with a solid explosive charge. The amount of charge used depends on the sensitivity of the mixture. A Shchelkin spiral, located just after the initiating section, shortens the transition to detonation in the largest tube. In the present study, lean mixtures of hydrogen/air are studied. The mixture composition is determined by the method of partial pressures of the different components in the evacuated tube. A bellows-type pump is used to circulate the gas inside the loop during 30 min to ensure proper mixing.

It is also found that the damping section causes the detonation to fail. Transition back to detonation in the subsequent smooth-walled tube may not occur within the length available for propagation. Thus, to promote

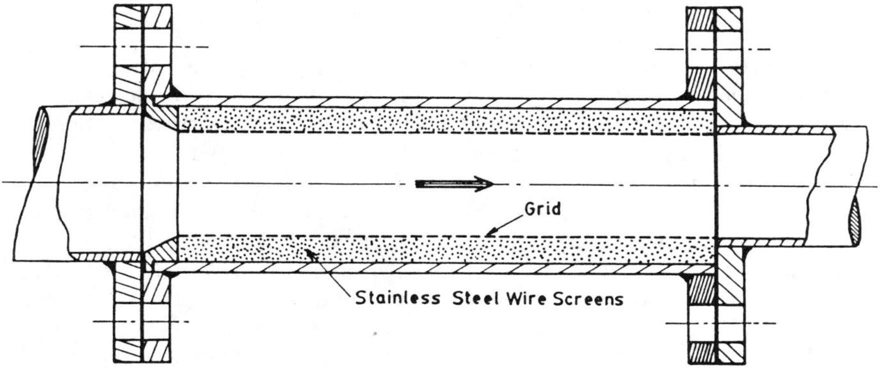

Fig. 2 Schematic view of the damping section.

transition a short rough-walled section with a length over diameter ratio of 3 is added to each tube following the damping section. As pointed out previously, large perturbations on the detonation front induced by obstacles of large blockage ratio may take a long time to decay. This will negate the effects of the damping section, whose purpose is to eliminate the large transverse perturbations generated by the abrupt area change and by the bend. Hence, in the present study only relatively small wall roughness is used instead of using obstacles or a Shchelkin spiral of large wire diameter. A thin liner, in which an internal screw thread of about 2.5-mm-deep is cut, is inserted into the smallest tubes to provide a rough-walled section to shorten the transition distance. For the largest tubes the rough-walled section consists of a length of ring obstacles. It is envisaged that the transverse perturbations induced by the grooves are weak and will neither cause large velocity fluctuations nor artificially support the wave, as in strong ignition.

All of the experiments are carried out at an initial pressure of 760 Torr and at ambient temperature. Diagnostics consisted of velocity measurements by a series of ionization probes spaced 0.5 m apart at the beginning and at the end of each tube in the loop. Smoke foils have been used in several experiments to confirm the occurrence

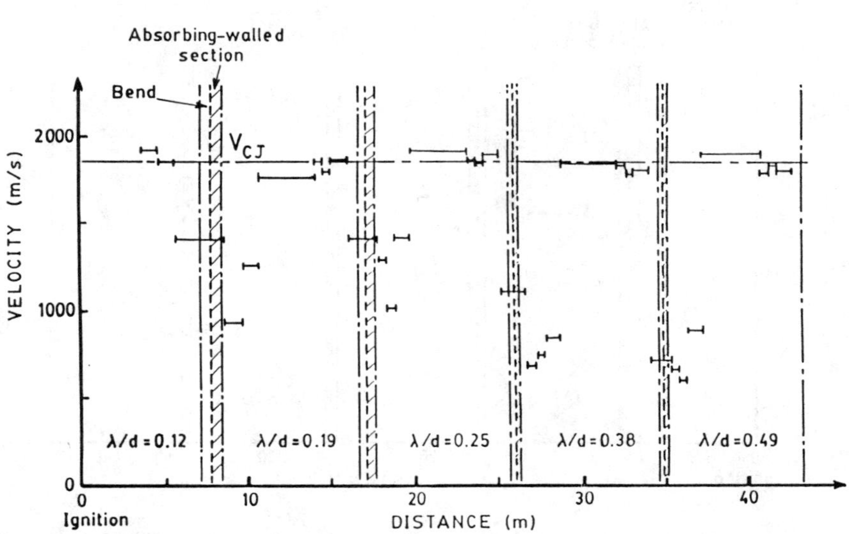

Fig. 3 Evolution of experimental detonation velocity with distance from ignition for a mixture concentration of 25 mol % hydrogen. Comparison with theoretical Chapman-Jouguet velocity V_{CJ}. Fiberglass is used as absorbing material and no rough-walled section is added.

of a detonation wave at the tube end. However, the cell size used is that one experimentally determined by Knystautas et al.[13] in the corresponding mixture. A schematic sketch of the detonation loop with the associated gas handling system and diagnostics is shown in Fig. 1.

Results and Discussion

Figure 3 shows the typical result for the velocity variation in the different tubes of the loop. The mixture contains a molar concentration of hydrogen in air equal to 25 percent (i.e., 25 mol % H_2). The corresponding cell size λ is equal to 18.5 mm for this mixture. Thus, λ/d = 0.12, 0.19, 0.25, 0.38, and 0.49 after the damping section of each tube from the larger to the smaller one. We note that the experimental detonation velocity is quite close to the normal Chapman-Jouguet (C-J) value before transmission through the first semicircular bend and damping section. Upon transmission the velocity drops to about 1000 m/s or less, but transition occurs readily, bringing the detonation back to its normal C-J value. In this experiment a layer of fiberglass insulation is used in the damping section, and no rough-walled section is added to

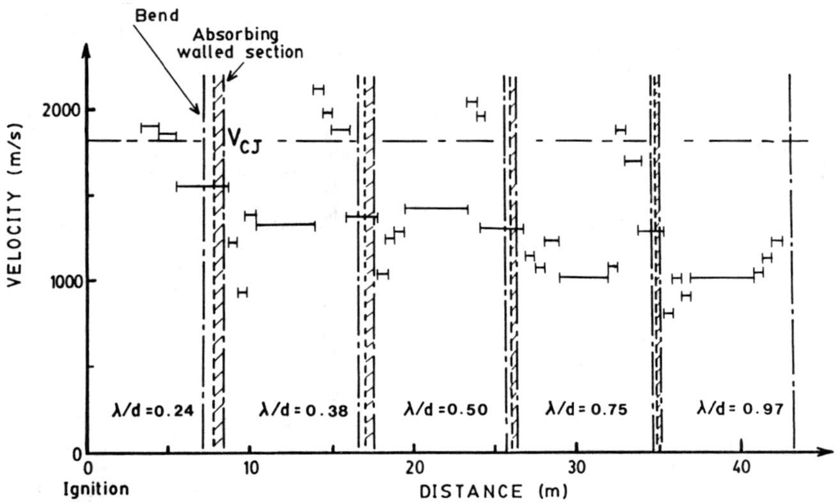

Fig. 4 Evolution of experimental detonation velocity with distance from ignition for a mixture concentration of 22 mol % hydrogen. Comparison with theoretical Chapman-Jouguet velocity V_{CJ}. Metallic screens are used as absorbing material and no rough-walled section is added.

promote transition. It is to be noted that the velocities measured throughout the bend and the damping section and in the middle part of the tubes are a mean value over a long distance.

In Fig. 4 the mixture composition is 22 mol % H_2 and λ = 37 mm. The corresponding values for λ/d = 0.24, 0.38, 0.50, 0.75, and 0.97 for all of the tubes used. Layers of stainless steel wire screens are used in the damping section in this experiment (and in those shown in the following figures), since it is found necessary to employ a stronger damping material. The fiberglass insulation is destroyed after a few experiments, necessitating its replacement. In this experiment no rough-walled section is used. It is noted that in this less sensitive mixture the onset of detonation is a more violent event, resulting in an overdriven detonation that subsequently decays, as shown in the tubes of d = 94 mm and 74 mm. In the last tube, $\lambda/d \leq 1$, and transition from flame to detonation is observed no more, as in the experiments of Knystautas et al.[8]

For mixtures with lower H_2 concentration, i.e., 20 mol % H_2, as shown in Fig. 5, transition to C-J detonation is not obtained even in tubes corresponding to a λ/d ratio as small as 0.57.

In order to shorten the transition distance and to observe the behavior of the wave for a longer time, a

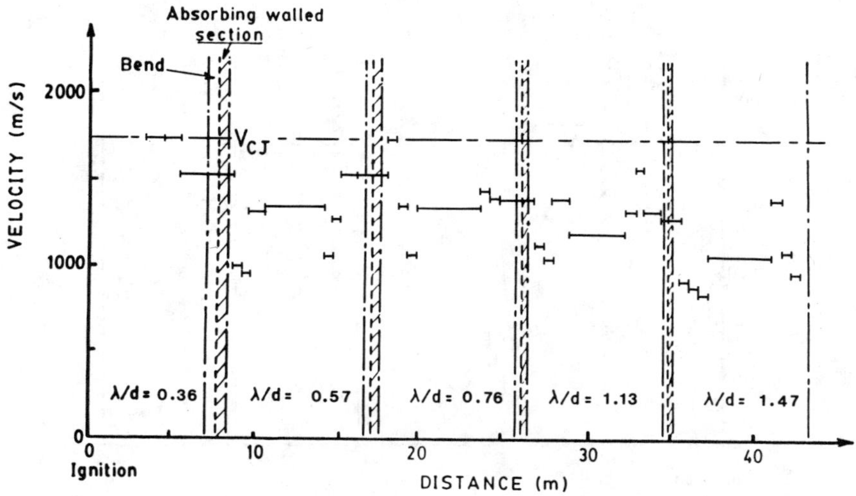

Fig. 5 Evolution of experimental detonation velocity with distance from ignition for a mixture concentration of 20 mol % hydrogen. Comparison with theoretical Chapman-Jouguet velocity V_{CJ}. Metallic screens are used as absorbing material and no rough-walled section is added.

2.50-m extension has been added to each tube and a rough-walled section installed after the damping section. Figure 6 gives results for the same mixture as shown in Fig. 5, i.e., 20 mol % H_2, except that there is a rough-walled section placed at about 2 m after the damping section. Transition to detonation readily occurs as the wave passes through the rough section for $\lambda/d = 0.57$, 0.75, and 1.12, and the velocity remains close to the normal C-J value. In the smallest tube ($\lambda/d = 1.46$), the rough-walled section makes the velocity increase, then large velocity fluctuations are observed. No self-sustained, steady C-J detonation wave can be obtained within the tube length. For the same mixture, i.e., 20 mol % H_2, if the rough-walled sections are positioned just after the damping sections, transition to detonation can be achieved in the smallest tube corresponding to $\lambda/d = 1.46$, as shown in Fig. 7.

This is in contradiction to the previous postulate[5] that $\lambda/d < 1$ for abrupt transition to occur. In the experiments of Knystautas et al.[8] where transition is not observed for $\lambda/d \geq 1$, a smooth tube is used in contrast to the present case, in which small transverse perturbations in the front are generated by the short rough section. Lee[5] also postulated that, for a slow mode of transition

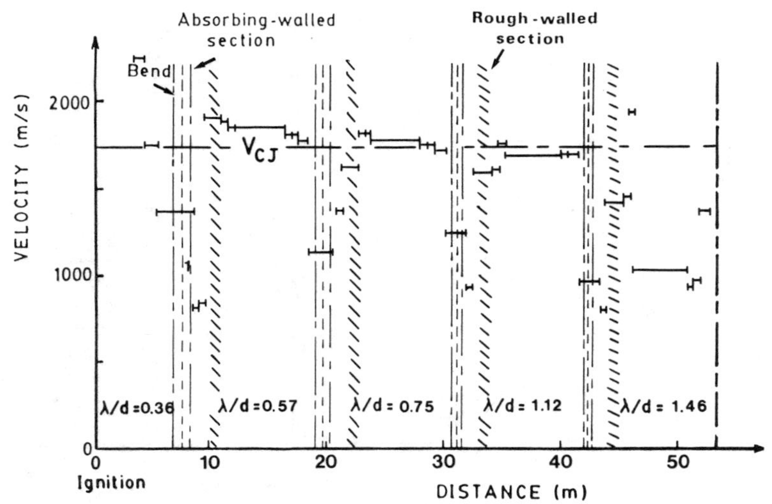

Fig. 6 Evolution of experimental detonation velocity with distance from ignition for a mixture concentration of 20 mol % hydrogen. Comparison with theoretical Chapman-Jouguet velocity V_{CJ}. Metallic screens are used as absorbing material and a rough-walled section is added 2 m after the damping section.

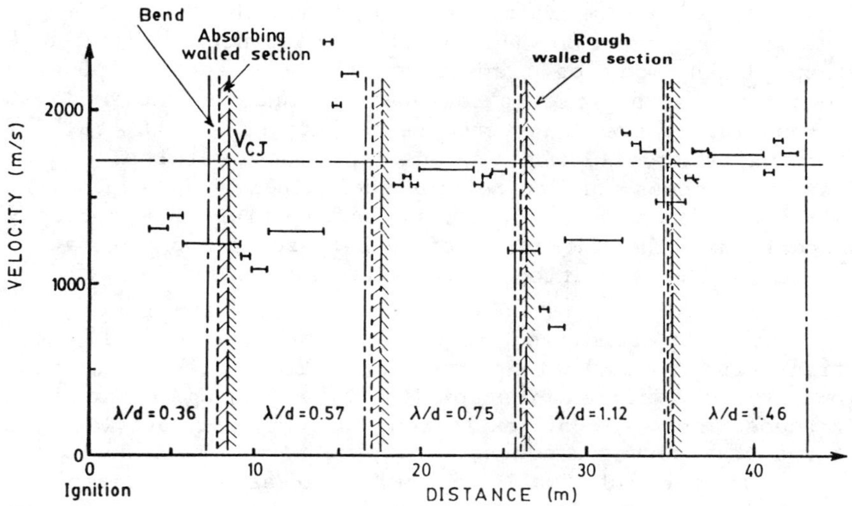

Fig. 7 Evolution of experimental detonation velocity with distance from ignition for a mixture concentration of 20 mol % hydrogen. Comparison with theoretical Chapman-Jouguet velocity V_{CJ}. Metallic screens are used as absorbing material and a rough-walled section is added just after the damping section.

to occur, $\lambda/d \leq \pi$. The slow mode of transition is due to the progressive growth and amplification of transverse waves, in contrast to the fast mode, where the onset of detonation is abrupt with a strong local explosion initiating the detonation and a retonation. Hence, in the present experiments, where transverse waves are generated by the rough section of the tube, transition probably occurs due to the amplification of these transverse waves. Thus, transition to detonation is observed for all the tubes used and, therefore, well within the limits as specified by the various criteria.

In Fig. 8, where the H_2 concentration is 17.75 mol % and λ = 117 mm, the λ/d ratio for the smallest tube is 3.29, just in excess of $\lambda/d = \pi$ for the slow mode to occur according to the criterion suggested by Lee.[5] However, it is found that transition is still observed.

Figure 9 gives results for a mixture with lower concentration, i.e., 16 mol % H_2 and λ = 253 mm. The λ/d ratio for the last three tubes are 3.39, 5.10, and 6.63, corresponding to d = 74, 49, and 38 mm, respectively. Transition to detonation is observed for λ/d = 3.39 and 5.10. This confirms that with some slight initial transverse perturbations transition can occur for $\lambda/d > \pi$.

However, for $\lambda/d > 2\pi$, transition is not observed.

It appears that the previous distinction between a fast abrupt mode of transition, where a local explosion center marks the onset of detonation, and the slow mode of detonation, where transverse perturbations amplify to form eventually the cellular detonation, is an arbitrary one. The two modes of transition definitely correspond to different phenomena, depending on initial and boundary conditions. The formation of a self-sustained detonation requires the formation of a cellular structure. If the initial condition is that of a weak ignition source and the boundary condition is that of a smooth wall where transverse perturbations are not formed, the transition must occur via an instability. The strong blast wave produced by a local explosion center will provide the transverse perturbation necessary to form a self-sustained wave. If the tube wall is rough and acoustic waves are emitted from it, these transverse acoustic waves form the basis for the development of the eventual self-sustained cellular detonation. If a very strong transverse perturbation is formed initially by a powerful ignition source, it is only a question of maintaining this strong wave complex from decaying by the chemical energy release to effect a detonation-like phenomenon. This is the case of the CH_4/air detonation observed by Wolanski et al.[11] in the

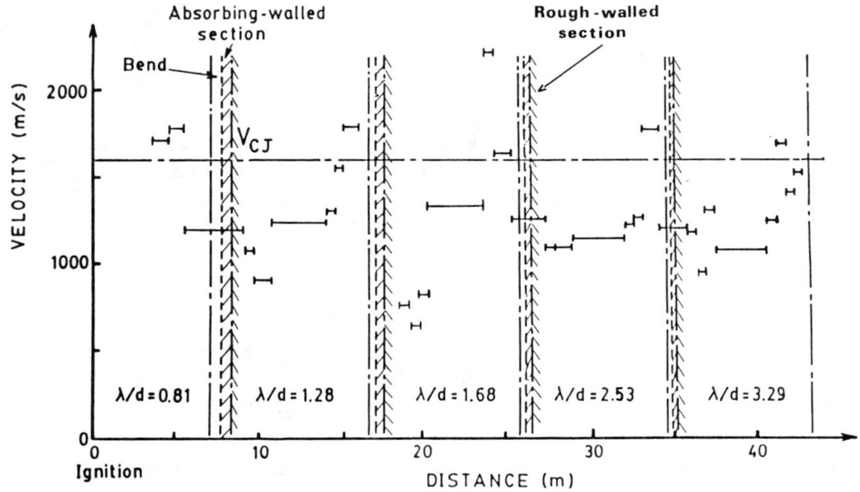

Fig. 8 Evolution of experimental detonation velocity with distance from ignition for a mixture concentration of 17.75 mol % hydrogen. Comparison with theoretical Chapman-Jouguet velocity V_{CJ}. Metallic screens are used as absorbing material and a rough-walled section is added just after the damping section.

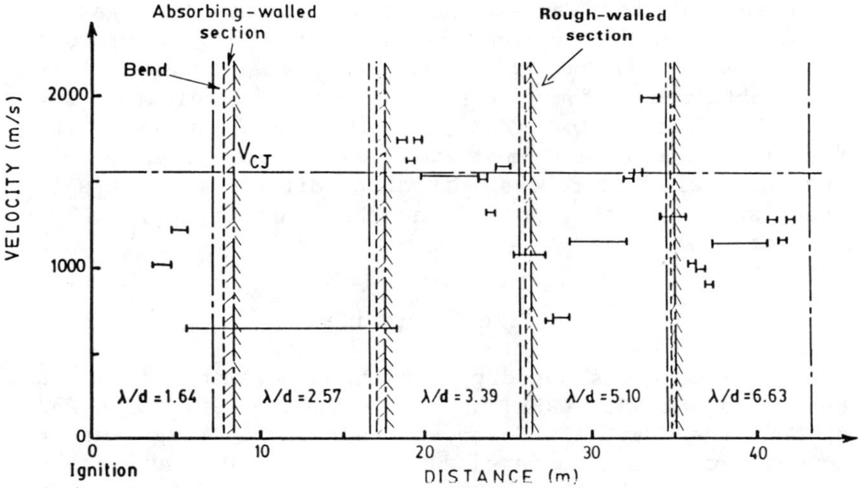

Fig. 9 Evolution of experimental detonation velocity with distance from ignition for a mixture concentration of 16 mol % hydrogen. Comparison with theoretical Chapman-Jouguet velocity V_{CJ}. Metallic screens are used as absorbing material and a rough-walled section is added just after the damping section.

5x5-cm tube. It should be noted that strong transverse perturbations are caused by strong igniters or by the volumetric explosion of shock heated gases. In the present experiments the rough-walled section produces only acoustic strength transverse waves as the shock propagates past it. Hence, the detonation formed is not an artificially induced detonation-like phenomenon with a strong igniter.

Conclusions

The present study elucidates on the numerous limit criteria that have been suggested previously. We should acknowledge that, although a detonation-like wave complex can be generated and propagate with relatively constant velocity (without noticeable decay over long distances), if a very powerful igniter is used initially to produce a strong transverse wave a true detonation is probably one that forms on its own accord via transition. Thus, the limit criterion is in fact the same as the transition criterion, as suggested by Lee.[5] The distinction between the two modes of transition is an artificial one, although they correspond to different identifiable phenomena. Both modes lead to the formation of the transverse waves that are essential for self-sustained detonation propagation. The criterion for detonation limits given previously, i.e., $\lambda/d \leq \pi$ is not found to be representative of the data obtained. The present results indicate that transition occurs for $\lambda/d \leq 2\pi$. It should be noted that extensive experiments must be carried out over a wide range of fuels as well as initial conditions to eventually establish firmly the $\lambda/d \leq 2\pi$ as the limit criterion for a circular tube.

Acknowledgments

This work was supported by the Department of National Defense of Canada [DRES] under Contract 01SG W7702-6-2521-A and the Natural Sciences and Engineering Research Council of Canada under NSERC Grants A-7091 and A-3347. One of the authors (G. Dupré) acknowledges support from canadian-french cooperation in the framework of CNRC/CNRS program.

References

[1]Sugimura, T., Fujiwara, T., and Lee, J.H., "Cellular Detonation-Instability and Substructure," 22nd Symposium (International) on Combustion, Univ. of Washington, Seattle, WA, 1988.

[2]Shchelkin, K., "Instability of Combustion and Detonation in Gases," Soviet Physics Uspekhi., Vol. 8, No. 5, New York, 1966, p. 780.

[3]Dove, J.E., and Wagner, H.G., "A Photographic Investigation of the Mechanism of Spinning Detonation," 8th Symposium (International) on Combustion, William and Wilkins, Baltimore, MD, 1962, pp. 589-600.

[4]Moen, I.O., Donato, M., Knystautas, R., and Lee, J.H., "The Influence of Confinement on the Propagation of Detonation near the Detonability Limits," 18th Symposium (International) on Combustion, The Combustion Institute, Pittsburgh, PA, 1981, pp. 1615-1622.

[5]Lee, J.H.S., "On the Transition from Deflagration to Detonation," in "Dynamics of Explosions," Progress in Astronautics and Aeronautics, edited by J.R. Bowen, J.-C. Leyer, and R.I. Soloukhin, AIAA, Vol. 106, New York, 1986, pp. 3-18.

[6]Dupré, G., Knystautas, R., and Lee, J.H., "Near-limit Propagation of Detonation in Tubes," in "Dynamics of Explosions", Progress in Astronautics and Aeronautics, edited by J.R. Bowen, J.-C. Leyer, and R.I. Soloukhin, AIAA, Vol. 106, New York, 1986, pp. 244-259.

[7]Denisov, Y.N., and Troshin, Y.K., "On the Mechanism of Detonative Combustion," 8th Symposium (International) on Combustion, William and Wilkins, Baltimore, MD, 1962, pp. 600-610.

[8] Knystautas, R., Lee, J.H., Péraldi, O., and Chan, C., "Transmission of a Flame from a Rough to a Smooth-walled Tube," in "Dynamics of Explosions," Progress in Astronautics and Aeronautics, edited by J.R. Bowen, J.-C. Leyer, and R.I. Soloukhin, AIAA, Vol. 106, New York, 1986, pp. 37-52.

[9] Manson, N., Private communication to J.H. Lee, 13th Symposium (International) on Combustion, Univ. of Utah, Salt Lake City, UT, 1971.

[10] Dupré, G., Péraldi, O., Lee, J.H., and Knystautas, R., "Propagation of Detonation Waves in an Acoustic Absorbing-walled Tube," in "Dynamics of Explosions", Progress in Astronautics and Aeronautics, edited by A.L. Kuhl, J.R. Bowen, J.-C. Leyer, and A. Borisov, AIAA, Vol. 114, Washington, DC, 1988, pp. 248-263.

[11] Wolanski, P., Kauffman, C.W., Sichel, M., and Nicholls, J.A., "Detonation of Methane-air Mixtures," 18th Symposium (International) on Combustion, The Combustion Institute, Pittsburgh, PA, 1981, pp. 1651-1661.

[12] Saint Cloud, J.P., "Contribution to the Study of the Propagation of Unstable Automonous Detonations in $C_3H_8 + 5\ O_2 + 10\ N_2$ Mixture," Master's Thesis, Univ. of Poitiers, Poitiers, France, 1971.

[13] Knystautas, R., Lee, J.H., and Guirao, C.M., "The Critical Tube Diameter for Detonation Failure in Hydrocarbon-air Mixtures," Combustion and Flame, Vol. 48, No. 1, 1982, pp. 63-83.

Effects of a Negative Step of Fuel Concentration on Critical Diameter of Diffraction of a Detonation

D. Desbordes*
Laboratoire d'Energétique et de Détonique, URA 193 au CNRS, ENSMA, Poitiers, France
and
A. Lannoy†
Electricité de France, Direction des Etudes et Recherches, Saint-Denis, France

Abstract

Diffraction of detonation wave from a pipe into a large volume, in which a compositional change (stoichio to lean) occurs stepwise just before the interface, was investigated. Laboratory-scale experiments were conducted with C_2H_2/O_2 and H_2/O_2 mixtures. The two explosive mixtures (stoichio and lean) were separated by a mylar film of 15 μm of thickness or by a removable plate. Deviations from values given by the impedance mismatching method are observed for the diffracting overdriven detonation characteristics (pressure and velocity). These deviations are explained by the bidimensional effects on the expansion behind the Chapman-Jouguet detonation wave propagating in the stoichiometric mixture contained in a pipe of small diameter (d = 52 mm).

The dynamic parameters of the diffracting detonation, namely the size λ of the cellular structure and the critical diameter of diffraction d_c that follows the classical d_c = 13 λ rule, are significantly reduced from their CJ values. Assuming that λ_{CJ} and λ are the sole scaling factors governing the detonability of a system, we have extrapolated the conclusions of the small-scale experiments to large-scale events. For instance, we have considered the problem of H_2-air detonations (involved in explosion hazards in nuclear power plants) where a reduction of about 2 is found on the critical diameter of diffraction in a lean

*Assistant Professor.
†Senior Research Engineer.

mixture of equivalent ratio of 0.4, when bidimensional effects are neglected.

Introduction

When a large amount of hydrocarbon or hydrogen is released and mixed with the air contained in rigid confinements of different shapes and volumes connected together, large compositional changes may arise in the different parts of the confinement. In such a situation, if ignition of the reactive mixture occurs the behavior of combustion depends significantly on the local initial and boundary conditions: therefore, the problem of the flame propagation regime is complex.

We have considered here a simple academic case that introduces the effects of a compositional change of the reactive mixture on the condition of propagation of a detonation. A near stoichiometric reactive mixture (equivalent ratio $\phi=1$) is contained in a long pipe of circular section of i.d. d connected with a large volume where a lean reactive mixture ($\phi < 1$) is located.

Because at stoichiometry the majority of reactive mixtures presents the greatest sensitivity to detonation (especially in confined situations through, for instance, Deflagration to Detonation Transition mechanisms enhanced by some existing obstacles in the tube), the detonation regime is assumed in the pipe, travelling at CJ conditions towards the large volume. The plane detonation, just before the diffraction into the large volume, encounters a compositional change. This compositional change is supposed to occur stepwise.

The diffraction of a detonation into a large volume in a homogeneous reactive mixture has been widely studied and appears now as a classical problem.[1,2] The situation considered here, when the combined effects of changes of geometry of propagation and composition occur, is, however, more complex. The condition of diffraction of the detonation depends strongly on the transmission of the self-sustained detonation from the stoichio to the lean reactive mixture. In a general way, just after such a compositional change the resulting detonation must be an overdriven wave and can be known a priori. The size λ of its tridimensional structure, obtained for a moderate degree of overdrive of the detonation, i.e., $D/D_{CJ} < 1.2$ (as expected for transmissions of a CJ detonation from $\phi = 1$ mixture to $\phi = 0.2 - 0.4$ mixtures), gives the size d_c of the critical diameter of diffraction by the classical relationship $d_c = 13\ \lambda$ as shown in Ref. 3 for overdriven detonation diffraction.

The purpose of the experimental investigation described in this study is to determine the critical conditions of diffraction of a detonation into a large volume where, just before the orifice of diffraction, compositional change occurs. We will focus especially on the problem of transmission of a detonation between two media of different compositions.

Laboratory-scale experiments are conducted with C_2H_2-O_2 mixtures. Characteristics of the transmitted quasisteady overdriven detonation wave obtained are compared to theory. Systematic lower experimental values are reasonably explained on the basis of the bidimensionnal effects behind a sonic wave.

The results obtained are extrapolated to large-scale detonation phenomena, as in H_2-air mixtures, assuming that the scaling factor of the detonability of a system is reduced to λ.

Schematization of the Problem

The determination of the critical diameter of diffraction with both change in geometry of propagation and proportion of fuel (CnHm or H_2) in the reactive mixtures needs for the first study presented here some convenient simplifications. The problem is presented as follows:

1) The stoichiometric mixture (ϕ = 1: mixture A), where the self-sustained detonation is produced, is confined in a rigid cylindrical tube closed at one end. This tube is connected at the other end to a bigger volume containing the lean mixture (ϕ < 1 mixture B). Its diameter is d and length L so that L > d.

2) Initial pressure and temperature in the two mixtures are the same, i.e., p_o = 1 bar and T_o = 293 K, respectively.

3) The detonation is generated in the mixture A, far from the interface with mixture B, so that it can be regarded as steady and plane when it encounters the interface.

4) The change of composition is assumed to be abrupt (negative step of concentration of fuel). The change of geometry occurs slightly later in the direction of propagation of detonation. See Figs. 1a-1b. (This is due to the experimental observation that there is some short reestablishment delay of the steady detonation in B.).

The problem is now reduced to two essential studies:

1) The determination of the detonation regime in mixture B, following the negative step of fuel concentration.

2) The evaluation of critical conditions that may yield the onset of the detonation of B undergoing an abrupt change

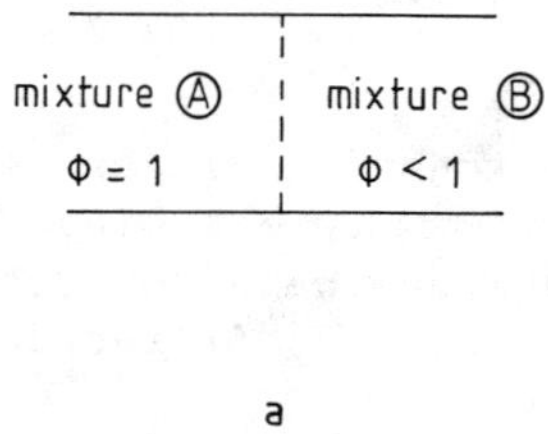

Fig. 1a Transmission of detonation from mixture A to B.

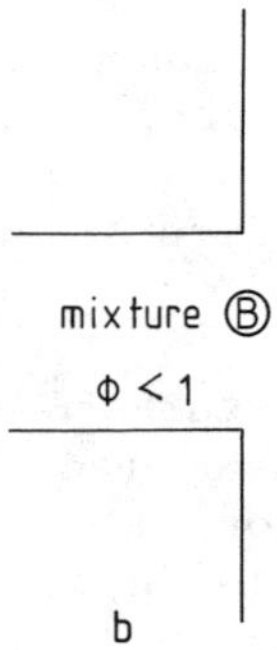

Fig. 1b Diffraction of detonation in B.

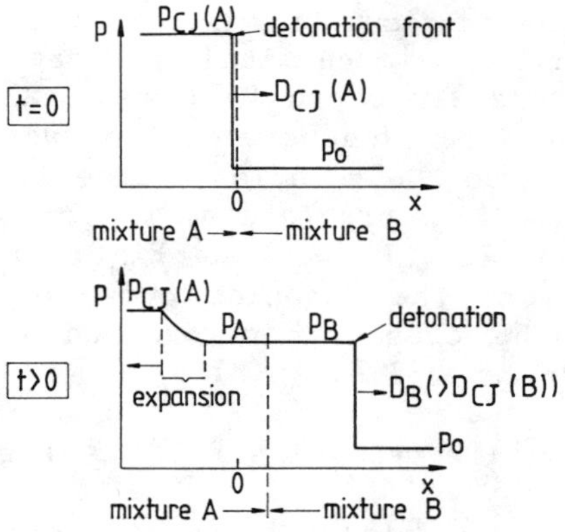

Fig. 2a Sketch of the transmission of a CJ detonation wave from the medium A to the medium B.

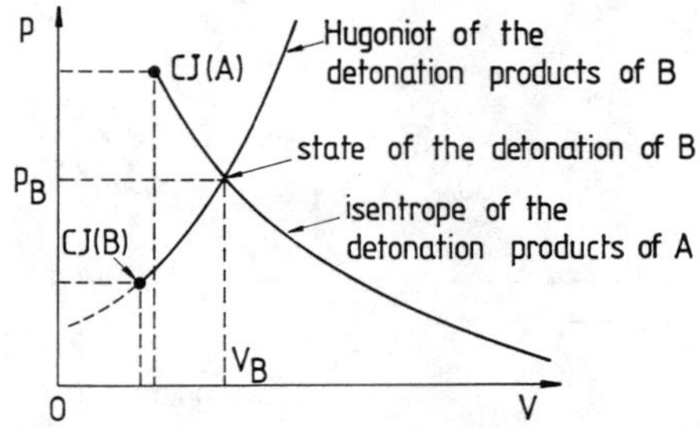

Fig. 2b Determination of the detonation of B in the p, V plane.

in its propagation geometry (plane to spherical diverging propagation).

Cell Size and Critical Diameter in Overdriven Detonation Conditions

When mixtures A and B are similar, the detonation propagates in a homogeneous medium, and the CJ condition is achieved. The detonation cell width is $\lambda_{CJ}(B)$, and when the detonation diffracts in free space the critical diameter of diffraction d_c is given by the classical empirical rule:

$$d_c(CJ) = 13\ \lambda_{CJ}(B)$$

In the case of a stoichiometric mixture in A and lean mixture in B, the CJ pressure p_{CJ} and particle velocity V_{CJ} of A exceed generally those of B.

The plane detonation in mixture A propagates at CJ conditions. Therefore, mechanical parameters of the new detonation in B, namely p and V, should satisfy continuity conditions on the interface between the two detonated media (see Figs. 2a-2b). So the solution of the detonation in B is given by p(B) and V(B), provided by the classical impedance mismatching method (see Ref. 4 and Fig. 2b) and belongs to the upper branch of the Hugoniot curve of the detonation products of B. The crossover detonation in B is then an overdriven steady wave: i.e., $p(B)/p_{CJ}(B) > 1$, $V(B)/V_{CJ}(B) > 1$ and $D(B) > D_{CJ}(B)$.

Consequently, the cell size $\lambda(B)$ is in general

$$\lambda(B) < \lambda_{CJ}(B)$$

As it has been recently shown in an overdriven detonation regime in gaseous mixtures (see Refs. 3, 5-7 in C_2H_2 and H_2

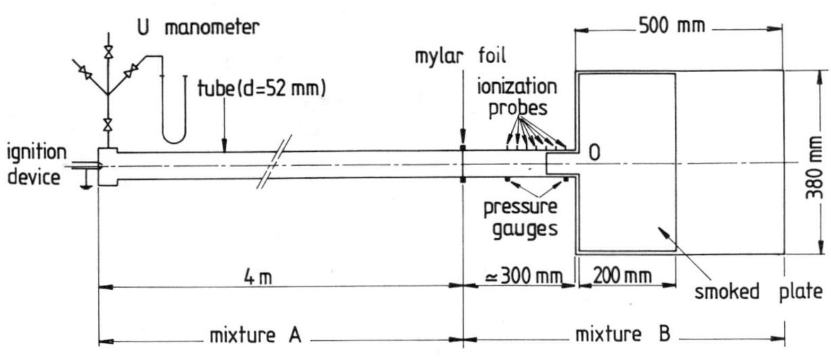

Fig. 3 Experimental device.

with O_2 or air), the magnitude of λ, depending on postshock conditions, varies as

$$\lambda \sim \rho_s^{-(n+1)} D \exp \frac{E_a}{RT_s}$$

where n is a constant parameter depending up on the given reactive mixture: ρ_s and T_s are the density and temperature, respectively, behind the shock wave propagating at a velocity D, assuming the frozen chemical composition; E_a is the apparent activation energy of the mixture; and R is the ideal gas constant.

The critical diameter of diffraction d_c of the plane overdriven detonation, if D/D_{CJ} takes moderate values (< 1.2), is still linked to the cell size by the classical rule $d_c = 13\lambda$.[3,6] The d_c value reported to that of the corresponding CJ state $d_c(CJ)$ is a sharp decreasing function of D/D_{CJ} and is expressed by

$$\frac{d_c}{d_c(CJ)} = \frac{\lambda}{\lambda_{CJ}}$$

Laboratory Tests

Objectives of Simulation

Experiments on explosive C_nH_m/air or H_2/air (H_2/air/steam) mixtures would need a very large-size tube and vessel owing to the cell width of the mixtures ranging from more than 10 mm to 1000 mm. For that reason and because the dynamic behavior of detonation shows in rough approximation a good similitude if one may refer to the size and the tridimensional structure, laboratory-scale tests were performed with C_2H_2-O_2 mixtures.

The aims of the experiments are:

1) To provide for each mixture B (different $\phi(B) < 1$) and for a given mixture A ($\phi(A) = 1$ or equivalent): a) the characteristics of the quasisteady overdriven detonation in mixture B by means of pressure p, velocity D, and cell size λ recordings; and b) the value of the critical diameter of diffraction d_c of that detonation.

2) To verify how far the aforementioned model may explain the experimental results.

3) To extend the observations of laboratory-scale tests to, for instance, the H_2-air systems involved in nuclear power plant explosion hazards.

To obtain more insight into the transmission phenomenon, by amplifying the difference between the CJ characteristics of the detonation of A and of B, other combinations of mixtures A and B are also used. For example, an overdriven detonation regime in B can be obtained with the same mixture used as A and B with $p_o(B) < p_o(A)$ or with different fuels in A and B.

Experimental Details

The experimental device is essentially that shown in Fig. 3. A thin mylar foil (thickness of 15 μm) separates the mixtures A and B. At the closed end of the d = 52 mm tube containing mixture A, the detonation is triggered in a direct way by means of an exploding wire. Three parameters of the detonation wave are measured on mixture side B just after the interface and just before the diffraction orifice: the pressure history p(t) by two KISTLER 603B pressure gauges at a distance of 150 mm, the velocity D by seven ionization probes at a distance of 25 mm, and the cell size λ by the smoked foil technique.

As mentioned previously, mixtures of C_2H_2-O_2 are used at room temperature. Mixture A is the equimolar mixture at $p_o(A)$ = 1 bar and corresponds to the maximum CJ detonation pressure and velocity. For mixtures B, the following are used:

1) Lean C_2H_2-O_2 mixtures up to $\phi = 0.3$.
2) Equimolar and stoichiometric C_2H_2-O_2 mixtures at initial pressure $p_o(B) < p_o(A)$.
3) A lean H_2-O_2 mixture.

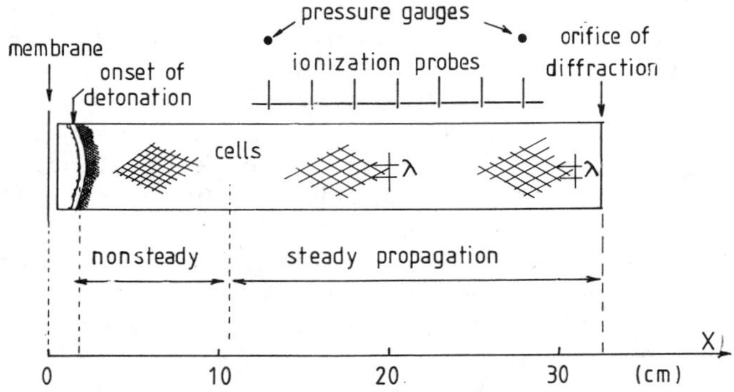

Fig. 4 Example of smoked foil records of the onset of detonation in mixture and its progressive stabilization.

Results and Analysis

Characteristics p_{st}, D_{st}, and λ_{st} of the steady self-sustained detonation wave of mixture B have been systematically measured as a function of ϕ(B). The two first parameters (thermomechanical) associated with the leading front, namely p_{st} and D_{st}, are very close to the CJ values. Thus, the third parameter (thermokinetical) λ_{st} is considered as λ_{CJ}(B).

The transmission of the detonation from the mixture A to the mixture B is followed with the help of the smoked foil technique. The essential features of the transmission appear in Fig. 4.

The detonation starts in B with a certain delay after the detonation of A has touched the membrane. This delay depends on the cellular structure size of B; the delay is short if λ is small. The onset of detonation is followed by a region where a very small cellular structure grows monotonically. About 10 cm after the membrane the detonation propagating in B reaches its stationary propagation characteristics, i.e., D, p, and λ remain constant.

Examples of pressure and cell recordings for steady self-sustained and overdriven detonation waves in B are given in Figs. 5a-5b respectively. As will be noted, the pressure histories are essentially different for these two regimes of detonation:

1) For the self-sustained detonation condition, the

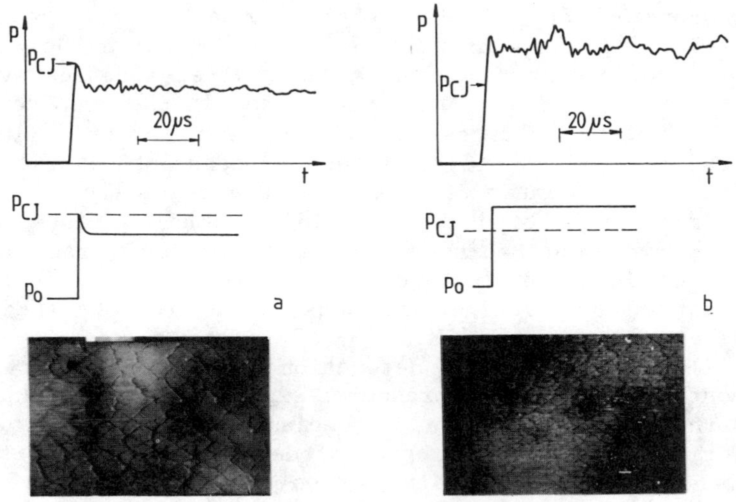

Fig. 5 Typical pressure and smoked foil records in: a) steady self-sustained, and b) steady overdriven detonation conditions.

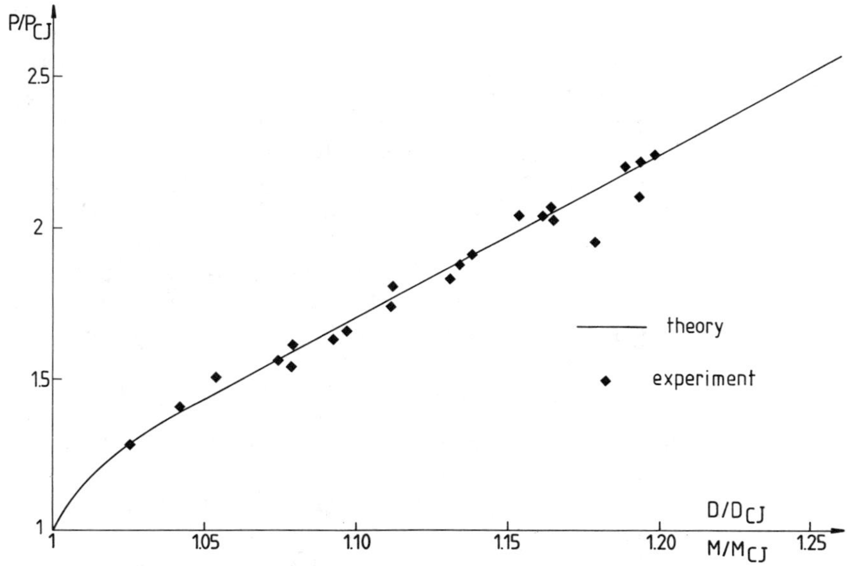

Fig. 6 Nondimensional detonation pressure p/p_{CJ} vs D/D_{CJ}.

first pressure jump is close to the CJ state (except for large cells systems). The pressure signal drops rapidly behind the front and in a few microseconds reaches a quasiconstant level that, for the detonation pipe used ($d \sim 50$ mm), takes the value of 0.7 - 0.75 p_{CJ}. This level is maintained for at least 100 μs after the leading front. The corresponding flow Mach number approaches the value of 1.25-1.30. As shown in Ref. 8, the bidimensional expansion due to confinement can produce, just after a sonic wave (as the CJ wave), a drastic and rapid drop in the properties of the detonation products. Then the characteristics of the supersonic flow behind a CJ wave propagating in a tube of finite size (of inner diameter d) are essentially sub-CJ.

As d tends to infinity, bidimensional effects should vanish, and flow behind the wave front could recover the CJ properties. For instance, in a pipe of d = 20 cm the constant sub-CJ level in the detonation products takes the value of .9 p_{CJ} and 1.1 M_{CJ}.[9]

2) For overdriven detonation conditions, the first discontinuity in the pressure signal is followed by a constant level pressure, especially as the degree of overdrive is high. This regime of detonation is supported by a plane constant velocity piston, and if walls of pipe induce bidimensional expansion their influence on the main flow characteristics behind the front is negligible because

the wave is subsonic (as a steady shock wave).

This discussion of the characteristics of the flow field behind a detonation wave is important for the following analysis concerning the transmission of detonation between the two mixtures.

Experimental values of overdriven detonation of B, i.e., p, D, and λ referred to its CJ values can give, with different accuracies, the real degree of overdrive of the detonation. Experimental values p/p_{CJ}, vs D/D_{CJ} and λ/λ_{CJ} vs D/D_{CJ} are reported in Figs. 6 and 7, respectively. Agreement with theoretical curves appears very good.

An a priori determination of the thermodynamical properties of the steady overdriven detonation of B, as pointed out previously, should be obtained by the classical impedance mismatching method. In the p, V plane are drawn two locii (cf., for instance, Figs. 8 and 9):

1) The first corresponds to the Rieman expansion of the detonation products of A, starting from CJ (A).

2) The second represents the Hugoniot branch of the detonation products of B.

The intersection of these two locii gives the solution of the detonation regime in B.

Experimental results of characteristics of detonation in B, defined by p(B) and $D(B)/D_{CJ}(B)$, are plotted in the p, V plane for a few tests. They are systematically below the theoretical prediction in such a way that, for instance, no difference from the CJ value appears in a detonation propagating in $\phi = 0.6$ (C_2H_2-O_2 mixture) if a CJ detonation in $\phi = 2.5$ is transmitted (see Fig. 8). This important deviation can be explained by the following considerations.

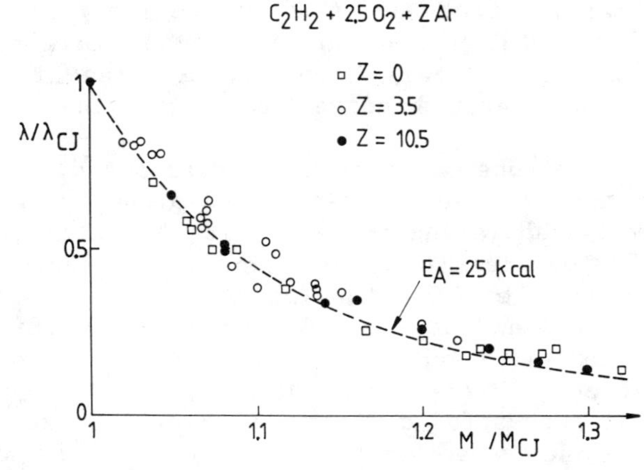

Fig. 7 Nondimensional detonation cell size λ/λ_{CJ} vs D/D_{CJ}.

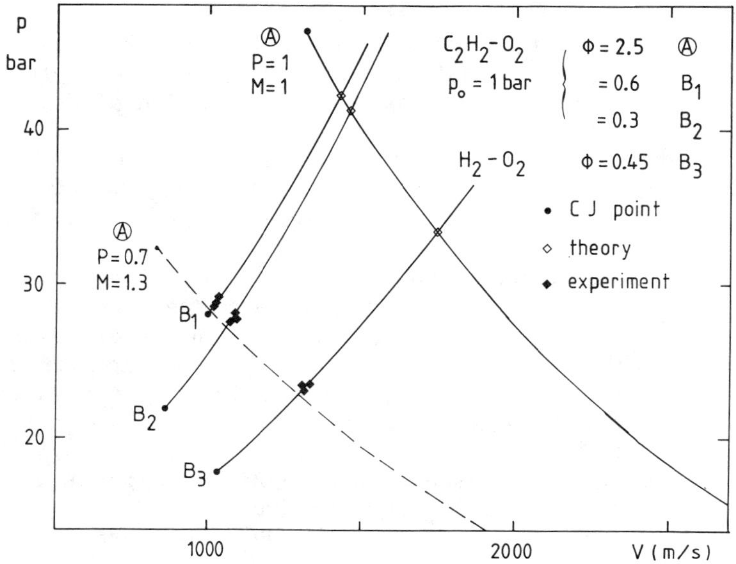

Fig. 8 Conditions of detonation in mixture B in the p, V plane for different compositions of mixtures A and B.

In fact, during the greatest part of the interaction between the two media studied here the steady overdriven detonation state of B is imposed by the quasiconstant sub-CJ state of long duration behind the C.J. wave of A and presented previously (i.e., p/p_{CJ} = 0.7 - 0.75 and M/M_{CJ} = 1.25 - 1.3). So in the p, V plane, a more realistic Rieman expansion locus of the detonation products of A, taking into account globally bidimensional effects of a real tube, can be drawn. Starting from the sub-CJ state ($P=p/p_{CJ}$ = 0.7 and $M=M/M_{CJ}$ = 1.3), this locus can now explain in a satisfying manner all the experimental results. Examples are given in Figs. 8-9, where dotted curves represent the new Rieman expansion locii.

Thus, for a tube of given inner diameter d, an a priori determination of the steady detonation propagating in B can be done. By extrapolating these results for large values of d, where bidimensional effects can be neglected, it is concluded that the sole knowledge of CJ conditions of mixtures A and B and correlated Rieman and Hugoniot curves may lead to a good prediction of the detonation characteristics of B, namely p, D, and λ.

Critical successful diffractions into a large volume of overdriven plane steady detonations follow the classical relationship of d_c and λ for D/D_{CJ} up to at least 1.16.

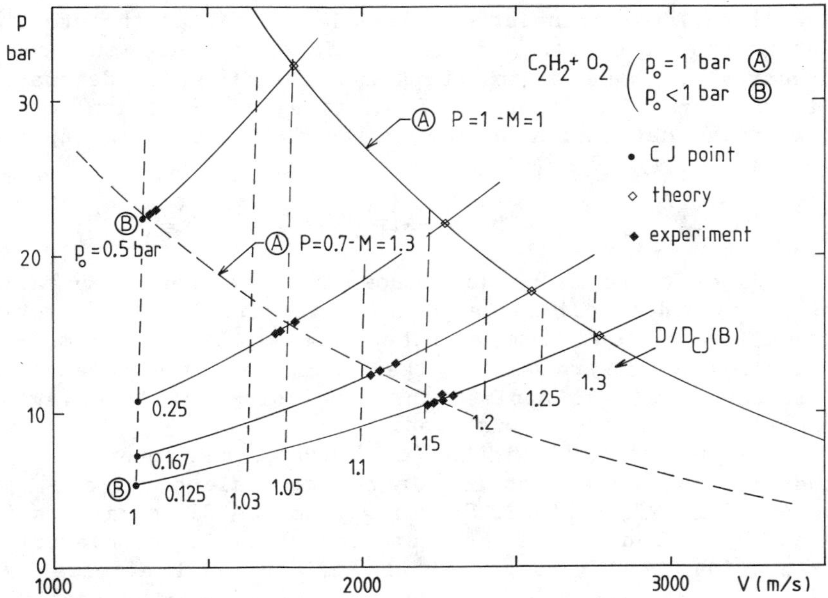

Fig. 9 Conditions of detonation in mixture B in the p, V plane for different initial pressures in mixtures A and B.

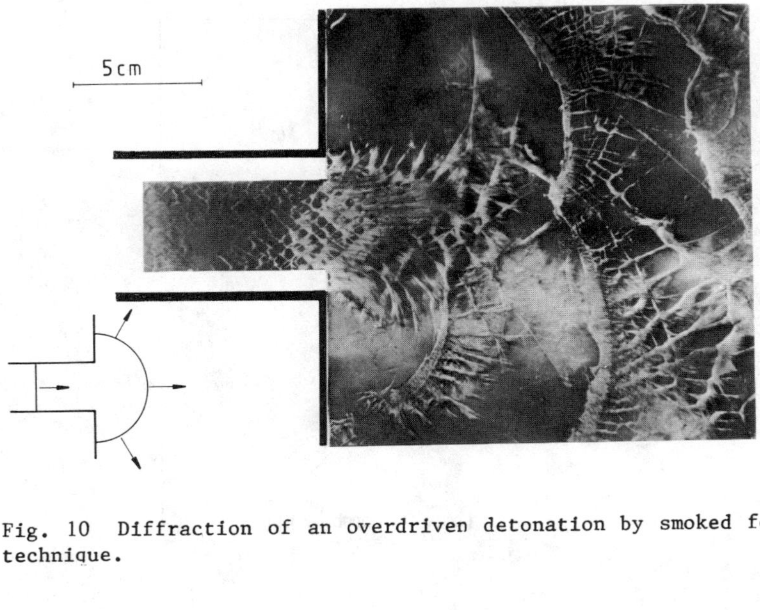

Fig. 10 Diffraction of an overdriven detonation by smoked foil technique.

Therefore, due to tne strong dependence of λ on the state of detonation (D/D_{CJ}...), critical diameters turn out to be drastically reduced in comparison to the CJ detonation diffraction. An example of critical diffraction of an overdriven detonation by the smoked foil method is reported in Fig. 10.

The H_2-Air Case

Since there is no doubt now that the dynamic behavior of gaseous detonations (especially the critical diffraction of a detonation from a pipe into a large volume) is essentially governed by the size of its elementary structure, we have applied the conclusions of the previous experiments to H_2-air mixtures.

Detonation cell widths λ_{CJ}[10] and CJ characteristics of that system, as a function of the equivalent ratio ϕ, are reported in Figs. 11-12. Starting from the CJ conditions for A and B, we can obtain the characteristics of the overdriven detonation in B (i.e., without bidimensional effects; see Fig. 13). For the H_2-air system, it is noteworthy that $p(B) = p_{CJ}(A)$ and $V(B) = V_{CJ}(A)$ whatever $\phi(B)$ is. So if we simply do the ratio $p_{CJ}(A)/p_{CJ}(B)$ or $V_{CJ}(A)/\ V_{CJ}(B)$, the overdriven degree of the detonation of B is known. The variation of the ratio λ/λ_{CJ} (since $\phi(A) = 1$ and $\phi(B) < 1$

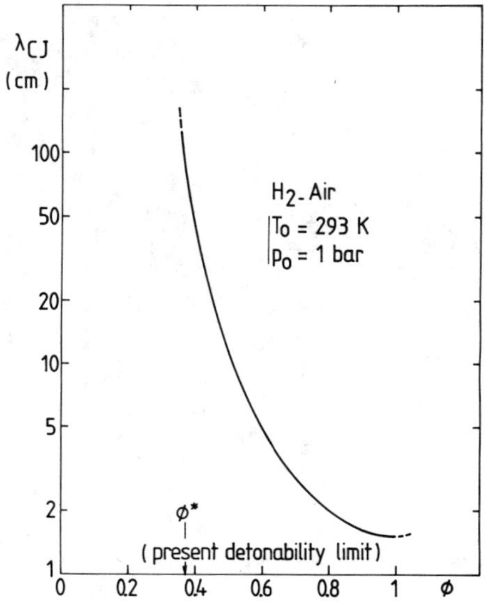

Fig. 11 Cell size λ_{CJ} vs equivalence ratio ϕ for H_2-air mixtures.

are known), i.e., $d_c/d_c(CJ)$, assuming an apparent activation energy value E_a = 19 kcal/mole of the mixture,[11] is reported as a function of $\phi(B)$ in Fig. 14, for $1 > \phi(B) > \phi^*$ where ϕ^* = 0.37 corresponds to the present detonability limit of the H_2-air mixture.[10] One observes that the lower the $\phi(B)$ value, the higher the reduction of the cell or the critical diameter size from their respective CJ values.

Consequently, it is possible to obtain for an H_2-air mixture as a function of $\phi(B)$ (< 1) (see Table 1) the magnitude of:

1) The normal (CJ detonation) critical diameter of diffraction given by $d_c(CJ) = 13\ \lambda_{CJ}$.

Table 1 Critical diameter of transmission of a detonation in CJ and in overdriven cases in H_2-Air mixtures as a function of ϕ

Equivalence ratio, ϕ	Critical diameter size $d_c(CJ)$, m	d_c (overdriven), m
1	0.20	0.20
0.6	0.65	0.55
0.5	1.6	1.1
0.4	6.6	3.2
0.37	16-17	6.5

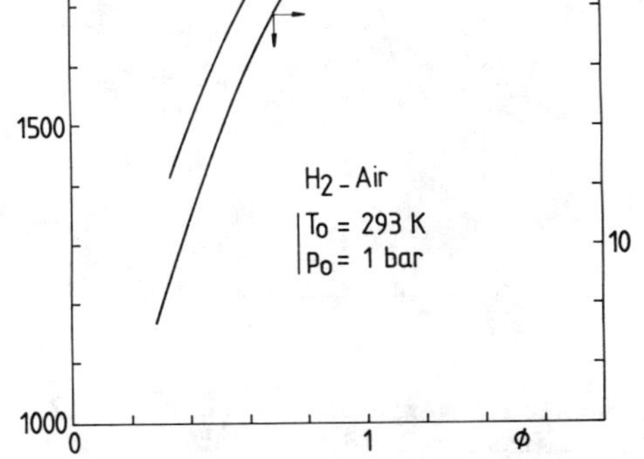

Fig. 12 CJ characteristics of H_2-air mixtures vs equivalence ratio ϕ.

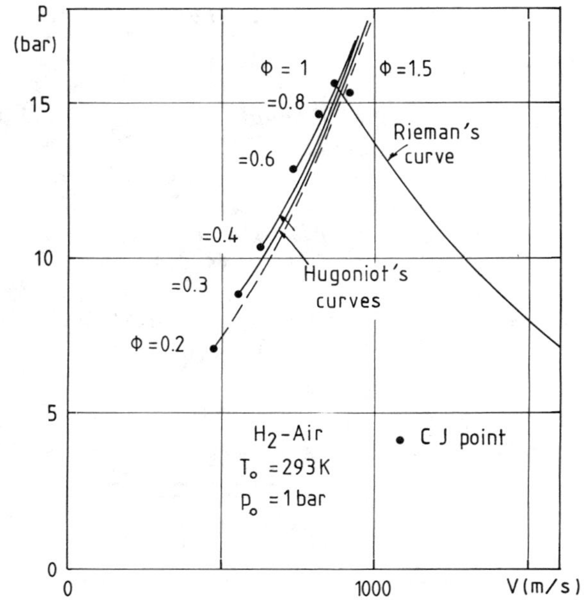

Fig. 13 Conditions of detonation in different lean H_2-air mixtures in the p, V plane by transmission of a CJ detonation in ϕ=1 H_2-air mixture.

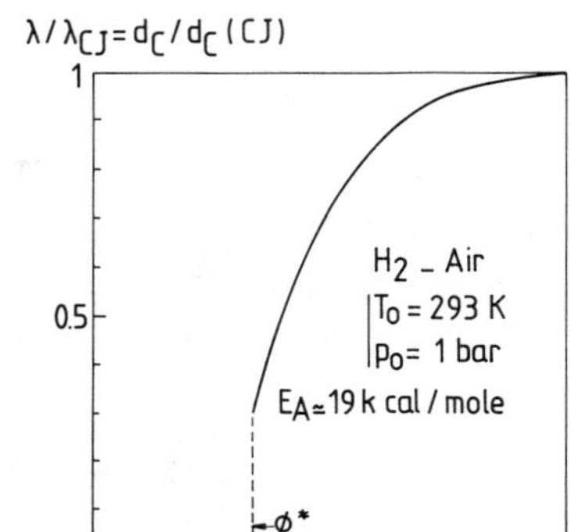

Fig. 14 Nondimensional variation of cell size λ/λ_{CJ} or critical diameter of transmission d_c/d_c(CJ) of a detonation in H_2-air mixtures as a function of equivalence ratio ϕ of the lean mixture.

2) The reduced (overdriven detonation) critical diameter of diffraction given by $d_c^c = 13\lambda$.

One may remark, for instance, that a reduction by a factor of 2 is obtained when $\phi(B) = 0.4$.

This extrapolation to H_2-air mixtures can be applied to H_2-air-steam mixtures without any significative difference on the ratio $d_c/d_c(CJ)$.[7]

Conclusion

This experimental investigation at laboratory-scale demonstrates that:

1) An overdriven steady detonation wave with velocity $D > D_{CJ}$ and a cell size $\lambda < \lambda_{CJ}$ can be created locally in a pipe by a rapid change of initial conditions of a reactive mixture (for instance, rapid drop of concentration of fuel in air).

2) This detonation can be transmitted from a duct to a large volume with a smaller critical diameter than that of the CJ detonation wave.

3) Likewise for small-scale results, where a CJ detonation propagating in a rigid tube in a stoichiometric mixture diffracts in a lean mixture contained in a large volume, critical diameters of diffraction have been determined in the case of H_2-air mixture. Particularly, when $\phi = 0.4$ in the lean mixture, calculations can predict a reduction by a factor of 2 of the critical diameter obtained for the CJ wave.

Somehow, it would appear that this study is pessimistic and not stricly valid because 1) it overestimates somewhat the condition of propagation of a detonation, and 2) it is based on the assumption that λ is the sole scaling parameter governing the detonability of a system. The limiting process of the transmission of a detonation from a mixture A to a less energetic mixture B has to be found in bidimensional effects in the detonation products of A. Besides, the interface between two mixtures of different composition is, actually, a time-dependent thickness of compositional gradient.[12]

Nevertheless, the conclusions of this study are important from a safety point of view and can be applied to other CnHm-air systems used, for example, in the chemical industry.

References

[1]Knystautas, R., Guirao, C., Lee, J.H., and Sulmistras, A., "Measurements of Cell Size Hydrocarbon-air Mixtures and Predictions of Critical Tube Diameter, Critical Initiation

Energy, and Detonability Limits," AIAA Progress in Astronautics and Aeronautics, Vol. 94, New York, NY, 1984, pp. 23-37.

[2]Lee, J.H., "Dynamic Parameters of Gaseous Detonations," Annual Review of Fluid Mechanics, Vol. 16, 1984, pp. 311-336.

[3]Desbordes, D., and Vachon, M., "Critical Diameter of Diffraction for Strong Plane Detonations," AIAA Progress in Astronautics and Aeronautics, Vol. 106, New York, NY, 1986, pp. 131-143.

[4]Berger, J., and Viard, J., "Physique des Explosifs Solides," Dunod, Paris, 1962.

[5]Huang, Z.W., and Xu, B., "Relation between Cell Size and Detonation Parameters in Overdriven Gaseous Detonations," Combustion and Flame, Vol. 67, No 2, 1987, pp. 95-98.

[6]Desbordes, D., "Transmission of Overdriven Plane Detonations: Critical Diameter as a Function of Cell Regularity and Size," AIAA Progress in Astronautics and Aeronautics, Vol. 114, Washington, DC, 1988, pp. 170-185.

[7]Desbordes, D., Brugier, J.L., and Guerraud, C., "Transition d'une Détonation Plane Forte dans les Mélanges H_2-air et H_2-air-H_2O Pauvres," Contract rept EDF/ENSMA (parts 1 and 2).

[8]Desbordes, D., Manson, N., and Brossard, J., "Influence of Walls on Pressure Behind Self-sustained Expanding Cylindrical and Plane Detonations in Gases," AIAA Progress in Astronautics and Aeronautics, Vol. 87, New York, NY, 1983, pp. 302-317.

[9]Desbordes, D., "Velocity and Pressure Field of Detonations in large Diameter Tubes," ENSMA Rept. Poitiers, France, 1983.

[10]Tieszen, S.R., Sherman, M.P., Benedick, W.B., Sherperd J.E., Knystautas, R., and Lee, J.H., "Detonation Cell Size Measurements in Hydrogen-air-steam Mixtures," AIAA Progress in Astronautics and Aeronautics, Vol. 106, New York, NY, 1986, pp. 205-219.

[11]Miyama, H., and Takeyama, T.J., "Kinetics of Hydrogen-oxygen Reaction in Shock Waves," Journal of Chemical Physics, Vol. 41, No 8, 1974, p. 2287.

[12]Donato, M., Donato, L., and Lee, J.H., "Transmission of Detonations Through Composition Gradients," Proceeding of the International Colloquium Berthelot-Vieille-Mallard et Le Chatelier, Vol. II, Bordeaux, France, 1981, pp. 467-472.

Oblique Detonation at Hypersonic Velocities

E. K. Dabora*
University of Connecticut, Storrs, Connecticut
and
D. Desbordes† and C. Guerraud‡
Laboratoire d'Energétique et de Détonique, URA 193 au CNRS, ENSMA, Poitiers, France
and
H. G. Wagner§
Max Planck Institute on Flow Research, Göttingen, Germany

Abstract

The technique used for establishing oblique detonation waves (ODW) by lateral expansion of the gas behind a normal detonation into a bounding reactive mixture is derived from the work of Dabora et al.[1] Normal driving Chapman-Jouguet high-velocity detonations (3000 m/s or more) are easily obtained with $CnHm/O_2/He$ systems. The bounding mixture is essentially of low CJ characteristics such as C_2H_2/O_2 diluted by 75% Ar. This mixture serves as a substitute, from a thermodynamical point of view, for CnHm-air mixtures (i.e., $D_{CJ} \cong 1800$ m/s, $p_{CJ}/p_o \cong 20$) and provides a much more regular cellular structure.

The laboratory-scale test section consists of a rectangular cross-sectional tube 2 cm x 3 cm, one narrow side of which is open but can be separated by a thin diaphragm of mylar from the oblique wave section (2 cm x 12 cm). The total observation length of the test section is 40 cm.

The experimental results using $C_2H_4 + 3\ O_2 + 8$ He mixture for the normal wave and $C_2H_2 + 2.5\ O_2 + 10.5$ Ar for the oblique wave side are described. By a convenient choice of the initial pressure of the two mixtures, different regimes of oblique detonation waves are observed. These regimes are characterized by pressure jump, cellular structure size, and detonation wave angle measurements. Streak photographs of the phenomenon are also given.

An a priori determination of characteristics and conditions for establishing ODW based on the detonation polar analysis is deduced.

---

*Professor, Department of Mechanical Engineering.

†Assistant Professor.

‡Engineer.

§Professor.

Introduction

It has been proposed recently that standing oblique detonation waves (SODW) can be used as an alternative to supersonic combustion vehicles (see, for example, Pratt et al.[2]). In other respects, ODW appear in premixed reactive mixtures as a chemical method for accelerating projectiles to ultrahigh velocities.[3] The work presented here is motivated by these applications with the aim of extending the range of initial conditions that have been attempted up to now.

In place of a very expensive hypersonic wind tunnel used for the study of SODW, a simple shock tube-like facility can be used to study ODW. The technique for establishing ODW is similar to that of Dabora et al.[1] An oblique shock (or detonation) wave is induced in a bounding inert (or reactive) gas by the lateral expansion of the gas behind a normal detonation wave propagating in a channel. This technique was also used by Liu et al.[4] to look into the diffraction and transmission of a detonation from one explosive layer to another.

As shown in Fig. 1, a well-established, normal, quasi-CJ detonation propagating in a rectangular rigid tube and meeting a yielding wall can produce an ODW in the bounding reactive gas. If the yielding wall is of negligible thickness, its deflection and formation of an oblique shock wave are immediate. Detonative combustion can follow with a certain delay, depending on the chemical kinetic parameters of the bounding mixture. The thickness and material of the wall define, for each detonation condition, the initial rate of expansion of the normal detonation products[1,5] and can noticeably modify the delay of onset of detonation in the bounding mixture.

To simulate hypersonic flow conditions, experiments are carried out with certain specific mixtures such as C_2H_4 - O_2 generally diluted by a large amount of helium, in order to obtain velocity at over 3000 m/s.

Bounding mixtures are stoichiometric C_2H_2-O_2 mixtures diluted by 75% of argon. These mixtures present CJ conditions close to those of CnHm-air mixtures. In addition, its tridimensional structure of very good regularity has a convenient size for laboratory simulation scale experiments (cell spacing in the range of 1-10 mm). Dependence laws on initial pressure and on degree of overdrive[6,7] are well defined.

The main purposes of this study are essentially: 1) to obtain ODW with convenient combination of initial pressure in both mixtures; and 2) to provide an a priori

Table 1 CJ properties of mixtures A and B

$C_2H_2 + 2.5\ O_2 + 10.5\ Ar$ (T_o = 298 K)							
p_o bar	a_o m/s	D_{CJ} m/s	a_{CJ} m/s	p_{CJ}/p_o	T_{CJ}/T_o	ρ_{CJ}/ρ_o	Γ_{CJ}
0.1	313	1692.6	934.6	20.44	10.69	1.811	1.173
0.167	313	1710.6	946.1	20.80	10.88	1.807	1.178
0.333	313	1735.5	962.5	21.35	11.27	1.803	1.187
$C_2H_4 + 3\ O_2 + 8\ He$ (T_o = 298 K)							
p_o bar	a_o m/s	D_{CJ} m/s	a_{CJ} m/s	p_{CJ}/p_o	T_{CJ}/T_o	ρ_{CJ}/ρ_o	Γ_{CJ}
0.5	540	3036.1	1668.7	22.79	11.36	1.819	1.167
1	540	3079.1	1696.0	23.36	11.71	1.816	1.174
1.5	540	3104.4	1712.3	23.69	11.93	1.813	1.178
2	540	3122.4	1724.0	23.92	12.08	1.811	1.181

determination of conditions for establishing ODW by detonation polar analysis.

Experimental Arrangement

For observing the interaction phenomenon between a primary detonating mixture (called A) and a bounding secondary explosive (called B), a special chamber (see Fig. 2) of 40 cm long and 2 cm thick has been constructed. It consists essentially in two parts: 1) a 2 cm x 3 cm rectangular cross-sectional tube containing the primary detonating mixture; and 2) a 2 cm x 12 cm rectangular chamber containing the secondary reactive mixture.

The two parts are separated by a 2 cm x 40 cm mylar membrane. The thickness of the membran in the range of 10 to 15 μm is chosen in order to maintain different initial pressure in the two reactive mixtures up to 2 bar.

The primary mixture is essentially $C_2H_4 + 3\ O_2 + 8$ He mixture used at room temperature and at different initial pressure (0.5 bar < p_o < 2 bar). This mixture is ignited at the end of a 3-m-long, 21-mm-i.d. tube by an electrical spark with the help of a Shchelkin spiral. In such a way, the detonation starts near the spark and the propagation is long enough to ensure a well-established, self-sustained detonation before entering the chamber. Two pressure gauges, IA and IIA (KISTLER 603 B), are located 25 cm one from another.

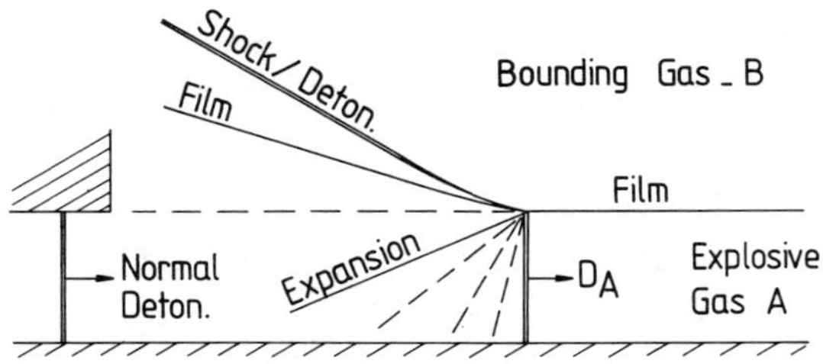

Fig. 1 Schematic of the formation of an oblique detonation by a normal detonation wave.

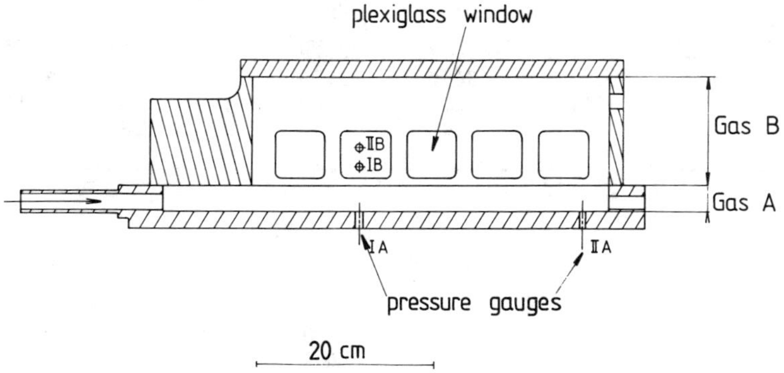

Fig. 2 The test section.

The secondary mixture is a $C_2H_2 - O_2$ stoechiometric mixture diluted by 75% in volume of argon. Two pressures gauges, IB and IIB (KISTLER 603B), are located vertically 2 and 4 cm from the membrane, on the side wall of the oblique part of the chamber. The other side of that chamber is equipped with plexiglass windows which allow direct luminosity photographic records of the phenomenon. The high-speed camera used is a THOMSON TSN 506.

On the wall are set smoked steel plates for cell size measurements and direction of normal propagation of the ODW. Thus, the ODW in B can be characterized experimentally by: 1) the angle of deflection of the wave θ_B; 2) the pressure jump across the wave p_B; and 3) the cell spacing λ_B.

First Analysis

We are looking only into the detonation regime of the combustion of the mixture B.

If a CJ detonation propagates in A, the detonation regime in B is assured if ODW angle θ (see Fig. 3a) takes a minimum value θ_{min}:

$$\theta_{min} = \sin^{-1}\left(\frac{D_{CJ}(B)}{D_{CJ}(A)}\right) \tag{1}$$

This condition is fullfilled if the expansion of the CJ detonation products of A is enough or, in other words, if the deflection angle δ is equal to or greater than δ_{min}. Three cases must be considered:

1) $\theta = \theta_{min}$, i.e., $D(B) = D_{CJ}(B)$. The detonation in B is a CJ wave.

2) $\theta > \theta_{min}$, i.e., $D(B) > D_{CJ}(B)$. The ODW in B is an overdriven steady wave.

3) $\theta < \theta_{min}$. No steady detonation wave can propagate in B.

The correlation between θ and δ is obtained by mass conservation across an oblique wave:

$$\delta = \theta - \tan^{-1}\left(\rho_o/\rho_1 \tan\theta\right) \tag{2}$$

where ρ_1/ρ_o represents the density ratio across the oblique wave.

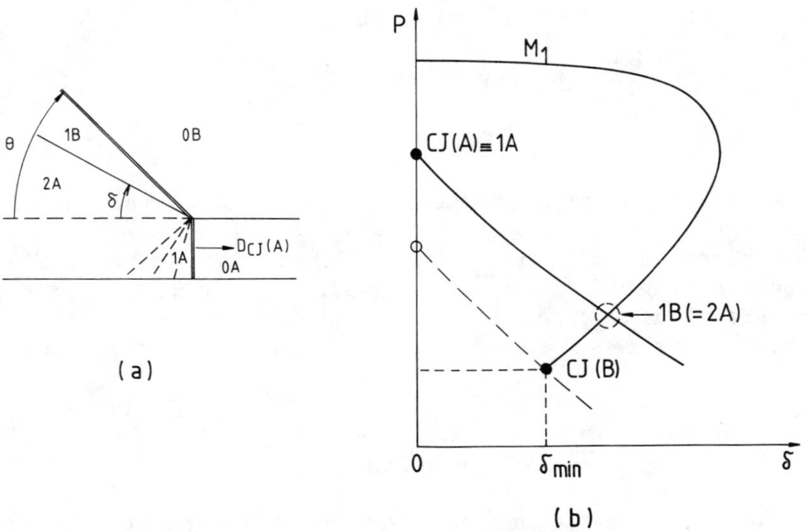

Fig. 3 Interaction in the: a) physical plane; and b) pressure deflection angle plane.

For a driving normal CJ detonation in A, the Mach number M_1 of the ODW in B is defined by

$$M_1 = \frac{D_{CJ}(A)}{a_o(B)} \tag{3}$$

where $a_o(B)$ is the sound velocity in fresh mixture B.

For a given M_1, the detonation polar of the ODW of B can be drawn in the P $(p_B/p_o(B))$, δ plane (Fig. 3b). We use for that the fact that an adimensional parameter of the Hugoniot of the detonation product (i.e., p/p_{CJ}, ρ/ρ_{CJ}, a/a_{CJ}, ...) calculated by the QUATUOR Code[8] can be expressed in first approximation by unique relations depending on M/M_{CJ}, regardless of the reactive mixture and initial conditions. The detonation polar of B is open at CJ point where $\delta = \delta_{min}$. It represents the locus of all thermomechanical conditions of detonation of B considering a given Mach number M_1.

The Prandtl-Meyer expansion of the CJ detonation products of A define another locus, starting from $P_{CJ}(A)$, $\delta = 0$, $M = 1$ in the P, δ plane (see Fig. 3b) and given by the classical relationships[9]

$$\delta = \left(\frac{\gamma + 1}{\gamma - 1}\right)^{1/2} \tan^{-1}\left(\left(\frac{\gamma - 1}{\gamma + 1}(M^2-1)\right)^{1/2}\right) - \cos^{-1}\left(M^{-1}\right) \tag{4}$$

and

$$P = P_{CJ}(A) \left(\frac{2+(\gamma - 1)M^2}{\gamma + 1}\right)^{\gamma/(1-\gamma)} \tag{5}$$

The intersection of these two locii gives the solution of the ODW of B.

Therefore, it is obvious that CJ or overdriven detonation regime of ODW of B requires a minimum of initial pressure of mixture A. The ratio of this threshold value of initial pressure of A to initial pressure of B is defined by π^*.

Results

We have limited our observation of interaction between the two explosives in varying the initial pressure ratio: $p_o(A) / p_o(B) = \pi$ from 2 ($= \pi^*$) to 20.

For mixtures used, the CJ detonation properties computed by the QUATUOR Code are listed in Table 1. Cell spacings of the mixture B, which supports the ODW, have been extensively measured in quasi-CJ conditions in function of p_o and in overdriven detonation conditions.

Consider the results[6] of C_2H_2-O_2-Ar systems, i.e.:

$$\lambda \propto M_s a_o \rho_s^{-1} \rho_o (O_2)^{-1} \exp E_A/RT_s \tag{6}$$

where s is the postshock conditions.

We have the following approximate relationship for $C_2H_2 + 2.5\ O_2 + 10.5$ Ar:

$$\lambda_{CJ}(\mathrm{mm}) \cong .777\ p_o(\mathrm{bar})^{-1.19} \tag{7}$$

and

$$\frac{\lambda}{\lambda_{CJ}} \cong \frac{D}{D_{CJ}} \exp \frac{E_A}{RT_{ZND}} \left(\left(\frac{D_{CJ}}{D}\right)^2 - 1\right) \tag{8}$$

where $E_A/RT_{ZND} \cong 5.2.$

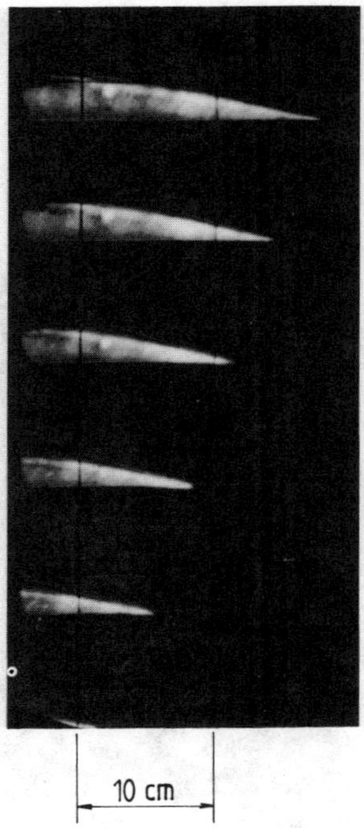

Fig. 4 Framing schlieren photos of oblique shock waves. Mixture A: 2 H_2 + O_2 + 2 He, p_o = 1 bar; Mixture B: Air, p_o = 1 bar; framing time: 10 μs/frame; exposure time: 0.5 μs.

For the mixture A, initial pressure varying from 0.5 to 2 bar, the normal detonation propagates close to the CJ velocity and is practically unaffected by expansion of the product because of its small structure size.

Few experiments have been carried out with air as a bounding mixture to investigate the interaction and to assure that film expands with a quasiconstant rate and that as a consequence the planeity of the oblique shock wave is good (see Fig. 4).

With a reactive bounding mixture, when direct detonation occurs we can distinguish two tendencies depending on relative value of ratio π to π^*:

1) For large value of π ($> 3\pi^*$), we observe quasi-immediate onset of the ODW as it is shown in Fig. 5. Cellular structure appears few millimeters (5-10 mm) from the initial position of the film. This structure, generally of small size in the present experiment, remains quasi-constant along the propagation except near the onset of ODW and maintains a nearly constant angle in the main direction of periodic collision of triple points. Thus, λ_B and θ_B can be deduced from the soot track technique.

2) For $\pi < 3\pi^*$, the onset of the detonation appears with a delay depending on the size of the structure of medium B. Smoked foil records presented in Fig. 6 show the

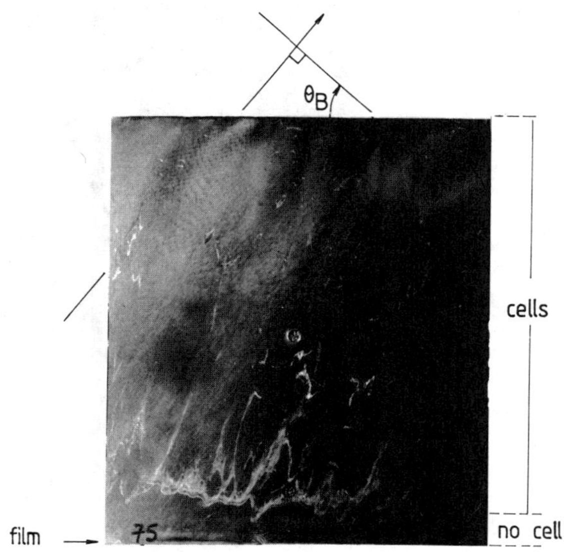

Fig. 5 Smoked foil record of ODW in mixture B for overdriven detonation condition. Mixture A: $C_2H_4 + 3\ O_2 + 8\ He$, $p_o = 1.666$ bar; Mixture B: $C_2H_2 + 2.5\ O_2 + 10.5\ Ar$, $p_o = 0.333$ bar.

amplitude of the quasihorizontal zone free of triple-point collisions when we are near the critical onset of ODW, i.e., when $\pi \cong \pi^*$. Local explosion appears with a large increase in cell size and direction of propagation before the stabilization of the phenomenon with constant λ_B and θ_B. This ODW is still driven by the expansion of the normal detonation products, but evolves to steady characteristics. Undoubtedly this delay substancially depends on the ratio π, the cell size of the steady ODW (λ_B), and the thickness of the mylar film.

It is obvious that for lower ratios of initial pressures, i.e., when $\pi < \pi^*$, no steady shock-flame propagation can be observed.

For these two kinds of onset of ODW in the secondary reactive mixture, typical pressure histories are given in Figs. 7-8 concerning normal and oblique detonation waves.

For the normal detonation in A, the amplitude of the first pressure spike $p_{st}A$ is slightly sub-CJ ($p_{st}A \sim 0.94 - 0.97\ p_{CJ}$). It is then followed by a sharp decrease corresponding to the expansion of the products. Figure 7 provides a comparison of pressure histories without expansion (with rigid confinment) and without losses, i.e., the Zeldovich-Taylor profile. Beyond this decrease the pressure maintains a quasiconstant level.

For the mixture B, when the onset of ODW is quasi-immediate the two pressures gauges give the same level of discontinuity p_B. This value maintains for at least a few

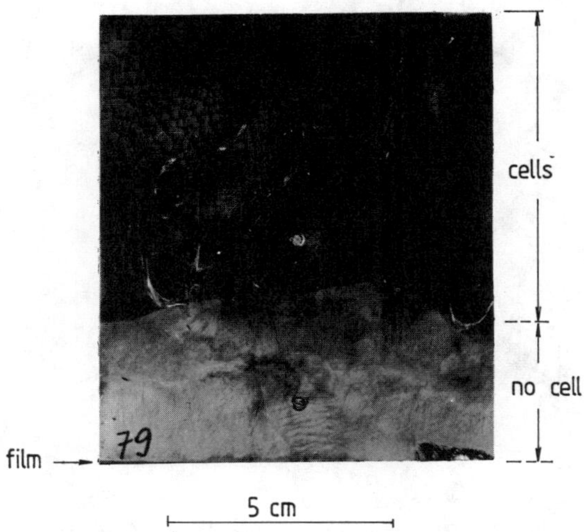

Fig. 6 Smoked foil record of ODW in mixture B for near CJ detonation condition. Mixture A: $C_2H_4 + 3\ O_2 + 8\ He$, $p_o = 0.666$ bar; Mixture B: $C_2H_2 + 2.5\ O_2 + 10.5\ Ar$, $p_o = 0.333$ bar.

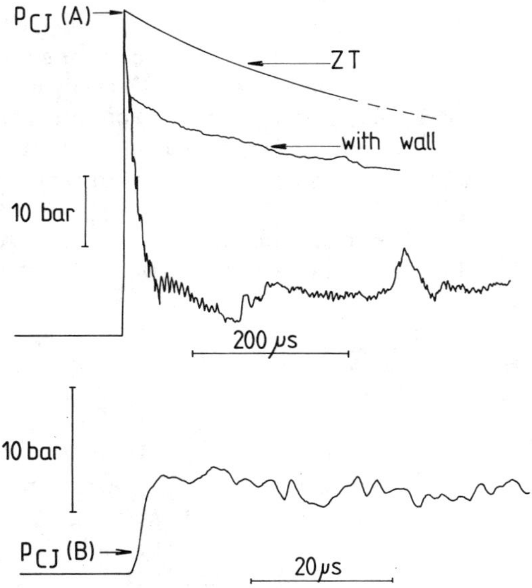

Fig. 7 Pressure histories during the interaction for overdriven ODW in mixture B: mixture A: C_2H_4 + 3 O_2 + 8 He, p_o = 2 bar; mixture A: C_2H_2 + 1.5 O_2 + 10.5 Ar, p_o = 0.1 bar.

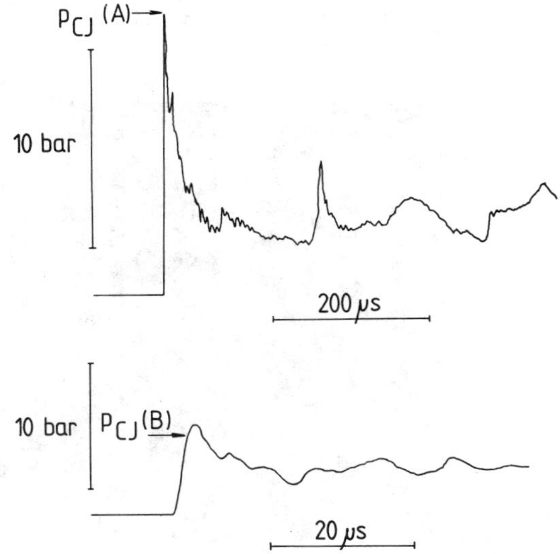

Fig. 8 Pressure histories during the interaction for near CJ ODW in mixture B: mixture A: C_2H_4 + 3 O_2 + 8 He, p_o = 0.666 bar; mixture A: C_2H_2 + 2.5 O_2 + 10.5 Ar, p_o = 0.333 bar.

10 μs, as it can be expected for overdriven detonation supported by a piston. Afterward, the signal decreases slowly. When onset of detonation is delayed, pressure histories of the first gauge and sometimes the second show rapid pressure fluctuations.

A typical streak schlieren record of steady ODW is given in Fig. 9. The detonation front appears clearly, immediately followed by the interface separating the two detonated media.

Discussion

The experimental values obtained for ODW in B, namely θ_B, λ_B, and p_B (see Table 2), provide independently, by comparison with the self-sustained detonation characteristics, λ_{st} and p_{st} ($\cong \lambda_{CJ}$ and p_{CJ}), the degree of overdrive of the detonation $D/D_{st}(B)$ with different accuracies. Cell size λ_B and pressure jump p_B are very sensitive to the regime of propagation of the detonation, especially near the CJ condition.[10] For a higher degree of overdrive, the pressure measurement p_B gives a more accurate definition of D/D_{st} (or $\sim D/D_{CJ}$).

Theoretical detonation polar of B and Prandtl-Meyer expansion of A are drawn in P, δ plane in Figs. 10-11, for a few cases of interactions.

Corresponding experimental results are plotted on these figures and are noticeably in good agreement with

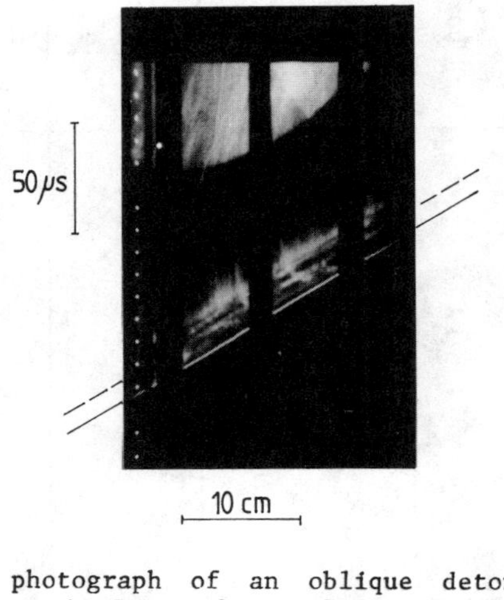

Fig. 9 Streak photograph of an oblique detonation (direct luminosity). Mixture A: $C_2H_4 + 3\ O_2 + 8\ He$, $p_o = 1.5$ bar; mixture B: $C_2H_2 + 2.5\ O_2 + 10.5\ Ar$, $p_o = 0.1$ bar.

Table 2 Conditions of interaction between mixture A and mixture B

Mixture A			Mixture B							
p_o bar	p_{st}/p_{CJ}	D_{st}/D_{CJ}	p_o bar	p_B bar	p_B/p_{CJ}	λ_B mm	θ_B °	D_{st}/D_{CJ}	λ_{CJ} mm	λ_B/λ_{CJ} %
2	0.97	0.97	0.1	7.94	3.88	.8-1	54	0.96	12	6.6-8.3
1.5	0.95	0.975	0.1	6.85	3.35	1-1.2	49	-	-	8.3-10
1.5	0.97	0.97	0.1	6.70	3.28	1-1.2	49	-	-	8.3-10
1.	0.95	0.96	0.1	5.60	2.74	1.5-1.8	44	-	-	12.5-15
1.	0.94	0.97	0.1	5.42	2.65	1.5-1.8	44	-	-	12.5-15
2	0.966	0.965	0.167	10.1	2.91	1-1.2	47	0.96	6.5	15.4-18.5
1	0.98	0.966	0.167	7.37	2.125	1.7-2.0	40	-	-	26.2-30.7
1	0.97	0.975	0.167	7.00	2.02	1.5-1.8	40	-	-	23-27.7
2	0.96	0.97	0.333	14.1	1.98	1:0	40	0.97	3	33
2	0.95	0.965	0.333	14.42	2.03	1.0	40	-	-	33
2	0.95	0.97	0.333	14.0	1.97	1.0	40	-	-	33
1.666	0.95	0.965	0.333	12.5	1.76	1.2	38	-	-	40
0.666	1.0	0.97	0.333	Detonation		3	?	-	-	100

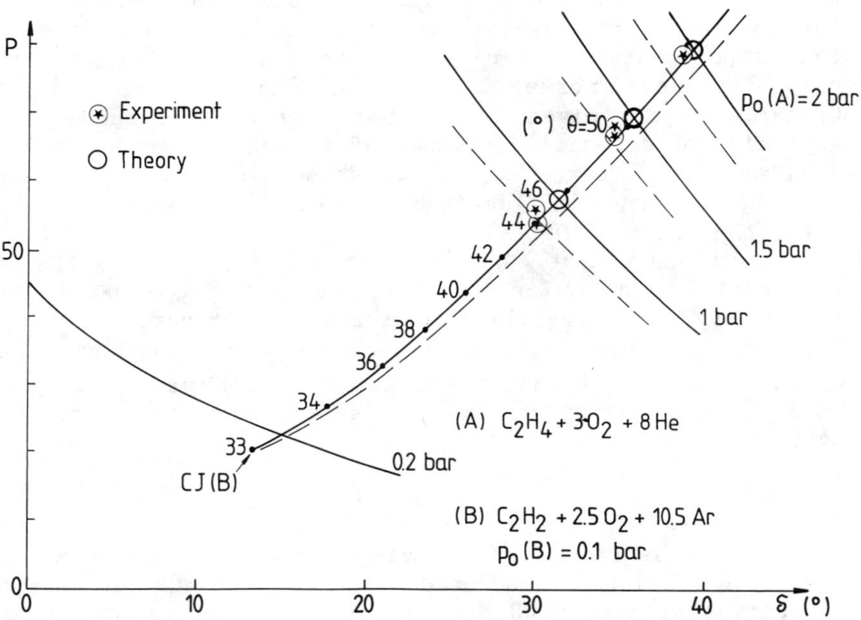

Fig. 10 P-δ plane of interaction. Mixture A: $C_2H_4 + 3\ O_2 + 8$ He at different initial pressures; mixture B: $C_2H_2 + 2.5\ O_2 + 10.5$ Ar, p_o = 0.1 bar.

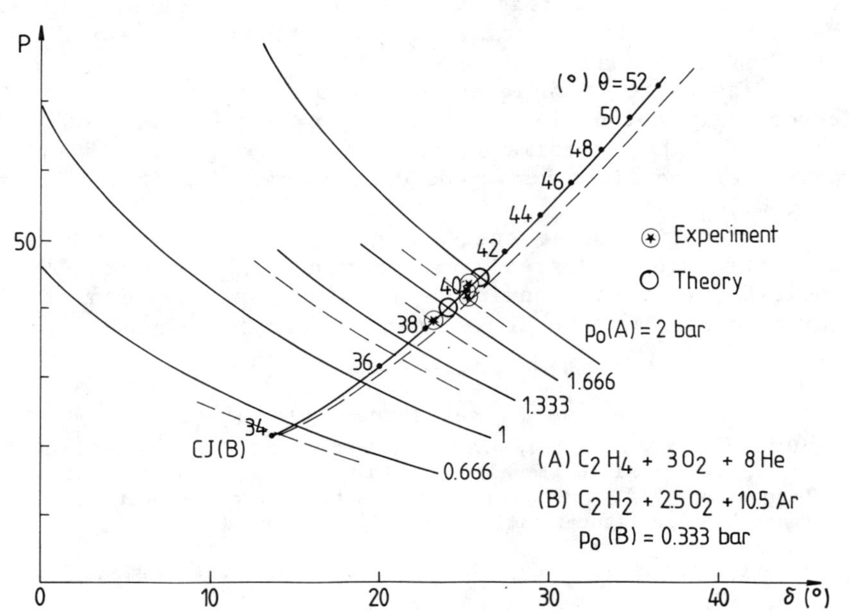

Fig. 11 P-δ plane of interaction. Mixture A: $C_2H_4 + 3\ O_2 + 8$ He at different initial pressures. Mixture B: $C_2H_2 + 2.5\ O_2 + 10.5$ Ar, p_o = 0.333 bar.

theoretical prediction. Experimental plots are, however, in the greatest part slightly lower than and are consecutive to the experimental, near sub-CJ characteristics of the self-sustained normal detonation propagating in A. Comparisons are given for dotted curves for Prandtl-Meyer expansion of detonation product of A starting from $\delta = 0$, $M = 1$, and $P = 0.9\ P_{CJ}(A)$ and for detonation polar of B with real Mach number M_1^{CJ} [defined by Eq. (3), where $0.96\ D_{CJ}$ takes the place of $D_{CJ}(A)$].

This agreement suggests that the non-negligible thickness of the mylar film separating the two media does not modify significantly the steady ODW properties in the bounding mixture. It modifies only the behavior of establishing the steady propagation of the wave (delay of onset and relaxation time to steady properties of the detonation).

Conclusions

A test section for observing the establishment of ODW has been designed and used. Mixtures of CJ detonation velocities at over 3000 m/s are used as a driver to create ODW in the bounding reactive mixture of low CJ detonation velocity ($\sim$ 1800 m/s) by the lateral expansion of detonation products. A convenient choice in initial pressure permits one to obtain different regimes of ODW travelling at high Mach numbers from normal CJ conditions to very highly overdriven detonations.

Diagnostic techniques used to characterize ODW are essentially detonation wave angle, pressure jump, and cell size of the tridimensional structure. All of these parameters permit a very good characterization of the regime of ODW.

Detonation polar analysis gives a very good prevision of interaction and steady propagation conditions of ODW. In addition, critical conditions of steady propagation of ODW corresponding to near CJ detonations can be deduced.

References

[1]Dabora, E.K., Nicholls, J.A., and Morrison, R.B., "The Influence of a Compressible Boundary on the Propagation of Gaseous Detonations," 10th Symposium (International) on Combustion, The Combustion Institute, Pittsburgh, PA, 1965, p. 817.

[2]Pratt, D.T., Humphrey, J.W., and Glenn, D.E., AIAA Paper 1987, pp. 1785.

[3]Hertzberg, A., Bruckner, A.P., and Bogdanoff, D.W., "Ram Accelerator: a New Chemical Method for Accelerating Projectiles

to Ultrahigh Velocities," AIAA Journal, Vol. 26, No 2, 1988, pp. 195-203.

[4]Liu, J.C., Liou, J.J., Sichel M., Kauffman, C.W., and Nicholls, J.A., "Diffraction and Transmission of a Detonation into a Bounding Explosive Layer," 21st Symposium (International) on Combustion, The Combustion Institute, Pittsburgh, PE, 1986, pp. 1639-1647.

[5]Murray, S.B., and Lee, J.H., "The Influence of Physical Boundaries and Gaseous Detonation Waves," Progress in Astronautics and Aeronautics: Dynamics of Explosions, Vol. 106, edited by J.R. Bowen, J.C. Leyer, and R.I. Soloukhin, New York, 1986, pp. 329-355.

[6]Desbordes, D., Brisot, D., et Guerraud, C., "Taille et Régularité de la Structure du Front de la Détonation dans les Gaz, Corrélation avec le Diamètre Critique de Transmission de la Détonation," (Size and Regularity of the Structure of Gaseous Detonation, Correlation with Critical Diameter of Transmission of the Detonation), Annales de Physique, France, Vol. 14, 1990, pp. 629-635.

[7]Desbordes, D., "Transmission of Overdriven Plane Detonations: Critical Diameter as a Function of Cell Regularity and Size," Progress in Astronautics and Aeronautics: Dynamics of Explosions, Vol. 114, edited by A.L. Kuhl, J.R. Bowen, J.C. Leyer, and A. Borisov, Washington, 1988, pp. 170-185.

[8]Heuzé, O., Bauer, P., Presles, H.N., and Brochet, C., "The Equations of State of Detonation Products and Their Incorporation into QUATUOR Code," 8th Symposium (International) on Detonations, Albuquerque, NM, 1985.

[9]Oppenheim, A.K., and Smolen, J.J., "Vector Polar Method for the Analysis of Wave Intersections," Rept No AS.66.16, College of Engineering, Univ. of California, Berkeley, CA, 1966.

[10]Desbordes, D., and Lannoy, A., "Effects of a Negative Step of Fuel Concentration on Critical Diameter of Diffraction of a Detonation," 12th ICDERS, Ann Arbor, Mich.; presented elsewhere in this volume.

Numerical Simulation of Layered Detonations

D. A. Jones*
Materials Research Laboratory, Melbourne, Australia
and
M. Sichel†
University of Michigan, Ann Arbor, Michigan
and
R. Guirguis‡
Naval Surface Warfare Center, White Oak, Maryland
and
E. S. Oran§
Naval Research Laboratory, Washington, DC

Abstract

Time-dependent numerical computations describing the interaction between a detonation propagating through a mixture of hydrogen and oxygen or an inert bounding layer are presented and compared to the results of shock-tube experiments. Three types of bounding materials are considered: the same mixture as in the propagating detonation, a leaner hydrogen-oxygen mixture, and an inert material. In each case, when the detonation first comes into contact with the bounding material, a blast wave expands into the bounding layer and hits a containing wall. Then, depending on the properties of the two materials and the state of the propagating detonation, the blast wave in the bounding material develops into one of a number of complex configurations. The numerical simulations are preformed using a two-dimensional time-dependent flux-corrected tranport algorithm and a calibrated parametric model of the chemical energy release. Simulations and experiments show qualitatively the same features, and the detailed output of the simulations provides important new information on the structures seen in the experiments.

Introduction

The interaction between an explosive charge and the bounding medium can have a significant influence both on the detonation propagating through

---

*Materials Research Laboratory, Melbourne, Australia.

† Department of Aerospace Engineering, University of Michigan, Ann Arbor, MI.

‡Detonation Physics Branch, Naval Surface Warfare Center, White Oak, MD.

§ Laboratory for Computational Physics and Fluid Dynamics, Naval Research Laboratory.

the charge and on the way energy is transmitted by the explosive to the surrounding medium. In the case of layered detonations, the bounding medium is an explosive with characteristics that may differ from those of the primary explosive. An important practical question in the design and use of such layered explosives is whether a suitable combination of explosives can enhance the effectiveness of the energy transmission compared to using a pure explosive with the same total energy.

When a detonation propagating through the primary explosive first comes into contact with the bounding or secondary explosive, there is an initial transient phase during which the basic interaction pattern becomes established. After that, there are a number of possible types of structures that can form in the bounding explosive, the exact nature of which depends on the specific properties of the primary and secondary explosives as well as on the degree of overdrive of the incident detonation. Figure 1, a schematic of a detonation propagating through a layered explosive,[1,2] shows the blast wave transmitted (the explosive bubble) and two of the simpler possibilites for subsequent structures: an oblique detonation with Mach reflection from the lower wall and an oblique shock with regular reflection.

The detonation interaction shown in Fig. 1 is important in a number of practical configurations. If the primary and bounding explosives are identical, the layered detonation pattern represents the diffraction that occurs when a detonation propagates past a step or an increase in cross-sectional area, a problem that also has been considered by Bartlma and Schroder.[3] If the wall of the upper shock tube is considered as an axis of symmetry, the layered detonation also can represent the propagation of a detonation from a smaller into a larger tube and then, as indicated by Matsui and Lee,[4] the diameter of the smaller tube must exceed a certain critical value if a detonation is to be initiated in the larger tube. If the wall of the lower channel is considered the axis of symmetry, the layered detonation simulates an inner explosive sheathed in an outer or primary explosive.

In the case of gas-phase explosives, the interface between the primary detonation products and the secondary products acts as a high-speed gaseous wedge that can induce an oblique detonation in the secondary explosive, a configuration that is one of the interaction patterns shown schematically in Fig. 1. Such oblique detonations are the basis of the oblique-detonation ramjet engine that has been proposed[5] as a supersonic combustion propulsion system. It is likely that reactive shock and detonation waves similar to those observed in layered detonations will, in fact, arise in supersonic propulsion systems such as the supersonic combustion ram jet.

In order to observe the complex interactions between detonating layers experimentally, it is necessary to use flow visualization methods in addition to various pressure and impulse measurements. Although such visualization methods are difficult if not virtually impossible to use in solid explosives, pulsed laser schlieren photography is an established method for studying complex structures in gaseous explosives. This technique has been combined with a specially designed shock tube to visualize transmitted shock and detonation structures in layered detonations for an extensive combination of primary and bounding explosives.[1,2]

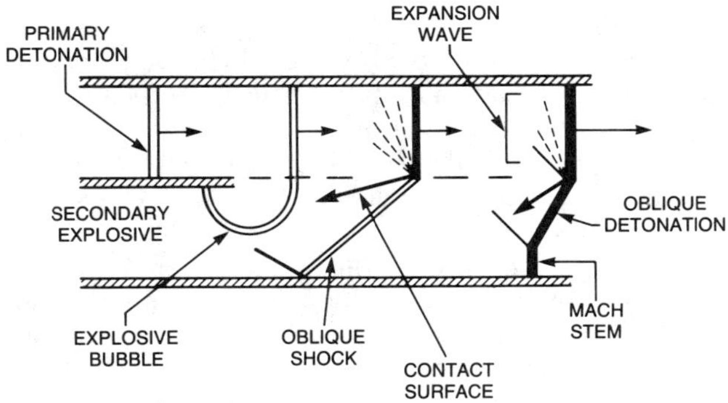

Fig. 1 Typical layered detonation interaction patterns.

In this paper, we present and compare experimental schlieren framing photographs with comparable numerical simulations of layered detonations. Previous analyses of layered detonations were restricted to the steady state using shock polars[1,6] and to quasisteady theory in the case of Mach reflections.[7] Both of these approaches produce results consistent with experiments for limited regions of the interaction zone; however, they do not describe the complex unsteady processes that actually occur. Here we simulate the layered detonations by solving the two-dimensional unsteady equations for a chemically reactive flow using the flux-corrected transport algorithm. The resulting interaction patterns are then compared to those observed for gaseous explosives under similar conditions. Although direct quantitative comparison was not possible because the simulated and experimental mixtures were not exactly the same, the simulations do reproduce the main interaction configurations observed in the experiments. The simulation then permits detailed exploration of the variation of other fluid dynamic variables that could not readily be measured and thus provides additional insight into the physics of the layered detonation interaction. We show that the numerical simulations can be used to interpret the experimental results and to extrapolate to conditions beyond those of the experiments.

The experimental investigations of layered detonations in gaseous explosives are described first, followed by a discussion of the steady-state shock-polar analysis. The numerical simulation is then described, and the results are compared with experimentally observed interaction patterns. Diagnostics from the simulation, which provide additional information about the physics of the interaction, are then presented and discussed. We conclude that the simulation captures most of the physical phenomena observed in the layered detonation experiments.

Experimental Measurements

A special detonation tube was used to obtain schlieren framing photographs of the interaction between a primary detonation and the bounding explosive mixture. This facility, described in detail in Refs. 1 and 2, is shown in Fig. 2 together with the associated optical instrumentation. It consists of two adjacent 1.6 cm^2 channels that are 3 m long. The primary explosive

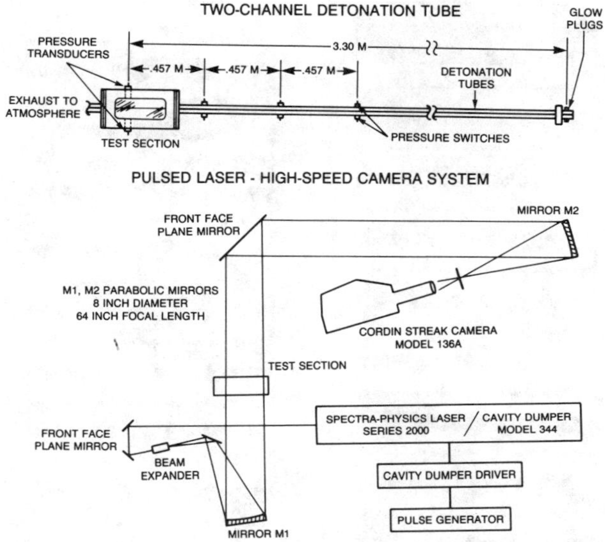

Fig. 2 Schematic diagram of layered detonation shock tube.

in the top channel is separated from the secondary (bottom) explosive by a very thin (15 – 50 Å) 1.6 × 15 cm^2 cellulose film. This film separates the two different explosives while creating minimal interference in the shock and detonation transmission process. The primary explosive is ignited by a glow plug 3 m upsteam of the test section to insure that the primary detonation is fully developed and stable. Pressure switches are used to monitor the velocity of the primary detonation, which has been observed to be very close to the theoretical Chapman-Jouguet value in all of the experiments. Two pressure transducers mounted on the top and bottom walls of the test section are used to sense the pressure history in both channels.

A system consisting of an argon-ion laser and a high-speed camera is used to obtain schlieren photographs of the interaction as the primary wave comes into contact with the bounding explosive in the test section. The laser pulse width of about 12 ns plays the role of a shutter so that the framing pictures are sharp and distinct. The delay between each photograph depends on the pulse repetition rate and was chosen to be 2 μs. It takes about 100 μs for the primary detonation to pass the test section so that typically 50 individual framing pictures are obtained for each run. Three sequences of framing photographs obtained with a stoichiometric H_2-O_2 mixture as the primary explosive and various secondary mixtures, taken from Refs. 1 and 6, are shown in Figs. 3–5 and are discussed later. Four marking wires, whose image appears on the framing photographs, are used to indicate various locations in the test section. From right to left, the first wire marks the start of the separating membrane, whereas the third is at the location of the two pressure transducers. The distances between the first and second and between the first and fourth wires are 5.08 and 10.16 cm, respectively.

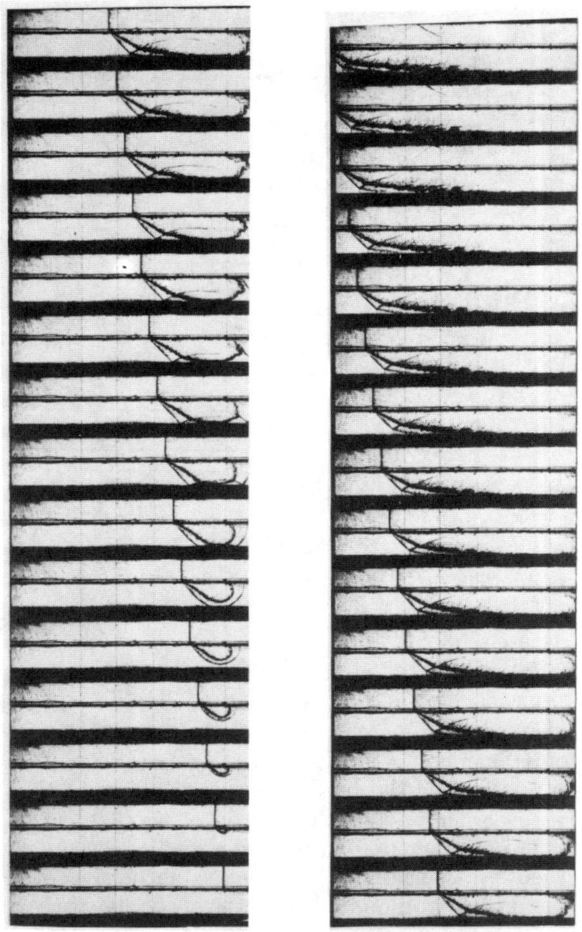

Fig. 3 Schlieren framing photographs, inert boundary. Primary mixture H_2:O_2/2:1; secondary mixture (inert) H_2:N_2/16:1.

Analysis and Discussion of Experimental Results

With stoichiometric H_2-O_2 as the primary explosive, Figs. 3–5 show the interactions that occur when the bounding mixtures are inert, stoichiometric, or lean, respectively. In each case, the interaction is initiated by an explosive bubble or blast wave generated by the high pressure behind the primary detonation. This wave propagates into the secondary mixture and is then reflected from the lower wall. However, after this initial transient phase, the interaction patterns are different for these three secondary mixtures.

Figure 3, the case of an inert bounding mixture, shows that the transmitted structure becomes an oblique shock, and the second half of the framing photographs show a transition from regular to Mach reflection. When

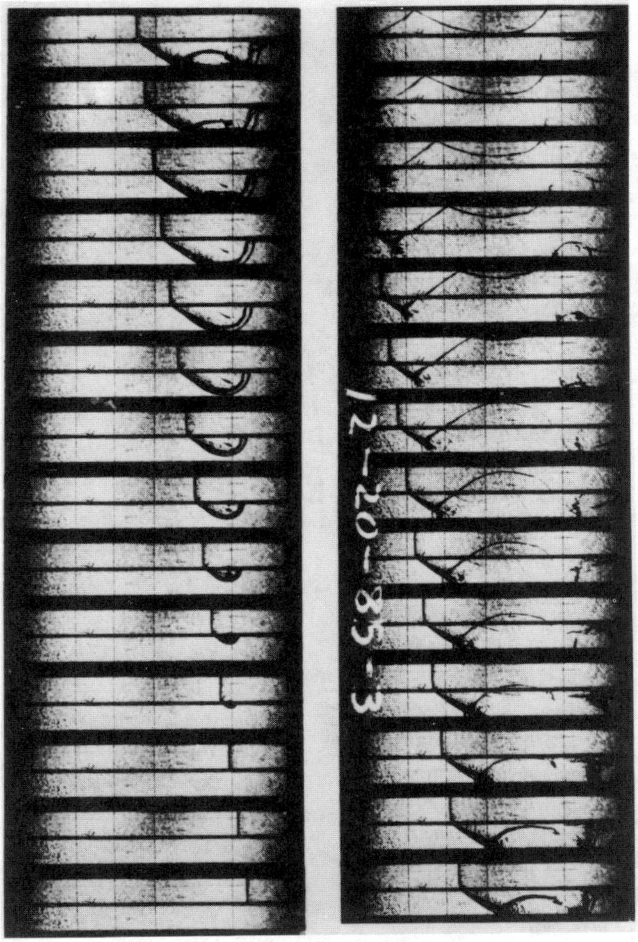

Fig. 4 Schlieren framing photographs, stoichiometric boundary. Primary Mixture H_2:O_2/2:1; secondary mixture H_2:O_2/2:1, $\phi = 1.0$.

both the primary and bounding mixtures are stoichiometric H_2–O_2, as shown in Fig. 4, an oblique shock again develops into a Mach reflection in the secondary explosive. In contrast to the case in which the bounding material is inert, the Mach stem is replaced by a normal accelerating detonation wave that catches up to the primary wave. Figure 5 shows that when the bounding mixture is lean, a complex pattern forms, consisting of an oblique detonation near the lower wall and an oblique shock wave nearer the primary wave.

Although a significant part of the interactions shown in Figs. 3–5 is unsteady, the interactions do appear to approach steady state in some cases and, even when this is not the case, the wave interactions can locally be considered as quasisteady. Thus, considerable information about the in-

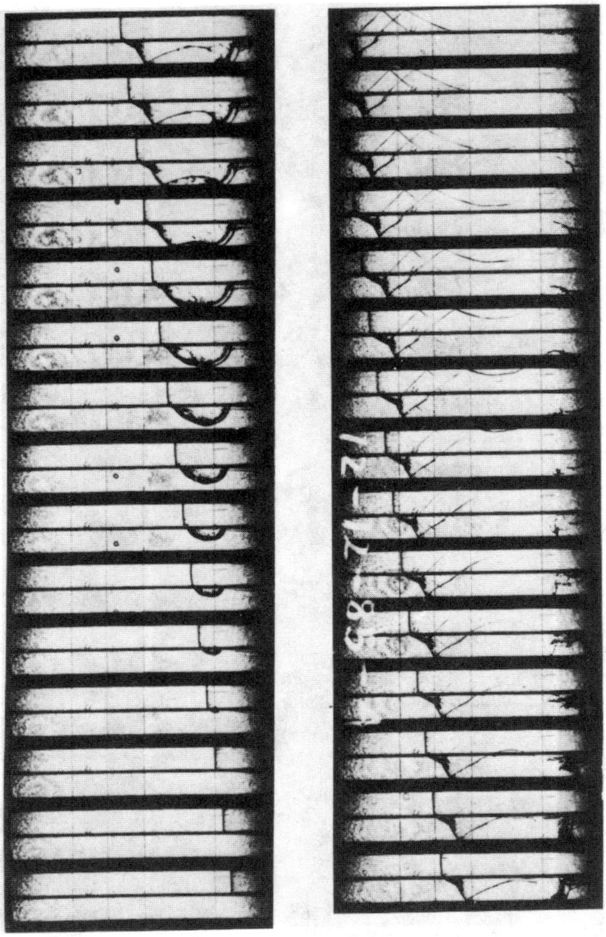

Fig. 5 Schlieren framing photographs, lean boundary. Primary mixture H_2:O_2/2:1; secondary mixture H_2:O_2/0.9:1, $\phi = 0.45$.

teraction can be obtained by analyzing the steady-state oblique shock and oblique detonation polars, especially in the plane of the pressure and flow-deflection angle ($p - \delta$ plane). A detailed analysis based on such polars has been conducted by Liou[6] and Fan et al.[7]

The interaction that occurs when the bounding medium is inert is shown in Fig. 6a in both the $p - \delta$ and physical planes, with each region on the physical plane represented by a point on the $p - \delta$ plane. In the case shown, an expansion wave behind the detonation connects the region behind the transmitted oblique shock and the high-pressure region at the Chapman-Jouguet (CJ) plane of the detonation. A contact surface or interface separates the combustion products behind the detonation and the inert gas that has been compressed by the oblique shock wave. Equality of direction and pressure at the interface are the conditions used to determine the strength of the expansion wave and the transmitted shock; however, the velocities on the two sides of the interface are generally not equal. Both the expansion and the incident and reflected oblique shock polars are shown in Fig. 6a and, since the reflected shock polar crosses the axis at $\delta = 0$, the oblique shock reflects regularly from the lower wall of the shock tube. If this polar does not cross the axis of the $p - \delta$ plane, a Mach reflection occurs at the shock-tube wall as shown in Fig. 6b.

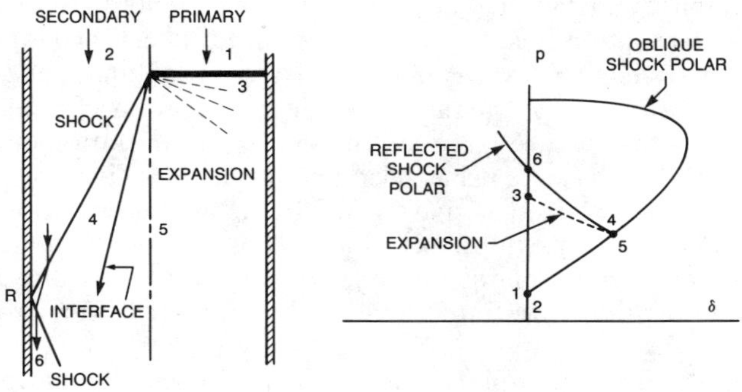

Fig. 6a Pressure-deflection angle and physical planes, oblique shock with regular reflection.

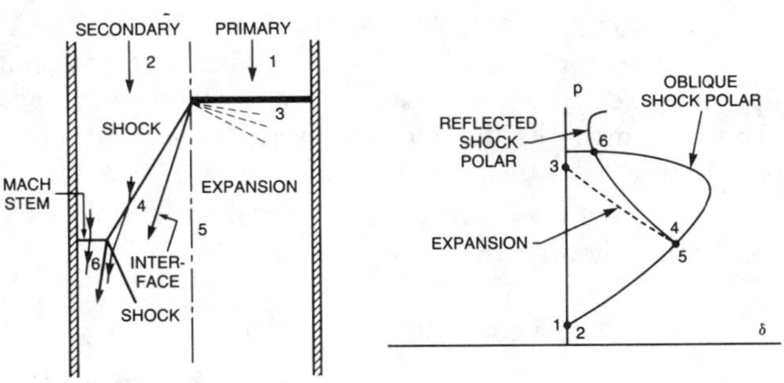

Fig. 6b Pressure-deflection angle and physical planes, oblique shock with Mach reflection.

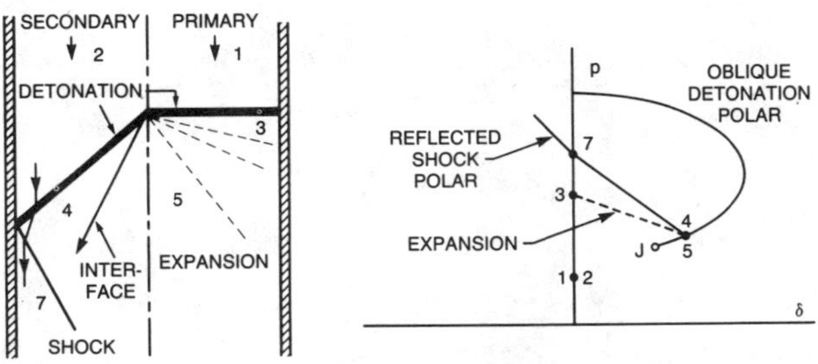

Fig. 6c Pressure-deflection angle and physical planes, oblique detonation with regular reflection.

An oblique detonation may be transmitted into the bounding medium if the bounding medium is reactive. The $p - \delta$ and physical planes corresponding to this case (Fig. 6c) show that the oblique detonation polar is open with point J corresponding to a CJ oblique detonation with sonic normal velocity at the CJ plane. Oblique detonations are generally overdriven, and the angle between the transmitted detonation and the oncoming flow is significantly greater than in the case of an oblique shock. In the case shown, the detonation reflects from the bounding wall as an oblique shock.

A complicating factor when the bounding medium is reactive is that an oblique detonation is not necessarily induced in the secondary explosive, so that the transmitted wave may be either an oblique shock or oblique detonation. Estimates of the induction length behind a transmitted oblique shock indicate that it is so large that oblique shock initiation of a detonation is unlikely.[2] In some cases, it has been found that the initiation of an oblique detonation is extremely sensitive to conditions at the interface between the primary and secondary explosive.[1] Thus, in Fig. 4, even though the bounding medium is reactive, an oblique detonation is not induced directly. Instead, a normal detonation is initiated behind the Mach reflection from the lower wall.

When the bounding mixture is lean (Fig. 5), the detonation velocity of the primary explosive is greater than that of the secondary explosive, so that any detonation induced in the secondary explosive will be overdriven. Then an oblique detonation, generated near the bottom of the lower shock tube, is connected by an oblique shock to the interface between the two media. According to the shock-polar analysis, both modes of interaction observed in this complex refraction are possible. It is interesting to speculate on the mechanism by which the oblique detonation is generated in this case. It is possible that instabilities due to the interface between the primary detonation products and the unreacted gas behind the oblique shock may provide the initiating disturbance.

Numerical Simulation

Numerical simulations of gas-phase detonations are based on solutions of the compressible, time-dependent, conservation equations for total mass density ρ, momentum density ρv, and energy E,

$$\frac{\partial \rho}{\partial t} = -\nabla \cdot \rho \mathbf{v} \tag{1}$$

$$\frac{\partial \rho \mathbf{v}}{\partial t} = -\nabla \cdot (\rho \mathbf{v}\mathbf{v}) - \nabla P \tag{2}$$

$$\frac{\partial E}{\partial t} = -\nabla \cdot (E\mathbf{v}) - \nabla \cdot (\mathbf{v}P) \tag{3}$$

where $\mathbf{v}$ is the fluid velocity and P the pressure. In a multispecies fluid in which chemical reactions affect transformations among the species, we also need individual species number densities $\{n_i\}$,

$$\frac{\partial n_i}{\partial t} = -\nabla \cdot n_i \mathbf{v} + Q_i - L_i n_i , \qquad i = 1, ..., N_s \tag{4}$$

where the $\{Q_i\}$ and $\{L_i\}$ are chemical production and loss terms, respectively, for species i. The effects of molecular diffusion have been omitted from these equations because they are generally insignificant on the time scales of interest for detonations. There is a constraint that defines the total number density N,

$$N = \sum_{i=1}^{N_s} n_i , \tag{5}$$

where N_s is the total number of different kinds of species present. The total energy is a sum of the kinetic and internal energy,

$$E = \frac{1}{2}\rho \mathbf{v} \cdot \mathbf{v} + \rho\epsilon \tag{6}$$

where ϵ is the specific internal energy.

An ideal-gas equation of state is used for the gas-phase calculations,

$$P = NkT = \rho RT \tag{7}$$

Also, we assume

$$\rho\epsilon = \sum_i \rho_i h_i - P = \rho h - P \tag{8}$$

$$h_i = h_{io} + \int_{T_o}^{T} c_{pi} \, dT \tag{9}$$

so that the properties of the individual species are taken into account. Here, k is Boltzmann's constant, $\{h_i\}$ the enthalpies of each species i, $\{h_{io}\}$ the heats of formation, and $\{c_{pi}\}$ the specific heats.

The detonation calculations described here were performed with a numerical model that solves the time-dependent two-dimensional Euler equations using the nonlinear, fully compressible flux-corrected transport (FCT) algorithm.[8,9] The FCT algorithm is an explicit, conservative, finite-volume method designed to insure that all conserved quantities remain monotonic and positive. It is particularly effective in maintaining steep gradients and accurate solutions in both supersonic and subsonic flow calculations. The use of this algorithm with direction and timestep splitting and a number of related calculations performed with it are described in some detail by Oran and Boris.[9] Specific applications to detonations are described in Refs. 10–14.

In all of the calculations described later, the full details of the chemical reactions are not included in the model. Instead, we use the induction parameter model that reproduces the essential feature of the chemical reaction and energy release process. In this model, three quantities are tabulated as a function of temperature, pressure, and stoichiometry: the chemical induction time, time of energy release, and final amount of energy released. These may be obtained by integrating the full set of elementary chemical reactions, as we have done for hydrogen-oxygen combustion in this paper, or they may be gathered from experimental data, as we have done previously for liquid nitromethane. Then a quantity called the induction parameter is defined and convected with the fluid in a Lagrangian manner. This parameter records the temperature history of a fluid element and, when the element is heated long enough, energy release is initiated. This model for including the properties of a chemical reaction mechanism in a numerical simulation was described originally by Oran et al.[10] and has been been developed further by Kailasanath et al.[11] and Guirguis et al.[12].

We simulate a planar detonation in a stoichiometric mixture of molecular hydrogen, oxygen, and nitrogen, H_2:O_2:Ar/ 2:1:7, propagating in a tube 1.6 cm high and 20 cm long. This tube rests on another tube, also 1.6 cm high, and is separated from it by a solid barrier that exists until the position 3.6 cm, when the contents of the upper and lower tubes come in contact. As in the experiments, the composition of the lower, stationary bounding mixture is varied: it can be an inert gas, the same mixture as the upper explosive, or a relatively rich or lean mixture. As already indicated, the chemical parameters for these mixtures are based on calculations using the detailed chemical reaction scheme for hydrogen and oxygen. In general, 500 computational cells span the x direction and 80 cells span the y direction, so that with an evenly spaced grid, $\Delta x = \Delta y = 0.04$ cm. The initial unperturbed mixtures are at standard temperature and pressure.

The detonation in the primary explosive was initiated by putting a fixed amount of energy into the first few cells and letting the detonation develop. This mode of initiation is similar to the direct intiation of a detonation using a high explosive charge as the initiator. The result is a blast wave that propagated into the reactive medium and decayed to a detonation propagating at the CJ velocity followed by a Taylor expansion wave.

Results of the Simulations

Figure 7 shows pressure contours from a numerical simulation in which a detonation, initially propagating in the stoichiometric argon-diluted mixture, comes in contact with a lean argon-diluted mixture with an equivalence ratio of 0.25. Elapsed times and time-step numbers are indicated on the figure, in which distance increases from left to right in contrast to the experimental framing pictures, in which distance increases from right to left. In this case, the initial detonation was initiated by a sudden energy release that creates a blast wave that transitions quickly into an overdriven detonation. When the detonation passes the barrier that separates the two mixtures, it is still overdriven and propagates with a velocity higher than the CJ value for that mixture. However, the structures formed do, in fact, look very much like

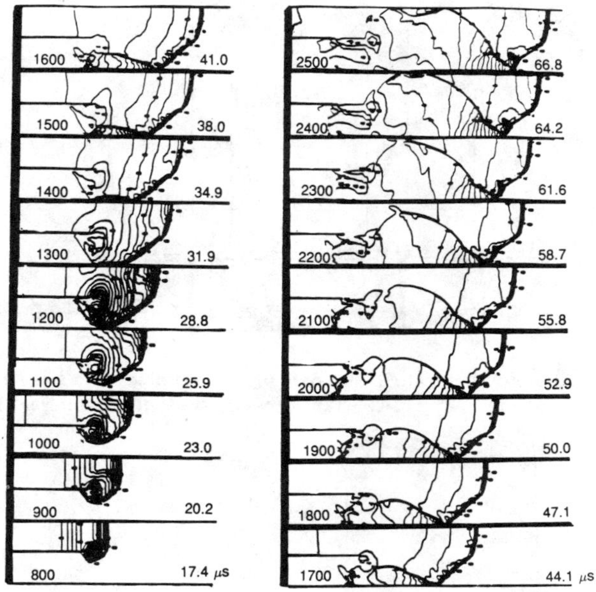

Fig. 7 Results of numerical simulation: constant pressure contours. Primary (upper) mixture H_2:O_2:Ar / 2:1:7; secondary (lower) mixture H_2:O_2:Ar / 0.5:1:7.

those shown in Fig. 5. Starting at 17.4 μs, there is a bubble or blast wave that grows and, at about 28.8 μs, it hits the bottom wall. The bubble is then reflected from the bottom wall and the reflection patterns look very much like the experimental patterns shown in Fig. 5. Between 28.8 μs and 34.9 μs, the interaction has evolved into an oblique shock transmitted into the secondary explosive. Then a further transition occurs near the lower wall resulting in an oblique detonation. By this time, the velocity of the primary detonation has reached a steady value of 1600 m/s. The pressure contours beyond 55.8 μs display the same combined pattern of an oblique shock and an oblique detonation observed in Fig. 5. Although quantitative comparison is not possible, this qualitative agreement indicates that the numerical simulation captures the main features of the interaction.

With the simulation, it is possible to study features of the flow that are not readily determined experimentally, and one such feature is the pressure variation within the flowfield. Figure 8 shows the pressure along a line through the middle of the bounding region at three different times in the evolution of the structure corresponding to Fig. 7. At 25.9 μs during the initial phase, the bubble has just formed and is moving downward into the lean mixture. The pressure profile of the leading front of the bubble is characteristic of a blast wave in that the pressure drops immediately behind the leading shock wave. The pressure rise across the shock moving back into the region underneath the barrier is also evident. At 34.9 μs and 47.1 μs, the pressure pattern is more complex because of the reflection at the lower

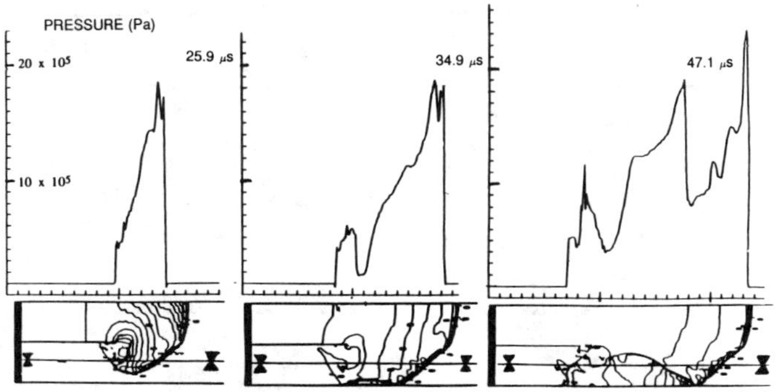

Fig. 8 Results of numerical simulation:: pressure profiles at a fixed distance above the lower shock-tube wall, and constant pressure contours at different times.

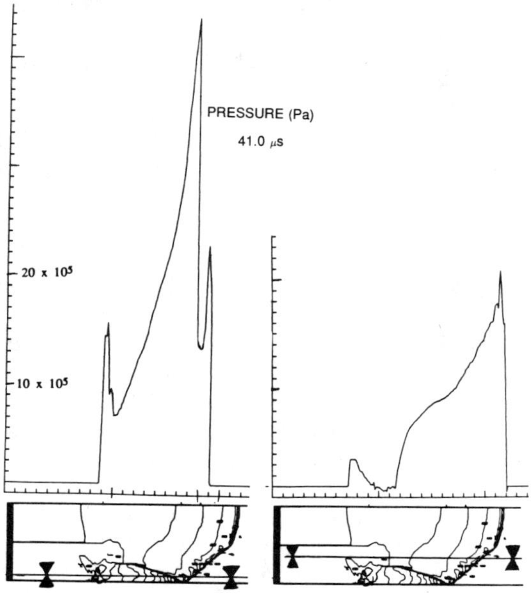

Fig. 9 Results of numerical simulation: pressure profiles at a fixed time at several levels above the lower shock-tube wall. ISTEP = 1600; time = 41.0 μs.

wall. The presence of the various reflected waves is particularly evident at 47.1 μs. A pressure probe in which the pressure is shown at a fixed instant of time (41.0 μs) is shown in Fig. 9 for lines that are 0.2 and 1.2 cm above the wall of the lower tube. It is interesting to note the very high pressure rise across the oblique shock reflection pattern near the lower shock-tube wall.

Figures 10 and 11 show the results of two computations in which both the upper and bounding materials are the stoichiometric argon-diluted mixture. The difference is that the bounding material in Fig. 10 is allowed to

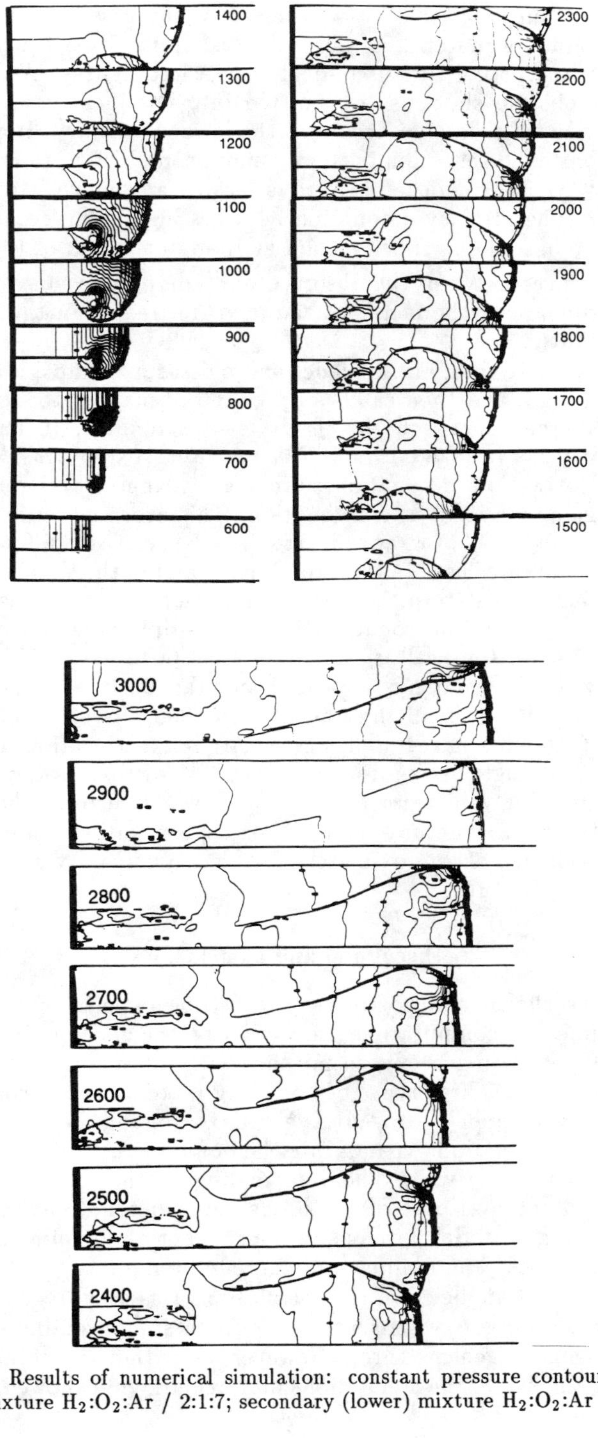

Fig. 10 Results of numerical simulation: constant pressure contours. Primary (upper) mixture H_2:O_2:Ar / 2:1:7; secondary (lower) mixture H_2:O_2:Ar / 2:1:7.

react whereas all reactions are suppressed in the case shown in Fig. 11 so that the bounding material is inert. The computational time-step number rather than the actual time is indicated on these figures. As in the experiments, an oblique shock is transmitted into the inert boundary, and an irregular or Mach reflection occurs at the lower boundary. However, in this case, the computed reflection pattern is more complicated than the observed one: a very complex shock structure is formed and moves steadily through the system. The primary detonation becomes highly curved, and a gradual transition to an oblique shock occurs at the later stages of the interaction. These differences between the observed and computed behavior are probably a consequence of the fact that the mixture treated in the simulation is relatively dilute.

In the reactive case, Fig. 10, a detonation starts behind a Mach reflection at the lower wall, and then catches up to the primary detonation. Because the only difference between the results shown in Figs. 10 and 11 is that the bounding gas is reactive in the latter case, it appears likely that the accelerating wave is a detonation as in Fig. 4, where the striations behind the wave again identify it as a detonation. This process occurs more rapidly than in the framing photographs in Fig. 4 because of the difference between the explosive mixures used in the experiments and in the simulation. By step 3000, the final time step in the calculation, the interaction does not appear to have reached any final configuration. The triple point formed when the bubble hits the bottom wall is moving upward (after step 1500) and at step 1900, a new triple point has appeared spontaneously above it and is also moving up and bounces off the wall at step 2300. By step 2000, the shock formed in the bounding region has gone through a transition to detonation. At step 3000, a single detonation spans the test section; however, it is likely that a triple point will begin propagating downward from the upper wall. The expected final structure in Fig. 10 is a propagating detonation with a detonation cell size of approximately 0.1 cm, significantly smaller than the total height of the tube.

Discussion and Conclusions

We have shown and compared shock-tube experiments and time-dependent numerical computations that describe the glancing interaction of a detonation in a hydrogen-oxygen mixture with a bounding layer of inert or explosive material. Three types of bounding materials were considered: the same mixture, a leaner hydrogen-oxygen mixture, and an inert material. In each case, the detonation expands into the bounding material as a blast wave that hits a containing wall. Then, depending on the properties of the two materials and the overdrive of the propagating detonation, the blast wave in the bounding material develops into a number of complex configurations consisting of shocks and oblique and curved detonations.

The explosive modeled in the calculations was a hydrogen-oxygen mixture diluted with argon, whereas the experiments were conducted with undiluted hydrogen-oxygen mixtures. However, even though the mixtures were not exactly the same, the simulations and experiments showed many of the

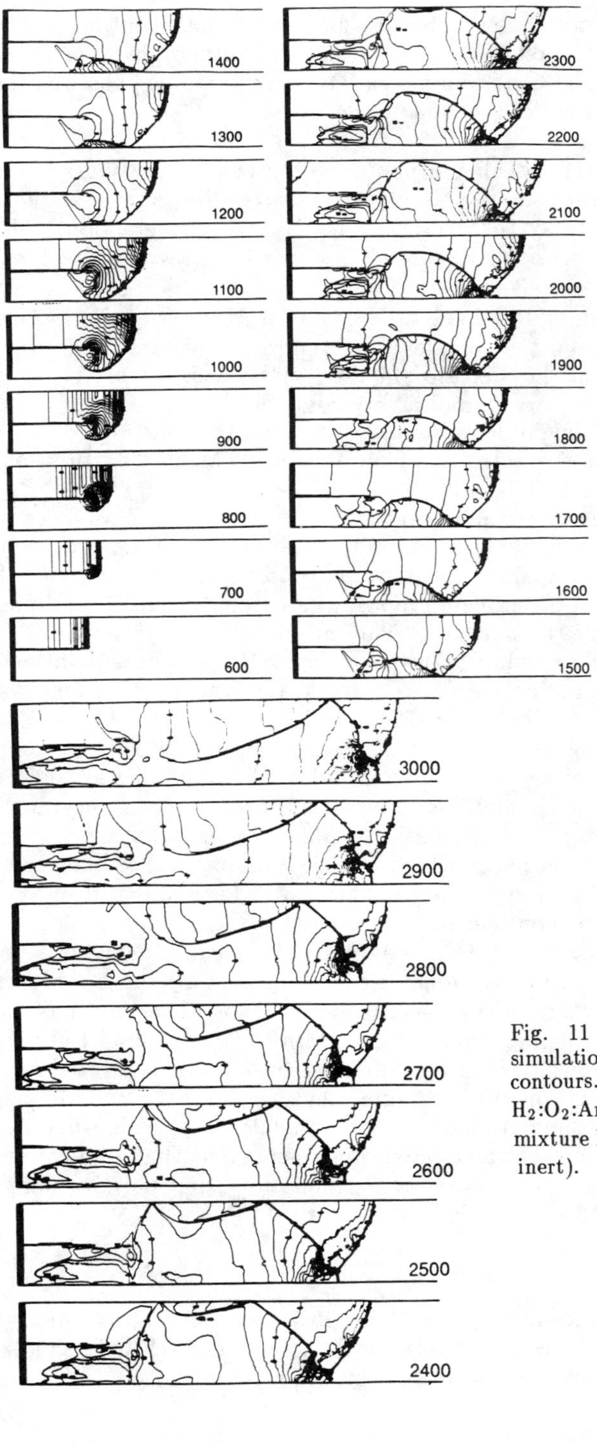

Fig. 11 Results of numerical simulation: constant pressure contours. Primary (upper) mixture H_2:O_2:Ar / 2:1:7; secondary (lower) mixture H_2:O_2:Ar / 2:1:7 (made inert).

the same shock and detonation structures, and the detailed output of the simulations provided interpretations of the schlieren photographs. The times after first contact with the bounding mixture at which various phases of the interaction took place were of the same order in both the experiments and the simulation.

It should be noted that schlieren photography senses density gradients whereas isobars are shown in the simulation results. Nevertheless, these representations both display the overall structure of the interaction. The pressure contours are insensitive to contact surfaces across which there is a density jump while pressure is constant.

When the bounding material is inert, regular or Mach reflection occurs at the lower boundary and the complexity of the resulting combined structure seems to depend on the mixture. In the simulations, the propagating detonation becomes very curved, and a very complex shock structure forms that moves steadily through the system. The undiluted mixtures used in the experiments also show a transition from regular to Mach reflection, but a relatively simpler structure results.

When the bounding material is identical to the primary explosive, complicated transient structures develop initially. Then if the primary detonation is strong enough to impart sufficient impulse to the unreacted material, the final structure would be a single propagating detonation with a cellular pattern characteristic of the material. The initial stages of this transition appears to be what is observed in both the experiments and the simulations. When the primary detonation is relatively weak, the whole detonation system could decay.

When the bounding explosive is lean, the experiments show a complex pattern consisting of an oblique detonation near the lower wall and an oblique shock wave. The simulations show structures that look very much like those seen in the experiments, again showing the propagating detonation connected to an oblique shock which, in turn, is connected to an oblique detonation. However, at the time the simulation was terminated, it did not appear that the configuration had reached a steady-state pattern.

Quantitative comparison of the simulation and the experiments was not possible since the explosive mixture modeled in the simulations was diluted with argon and the mixture in the experiments was not. This choice of explosive was dictated by the fact that the induction parameter model used in the simulation had been calibrated and carefully tested for the diluted mixture. However, there was qualitative agreement between the experiments and the corresponding simulations, and the simulations provided detailed insights into the complex shock and detonation phenomena that are involved in these interactions. It is anticipated that simulations that permit quantitative comparisons will be available in the future.

Acknowledgments

This work was sponsored by the Naval Research Laboratory through the Office of Naval Research and by the U.S. Army Research Office under Grant DAAG29-83-K-009, Dr. David M. Mann, project monitor.

References

[1] Liu, J.C., Sichel, M., and Kauffman, C.W., "The Lateral Interaction of Detonating and Detonable Gaseous Mixtures," *Progress in Astronautics and Aeronautics*, 1988, Vol. 114, pp. 264–283.

[2] Liu, J.C., Liou, J.J., Sichel, M., Kauffman,C.W., and Nicholls, J.A., "Diffraction and Transmission of a Detonation into a Bounding Explosive Layer," *Proceedings of the 21st Symposium (International) on Combustion*, The Combustion Institute, Pittsburgh, 1987, pp. 1659–1668.

[3] Bartlma, F., and Schroder, K., "The Diffraction of a Detonation Wave at a Convex Corner," presented at *10th International Colloquium on Dynamics of Explosions and Reactive Systems*, Berkeley, 1985.

[4] Matsui, H., and Lee, J.H., "On the Measure of the Relative Detonation Hazards of Gaseous Fuel-Oxygen and Air Mixtures," *Proceedings of the 21st Symposium (International) on Combustion)*, The Combustion Institute, Pittsburgh, 1989, pp. 1269–1280.

[5] Sheng, Y., and Sislian, J.P., "A Model of a Hypersonic Two-Dimensional Oblique Detonation Ram Jet," University of Toronto Institute of Aerospace Studies, Tech. Note 257, 1985.

[6] Liou, J.J., "Analysis of the Wave Interaction Between a Propagating Detonation and a Bounding Explosive Layer," Ph.D. Thesis, Department of Aerospace Engineering, The University of Michigan, Ann Arbor, Michigan, MI, 1986.

[7] Fan, B.C., Sichel, M., and Kauffman, C.W., "Analysis of Oblique Shock – Detonation Wave Interactions in the Supersonic Flow of a Combustible Medium," AIAA Paper AIAA-88-0441, Jan. 1988.

[8] Boris, J.P., and Book, D.L., "Solution of the Continuity Equation by the Method of Flux-Corrected Transport," *Methods in Computational Physics*, Vol. 16, 1976, pp. 85–129.

[9] Oran, E.S., and Boris, J.P., *Numerical Simulation of Reactive Flow*, Chap. 8, Elsevier, New Yori, 1987.

[10] Oran, E., Boris, J.P., Young, T.R., and Picone, J.M., "Numerical Simulations of Detonations in Hydrogen-Air and Methane-Air Mixtures," *Proceedings of the 18th Symposium (International) on Combustion*, The Combustion Institute, Pittsburgh, PA, 1981, pp. 1641–1649.

[11] Kailasanath, K., Oran, E.S., and Boris, J.P., "Determination of Detonation Cell Size and the Role of Transverse Waves in Two-Dimensional Detonations," *Combustion and Flame*, Vol. 61, 1985, pp. 199–209.

[12] Guirguis, R., Oran, E.S., and Kailasanath, K., "Numerical Simulations of the Cellular Structure of Detonations in Liquid Nitromethane — Regularity of the Cell Structure," *Combustion and Flame*, Vol. 65, 1986, pp. 339–366.

[13] Guirguis, R., Oran, E.S., and Kailasanath, K., "The Effect of Energy Release on the Regularity of Detonation Cells in Liquid Nitromethane," *Proceedings of the 21st Symposium (International) on Combustion)*, The Combustion Institute, Pittsburgh, 1987, pp. 1659–1668,

[14] Oran, E.S., Kailasanath, K., and Guirguis, R.H., "Numerical Simulations of the Development and Structure of Detonations, *Progress in Astronautics and Aeronautics*, 1988, Vol. 114, pp. 155–169.

Chapter III. Nonideal Detonations and Boundary Effects

Photographic Study of the Structure and Propagation Mechanisms of Quasidetonations in Rough Tubes

A. Teodorczyk,* J. H. S. Lee,† and R. Knystautas†
McGill University, Montreal, Quebec, Canada

Abstract

The structure and propagation mechanisms of quasidetonations in an obstacle-filled two-dimensional channel have been studied. Unlike previous photographic observations, the present study used a narrow channel (of the order of a detonation cell size) to minimize three-dimensional effects. This simplifies the complex wave processes and permits the fundamental mechanisms to be identified. High-speed framing schlieren photography (typically 500,000 frames/s) with a Barr and Stroud camera was used to observe detonation processes in a 16 mm x 57 mm x 1.2 m long detonation channel. Stoichiometric mixtures of H_2 and O_2 in the range of initial pressures 50 Torr $\leq p_o \leq$ 150 Torr were studied. In addition to photographic observation, average wave speeds were measured by using photodiodes and pressure by piezoelectric transducers. The photographic observation clearly identifies the propagation mechanism of quasidetonation to be one of autoignition by shock reflections. Three main shock reflection processes are observed: 1) regular (or Mach) reflection of the diffracted shock from the bottom wall of the channel; 2) normal reflection of the Mach stem from the front face of the obstacle; and 3) normal (or Mach) reflection of the transverse shock from the top wall of the channel. Depending on the obstacle height, spacing, and distance to the top wall, one or more of these three reinitiation mechanisms can occur. Furthermore, reflections of two transverse waves (as in the normal reinitiation process of a cellular detonation) are also possible in place of reflection from the solid top (or bottom) wall of the

*Visiting Professor, Department of Mechanical Engineering. Permanently, Professor, Warsaw University of Technology, Warszawa, Poland.

†Professor, Department of Mechanical Engineering.

channel. The obstacle plays an important role in attenuating the detonation by diffraction, whereas the expansion waves cause the detonation to fail by separating the reaction zone from the shock. Thus the obstacles play a dual role: a positive role of reinitiation by providing a surface for shock reflection and a negative role of attenuation by diffraction. On a global basis, obstacles have a negative effect on the overall average propagation velocity of the detonation. Quasidetonation is only possible on a scale larger than that of a cell size λ. The present study clearly demonstrates the multidimensional nature of cellular detonation processes and the essential role played by the transverse waves not only in the reinitiation process, but also in the feedback process of the chemical energy release to drive the leading shock front. Therefore, it may be concluded that the quasi-one-dimensional model of Zel'dovich of heat and momentum losses does not describe the propagation mechanism of this quasidetonation.

Introduction

The propagation of quasidetonations in an obstacle environment is of great practical interest. Numerous studies have been carried out in past decades on this problem, starting with the pioneering works of Laffitte[1] and Shchelkin.[2]

Theoretical studies have also been carried out, including quasi-one-dimensional analytical models[3-5] and complex numerical simulations of two-dimensional shock-wave processes in an obstacle field.[6] Despite all of these studies, the basic mechanisms responsible for the propagation of quasidetonations are not understood. Previous experimental observations via high-speed photography of the self-luminous events[7] or streak schlieren photography[8] have yielded little information on the structure and propagation mechanisms. Recently, results of high-speed schlieren photography of quasidetonations have been reported.[9,10] However, these studies indicate an extremely complex structure of three-dimensional interacting shocks, extended reaction zones and density interfaces. It is very difficult to understand these complex three-dimensional structures and identify the mechanisms involved. Earlier studies[9,10] have revealed that shock reflections resulting in autoignition provide an important reinitiation mechanism. However, the channels used in these experiments were wide (of dimension of many cell sizes), and, thus, three-dimensional events tend to obscure the basic mechanisms involved. The present paper reports results of a photographic study of the structure of quasidetonation in a narrow channel. Preliminary studies[11-14] have

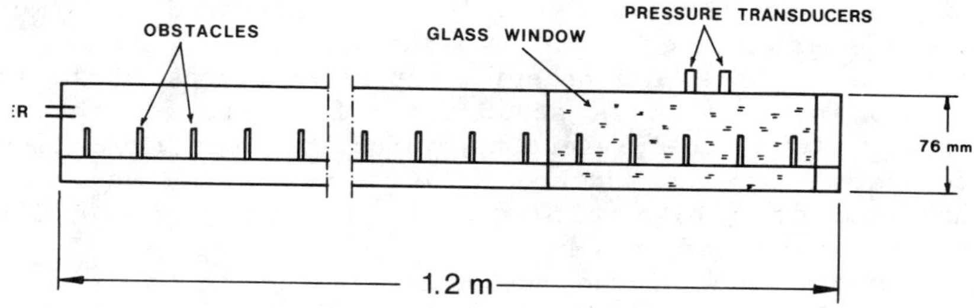

Fig. 1 Experimental apparatus.

indicated that the minimum width must be at least greater than $\lambda/2$ for stable detonation propagation. This limit condition corresponds to having a single-headed spin wave (or one "slapping wave") in the minimum dimension of the rectangular channel used. In the present study, the channel width was adjusted to the range $\lambda \lesssim w \lesssim 1.5\ \lambda$ and, hence, three-dimensional effects (in the line of sight) are minimized. This permits the complex structure of quasidetonations to be analyzed more readily.

Experimental Details

The experimental apparatus used in the present study (Fig. 1) consists of a long channel of rectangular cross section. The channel itself was constructed of steel beams as the top and bottom walls, with metal plates and optical-quality glass panes forming the side walls. The glass was bonded to the side edges of the steel beams with epoxy glue, which also assured that the resulting chamber was vacuum-tight. The glass side walls provide a see-through capability for 60 cm of the channel length (in its end part). The inner dimensions of the rectangular channel are 76 mm high x 16 mm wide in cross section by 1.2 m long. The obstacle field was constructed of short lengths of an aluminum flat bar attached periodically at equal distances on a 16 x 19 mm cross-section aluminum bar. Once assembled, the obstacle array was slid into the chamber along the lower steel channel wall and attached to it once in place. After assembly, each obstacle is 25.4 mm high, 16 mm wide, and 1.6 mm thick. The height d of the unobstructed passage is always the same, d = 31.6 mm. For a particular obstacle array, the interobstacle distance L is constant. Three spacing configurations were studied in the present experiments, with L/D = 0.80, 1.61, and 3.22.

In any given experiment, the channel was first evacuated to levels better than 0.01 Torr. The apparatus was then filled to the predetermined initial pressure level from a high-pressure reservoir containing the prepared stoichiometric H_2-O_2 mixture. The experiments in the channel were carried out at subatmospheric initial pressures in the range 50-150 Torr, which is about the safe limit for the glass side walls of the present apparatus. Direct initiation of detonation in the mixture at one end of the channel was achieved by a high-voltage exploding wire using a 100-μF capacitor charged to 4 kV.

Observation of the combustion phenomena was via high-speed schlieren photography. The schlieren system was of the double-pass type, and the main mirror delineating the field of view was a 30-cm-diam spherical mirror of high optical quality. The high-speed camera used to record the events was a Barr and Stroud rotating mirror type. Through an array of 30 prisms, the image of the event was projected on a stationary piece of 35-mm film (TIMAX-400), typically 75 cm long. This provides 30 frames of single-event photography up to a maximum framing rate corresponding to about 2 million frames/s. In the present experiment, the camera was run close to 500,000 frames/s, which corresponds to about a 2-μs between each frame. The camera was illuminated from a linear xenon flash lamp (Xenon Corporation FPA 8100C). The duration of the illuminating flash is about 1 ms at half peak intensity level. Synchronization of events (ignition and illumination flash) was via two BNC Corporation model 7010 digital delay generators. The initiation of events was manual followed by a pulse from a photodiode associated with the rotating mirror in the camera.

Velocity measurements were obtained by spacing a number of photodiodes (type TIL 78 N-P-N silicon phototransistor) along the glass side wall of the channel. The distribution of the diodes was at least 10 obstacle spacings apart so that the velocities reported are average values over a number of obstacles. Pressure measurements were obtained by calibrated fast-response piezoelectric transducers (PCB 113A24, 5 mV/psi). Two transducers were used located at the top wall of the channel in its see-through section.

Results and Discussion

Figure 2 shows the quasidetonation velocity as a function of initial pressure for the mixture studied. Also shown are the corresponding deflagration velocities in the "choking regime" prior to transition to the quasidetonation regime. It is evident that transition to detonation occurs when both characteristic dimensions d and L accommodate several detonation cell sizes λ. Comparison with the results of previous studies in wider channels[9] shows that, for similar values of d/λ and L/λ, the quasidetonation as well

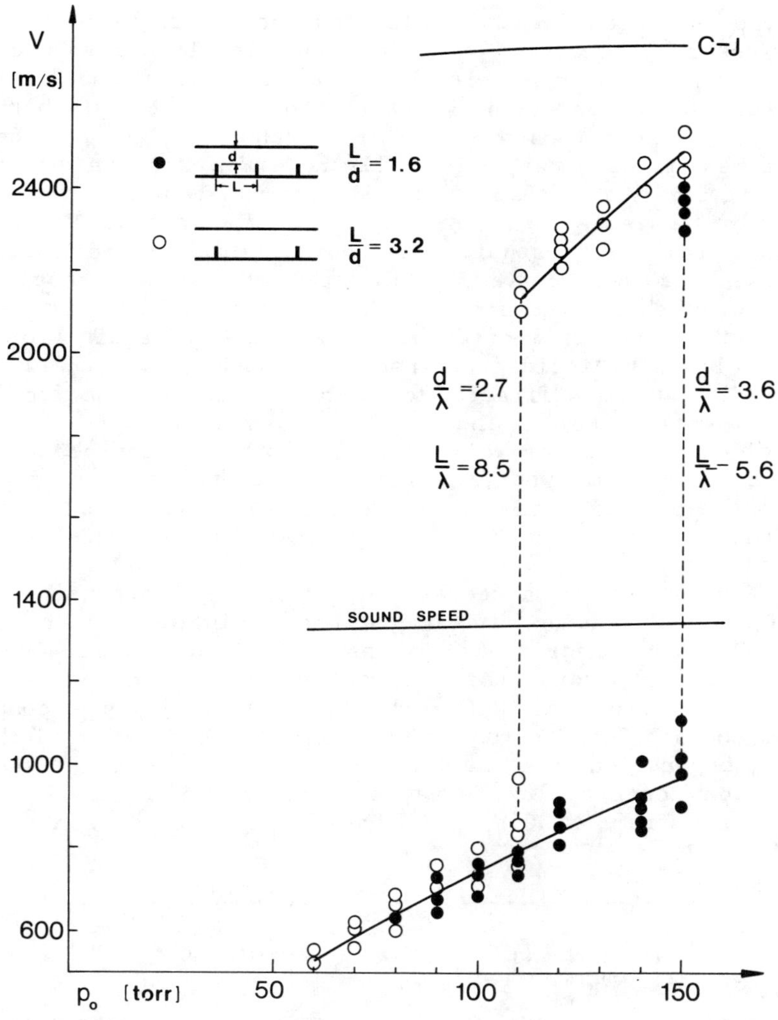

Fig. 2 Variation of average propagation velocity with mixture sensitivity (initial pressure p_o for $2H_2 + O_2$ mixture).

as deflagration velocities are lower in the narrow channel. This indicates the three-dimensional nature of the quasidetonation and fast deflagration processes, which is also reflected in the propagation velocity of the wave itself. That is, the added dimension provides for transverse wave collisions and reinitiation to offset the attenuating effect of the obstacles.

In general, the obstacles and walls (top and bottom walls in the present case of a two-dimensional geometry) provide surfaces for the reflection and diffraction of shock and detonation waves. To understand clearly the subsequent photographic records, it is best to review the salient features of a normal shock interacting with a simple barrier for an obstacle in a two-dimensional channel. A sequence of sketches illustrates this well-known interaction process. In Fig. 3a, a normal shock is shown advancing into the barrier obstacle. Figure 3b shows the normal reflected shock from the obstacle, the normal transmitted incident shock, and the curved cylindrical diffracted shock and reflected shock. Figure 3c shows the wave processes at a later time when reflected shock has reflected again from the top wall. The incident transmitted shock is now completely curved, and the diffracted wave is seen to reflect from the bottom wall as well. The angle is shown to be less than the critical so that the reflection is normal. Figure 3d shows later on in time when the normal reflection at the bottom wall has turned into a Mach reflection. In a combustible mixture, a reaction front will trail behind the leading shock complex.

Figure 4 shows a sequence of schlieren photographs when a normal detonation wave encounters a single barrier. The first frame clearly shows the reflected shocks in the combustion products, the transmitted and diffracted leading detonation front in which the reaction zone is decoupled from the shock and normal reflection of the diffracted shock just begins on the bottom wall. The decoupled shock-reaction zone complex decays continuously and its velocity decreases from ~2000 m/s at the second frame to about 1200

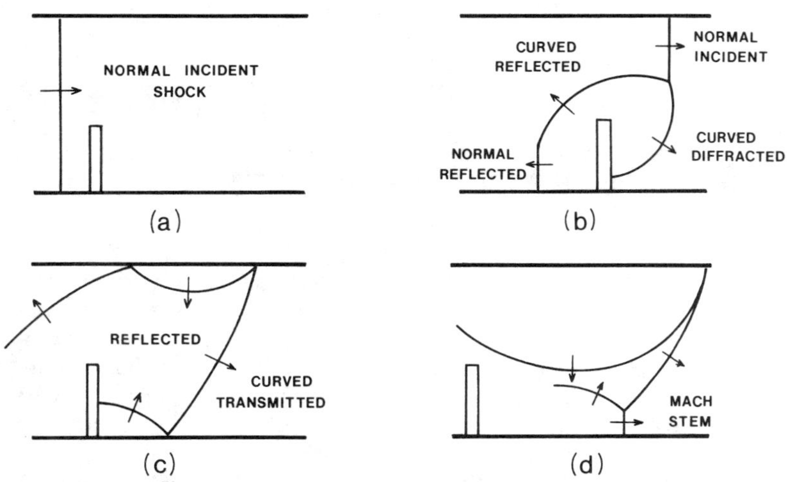

Fig. 3 Four subsequent stages of normal shock-wave interaction with single obstacle in a rectangular channel.

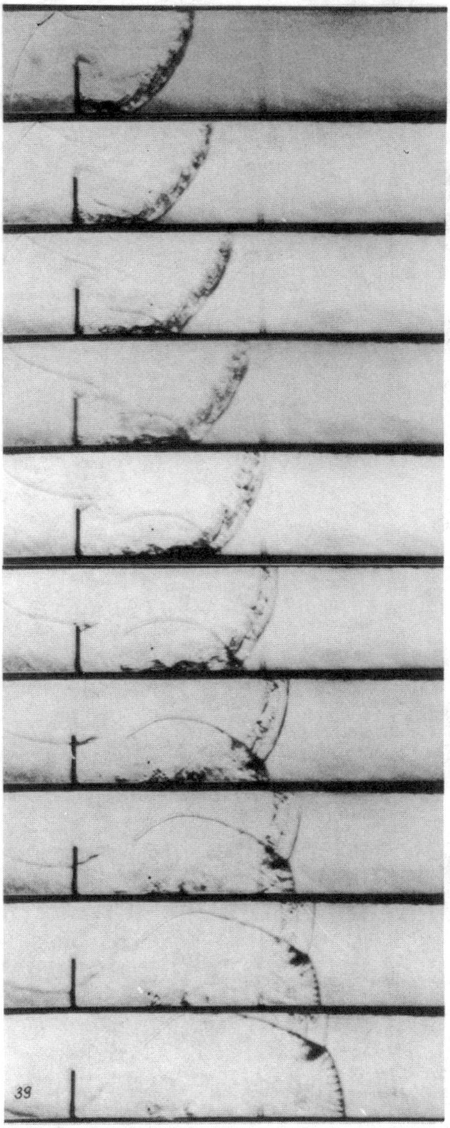

Fig. 4 Interaction of detonation wave with single obstacle in $2H_2 + O_2$ mixture. Detonation reinitiation in Mach reflection zone at bottom wall; p_o = 140 Torr, time between frames Δt = 6 μs.

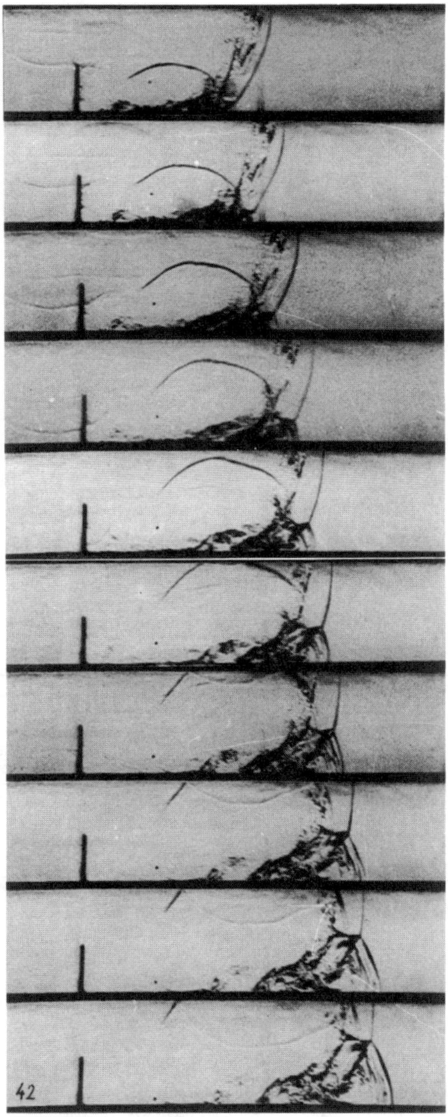

Fig. 5 Interaction of detonation wave with single obstacle in $2H_2 + O_2$ mixture. No detonation reinitiation; p_o = 90 Torr, Δt = 6.1 μs.

m/s at last frame. The reaction zone is seen to decouple progressively from the shock similar to the subcritical direct initiation regime when initiation energy for a cylindrical detonation is less than the critical value required for direct initiation.[15] In the sixth frame of Fig. 4, the shock angle at the bottom wall is approximately 58 deg, marking the transition from regular to Mach reflection. Since the Mach stem is sufficiently strong, autoignition and reinitiation to a multiheaded cellular detonation occurs in the Mach stem. The reinitiated Mach stem then propagates at ~2400 m/s, which corresponds to self-sustained detonation with its characteristic fine cellular structure. The decoupled deflagration near the top wall continues to attenuate with the reaction zone now completely separated from the leading shock front as the reinitiated cellular detonation from the bottom wall sweeps upward. Without further encounters with another obstruction, a multiheaded cellular wave then continues to propagate along the channel. If, however, another obstacle is encountered, the detonation will diffract around it and the expansion waves will cause the detonation to fail. The reinitiation process then repeats itself. Figure 4 shows one reinitiation mechanism that is possible when the Mach reflection of the diffracted shock from the bottom wall is sufficiently strong for autoignition. In Fig. 5, the conditions are such that the Mach stem is not strong enough to cause reinitiation (leading shock propagates with velocity 1000 m/s). The turbulence in the shear layer and the wall jet associated with the triple point cause intense combustion activity in the region of the reflected shock. However, the Mach stem is clearly a shock front and the reaction zone is not coupled to it as compared with the previous case shown in Fig. 4.

In Fig. 6, the effect of obstacle spacing is demonstrated clearly. As in the previous case shown in Fig. 4, reinitiation occurs at the Mach stem on the bottom wall. However, prior to complete reinitiation of the decoupled wave by the upward-growing detonation, reflection and, subsequently, diffraction of the detonation occur again by encountering another obstacle. In the sixth frame of Fig. 6, the curved, diffracted, and reflected shock with a reaction zone close behind is clearly evident. However, as this cylindrical deflagration expands, progressive decoupling of the reaction zone occurs. In the last frame, this entire shock and reaction zone complex is decoupled. The destructive role played by obstacles in causing the failure of detonation is now clear. With higher obstacle density (i.e., more obstacles per unit length), more frequent attenuations by diffraction of the reinitiated detonation

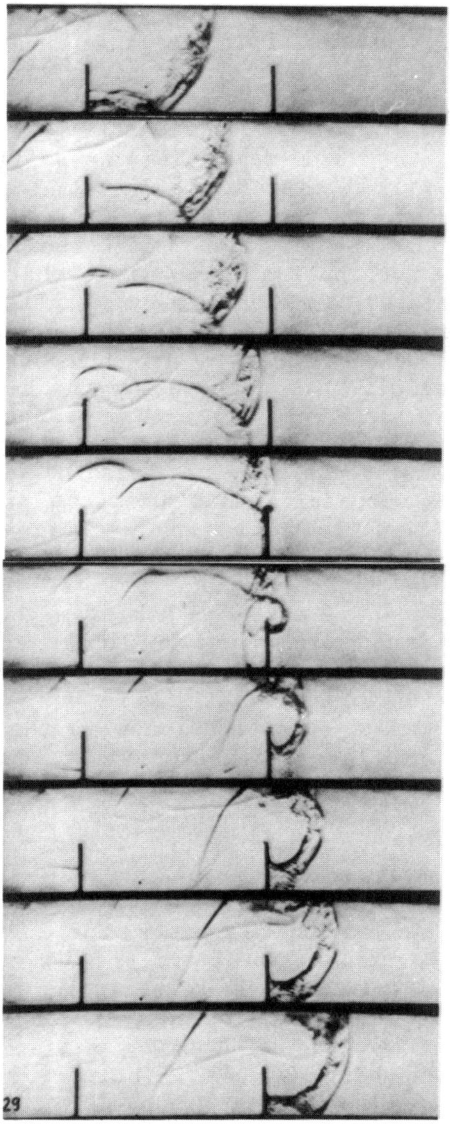

Fig. 6 Propagation of quasidetonation in obstacle array in $2H_2$ + O_2 mixture. Detonation reinitiation in Mach reflection zone at bottom wall; p_o = 140 Torr, Δt = 6 μs.

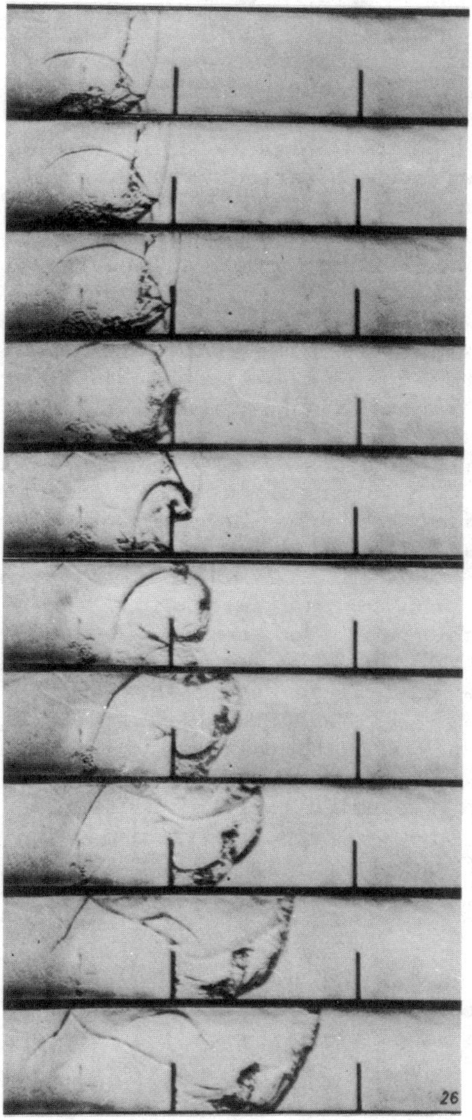

Fig. 7 Propagation of quasidetonation in obstacle array in $2H_2$ + O_2 mixture. Detonation reinitiation by normal Mach stem reflection from the obstacle with subsequent enhancement via reflection from top wall; p_o = 120 Torr, Δt = 6.1 μs.

occur. This explains the decrease in the "averaged" velocity of quasidetonations with increasing obstacle density reported previously.[5]

As shown in Fig. 5, the Mach stem at the bottom wall fails to cause reinitiation, and the reaction zone is not coupled to it. However, when this Mach stem encounters another obstacle, the normal reflection of the Mach stem from the front of the obstacle then may cause reinitiation. Figure 7 shows such a reflected Mach stem initiation. In the first two frames of Fig. 7, the incident shock, Mach stem, reflected shock complex followed by a reaction zone is clearly evident. After normal reflection of the Mach stem from the obstacle, an autoexplosion and reinitiation occur. The part of the detonation that transmits through the reaction zone now propagates into the combustion products; thus, it becomes a shock. However, the part of the detonation that diffracts around the obstacle and propagates in the unburned mixture between the leading shock front and the reaction zone develops into a cylindrical detonation. However, as shown in the sixth and seventh frames, the sharp expansion around the obstacle causes the lower part of the detonation to fail, whereas the upper part continues as a normal multiheaded cellular wave. Reflection from the top wall further enhances the detonation, and, in the last two frames, the self-sustained cellular detonation near the top wall eventually will sweep down to reinitiate the decoupled complex of the diffracted wave. Thus, this mode of reinitiation is due to the reflected Mach stem from the obstacle. The obstacle now acts to play a positive role in assisting the reinitiation process, while simultaneously playing a destructive role causing failure by diffraction.

Figure 8 shows a similar reinitiation process due to reflection of the Mach stem from the obstacle. The rapid expansion of the reflected shock due to autoexplosion clearly is evident by comparing frames 2 and 3. In the present case, the diffraction causes the reinitiated detonation to fail, and it becomes a cylindrical deflagration with progressive decoupling of the reaction zone from the shock front (fifth and sixth frames). However, the normal reflection of the leading shock from the top wall causes a reinitiation process. The cellular detonation subsequently sweeps down to engulf the entire decoupled front. The last two frames show a cellular detonation over the entire front, and failure recommences when the detonation diffracts over the obstacle again in the last frame. The accompanying pressure records for Fig. 8 are also shown, and the location of the pressure transducers on the channel are indicated by PT1 and PT2. The first transducer records the large pressure rise at the moment of autoexplosion at the top wall caused

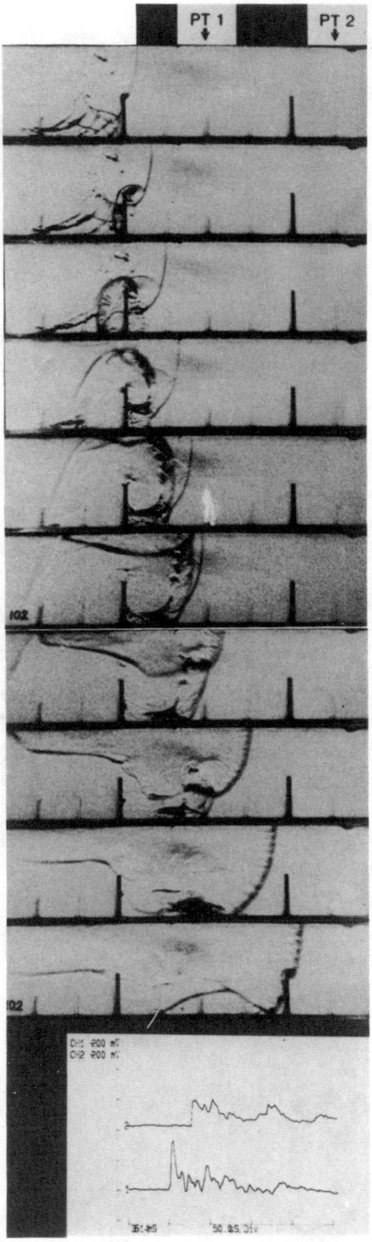

Fig. 8 Propagation of quasidetonation in obstacle array in $2H_2$ + O_2 mixture; p_o = 120 Torr, Δt = 6 μs.

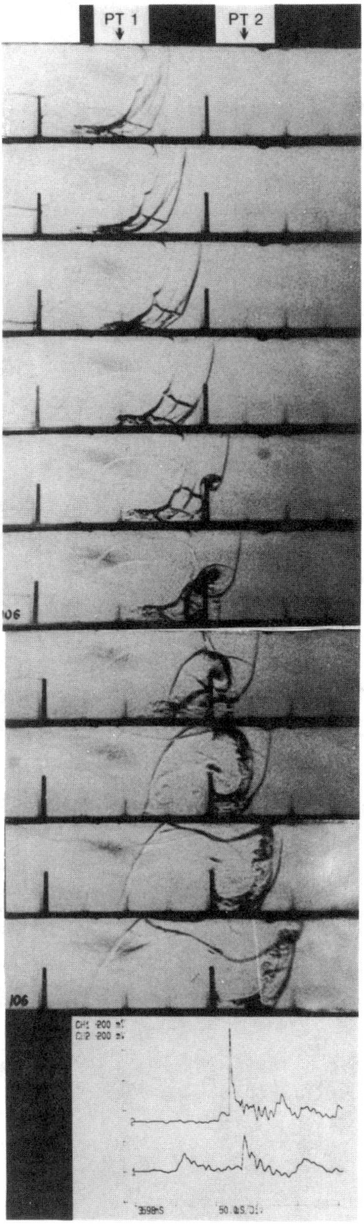

Fig. 9 Propagation of quasidetonation in obstacle array in $2H_2$ + O_2 mixture; p_o = 120 Torr, Δt = 6 μs.

by normal reflection of the cylindrical shock wave. The second transducer registers the cellular detonation pressure prior to its destruction by diffraction.

A similar reinitiation by Mach stem reflection from the obstacle--failure of the reinitiated wave and reinitiation again by reflection from the top wall--is clearly demonstrated by the case shown in Fig. 9. Of interest, are accompanying pressure traces from the two transducers. The first transducer records a series of weak shock waves typical of a fast deflagration driven by a turbulent reaction zone. The second peak in the pressure record of PT1 clearly identifies it as due to the reflected shock from the obstacle. The pressure trace from PT2 records the failing cylindrical deflagration with the shock slightly decoupled from the reaction front. However, when the shock reflects from the top wall, an extremely high pressure peak typical of a transition process is recorded. The subsequent overdriven detonation near the top wall is clearly evident in the last frame.

Depending on the obstacle height and spacing, as well as on the vertical height of the channel, one or more of the above-mentioned reinitiation mechanisms can occur. Furthermore, if a sufficiently strong transverse shock wave coming down from the top were to collide with one propagating upward, reinitiation can also occur similar to a reflection from the solid boundary of the top wall (Fig. 9). Thus, for arbitrary obstacle configurations and channel size, the number of waves and their interactions within the structure of the quasidetonation can be very complicated. It is doubtful that a general description of the structure of a quasidetonation can be obtained. However, it may be said that the propagation mechanism is one of continuous reinitiation and attenuation (or failure) by diffraction around the obstacles. This mechanism essentially is identical to that of a normal detonation, where reinitiation occurs when transverse waves collide and the reinitiated wave fails between collisions. In quasidetonations, the reinitiation is controlled by the obstacles. In a normal cellular detonation, resonance between the nonlinear chemical and gasdynamic processes controls the frequency of the reinitiations (in cell size). Numerous photographic records have been taken for different obstacle configurations and mixture sensitivity (i.e., cell size). As mentioned previously, the structure is much more complex, but the three modes of reinitiation described are found to play the central role.

Concluding Remarks

By taking high-speed schlieren photographs of quasidetonations in a narrow channel to minimize three-dimensional effects, it has been possible to reveal the basic mechanisms of propagation of the so-called quasidetonation wave. As in a normal detonation, continuous periodic reinitiation by shock reflections constitutes the principal mechanism of propagation. Diffraction by the obstacles results in failure of the detonation by causing the diffracted shock to decouple from the reaction zone. As in normal detonation, quasidetonation is also a cyclic process with the characteristic length scale controlled by the obstacle spacing. Actually, the spacing and height are linked in a complex fashion and control the reinitiation mode. It is evident that quasidetonations can occur only on a scale larger than the cell size of the mixture. Experiments have indicated that the quasidetonation regime occurs only when the obstacle spacing, height, and distance to the top wall exceed the cell size λ of the mixture. Thus, the presence of the obstacles in a global sense causes a destructive effect by attenuating an otherwise normal Chapman-Jouguet detonation in the channel. It is also understandable why in a near-limit mixture, when the tube dimension is of the order of a cell size, obstacles lead to a failure of the detonation wave, as observed by Saint-Cloud.[16] It may be concluded from the present study that the quasi-one-dimensional models, which serve as the basis for numerous theoretical studies, cannot provide a correct description of the quasidetonation process even on a qualitative basis. The concept of momentum and heat losses in a quasi-one-dimensional reactive flow simply cannot simulate the three initiation mechanisms via Mach stem reflection or reflected transverse wave ignitions. The two-dimensional process of shock or detonation diffraction and attenuation causing failure also cannot be modeled adequately by momentum losses in a quasi-one-dimensional flow. Similar to the one-dimensional ZDN model, which has failed to describe the propagation mechanism of a normal cellular detonation, these quasi-one-dimensional models also fail to describe quasidetonations. The implications are that all detonation processes are transient and multidimensional. The multidimensional nature of the detonation process is demonstrated further by the fact that, if the transverse shocks are damped by the use of porous walls, quasidetonations just as normal cellular detonations also fail to exist.[9,17] The use of porous walls in an obstacle field eliminates reinitiation by normal or Mach reflections from the top and bottom walls. However, reinitiation via the normal reflections from the obstacle is still possible. Thus, it seems that the (at least) two-dimensional requirement for the detonation process has more to do with just the reinitiation process. The positive feedback of the chemical energy release to the leading shock also appears to require the transverse shock wave rather than a one-dimensional push from behind.

Acknowledgments

This work was supported by the Department of National Defense of Canada (DRES) under contract 01SG W7702-6-2521-A and the Natural Sciences and Engineering Research Council of Canada under NSERC Grants A-7091 and A-3347.

References

[1]Laffitte, P., Compt. Rend., Vol. 186, No. 1, 1928, p. 95.

[2]Shchelkin, I. I., "Effect of Roughness of the Surface in a Tube on Origination and Propagation of Detonation in Gases," Journal of Experimental and Theoretical Physics (USSR), Vol. 10, 1940, p. 823.

[3]Gelfand, B. E., Frolov, S. M., and Polenov, A. N., "Specific Features of Detonation in Systems with Losses of an Arbitrary Type," Archivum Combustionis, Vol. 7, No. 1-2, 1987, pp. 197-214.

[4]Wolanski, P., Lee, D., Sichel, M., Kauffman, C. W., and Nicholls, J. A., "The Structure of Dust Detonations," Progress in Astronautics and Aeronautics: Dynamics of Shock Waves, Explosions and Detonations, Vol. 94, edited by J. R. Bowen, N. Manson, A. K. Oppenheim, and R. I. Soloukhin, AIAA, NY, 1989, pp. 241-263.

[5]Gu, L. S., Knystautas, R., and Lee, J. H., "Influence of Obstacle Spacing on the Propagation of Quasidetonation," Progress in Astronautics and Aeronautics: Dynamics of Explosions, Vol. 114, edited by A. L. Kuhl, J. R. Bowen, J. C. Leyer, and A. Borisov, AIAA, NY, 1988, pp. 232-247.

[6]Kuhl, A. L., "Numerical Calculations of Shock Decay in Rib Walled Ducts," TRW, Rept. 78, 4735.9-14, Sept. 1976.

[7]Wagner, H. G., "Some Experiments About Flame Acceleration," First International Meeting on Fuel Explosion, University of Waterloo Press, SM Study 16, Montreal, 1981, pp. 77-99.

[8]Brochet, G., "Contribution à L'étude des detonations instables dans les mélanges gazeux," Thèse doctorat, Univ. de Poitiers, France, Chap. VII, 1966, pp. 107-121.

[9]Teodorczyk, A., Lee, J. H. S., and Knystautas, R., "Propagation Mechanism of Quasidetonations," 22nd Symposium (International) on Combustion, The Combustion Inst., Pittsburgh, PA, 1988, pp. 1723-1731.

[10]Chan, C. K., and Greig, D. R., "The Structures of Fast Deflagrations and Quasidetonations," 22nd Symposium (International) on Combustion, The Combustion Inst., Pittsburgh, PA, 1988.

[11]Lee, J. H. S. "On the Transition from Deflagration to Detonation," Progress in Astronautics and Aeronautics: Dynamics of Explosion, Vol. 108, edited by J. R. Bowen, J. C. Leyer, and R. I. Soloukhin, AIAA, NY, 1986, pp. 3-18.

[12]Vasiliev, A. A., "Geometric Limits of Gas Detonation Propagation," Fizika Goreniya i Vzryva, Vol. 18, 1982, pp. 132-136.

[13]Dove, J. E., and Wagner, H. Gg., "A Photographic Investigation of the Mechanisms of Spinning Detonation," 8th Symposium (International) on Combustion, Academic Press, New York, 1961, pp. 589-600.

[14]Moen, I. O., Donato, M., Knystautas, R., and Lee, J. H., "The Influence of Confinement on the Propagation of Detonations near the Detonability Limit," 18th Symposium (International) on Combustion, The Combustion Inst., Pittsburgh, PA, 1981, pp. 1615-1622.

[15]Bach, G. G., Knystautas, R., and Lee, J. H., 12th Symposium (International) on Combustion, The Combustion Inst., Pittsburgh, PA, 1969, p, 853.

[16]Saint-Cloud, J. P., Guerraund, C., Brochet, C., Manson, N., "Quelques particularites des detonations très instables dans les mélanges gazeux," Astronautica Acta, Vol. 17, 1972, pp. 487-498.

[17]Dupré, G., Peraldi, O., Lee, J. H., and Knystautas, R., "Propagation of Detonation Waves in an Acoustic Absorbing-walled tube," Progress in Astronautics and Aeronautics: Dynamics of Explosions, Vol. 114, edited by A. L. Kuhl, J. R. Bowen, J. C. Leyer, and A. Borisov, AIAA, NY, 1988, pp. 248-263.

Calculation of the Velocity of Gaseous Detonation in a Rough Tube Based on Measurements of Shock Wave Attenuation

S. M. Frolov, B. E. Gelfand, and S. P. Medvedev
USSR Academy of Sciences, Moscow, USSR

Abstract

When considering the mechanism of gaseous quasidetonation in rough tubes, one introduces drag and heat transfer coefficients into governing equations describing unsteady combustible gas flow with shock waves. These quantities are usually taken from steady unreacted flow measurements in the form of empirical correlations. Since quasidetonation is a topic of considerable practical importance, one needs a detailed, quantitative theoretical and experimental substantiation of the approach. For this purpose we have developed a simple approximate analytical theory to predict the rate of attenuation of a planar shock wave propagating through a gas in a tube with rough walls. It involves the following basic assumptions: 1) the flow is one-dimensional; 2) heat transfer to the walls is neglected; and 3) shock wave velocity is independent of conditions behind it (this is known as Whitham's approach) and depends only on tube geometry, distance traveled by the shock wave in a section with obstacles, and its initial Mach number.

Comparison of the solution with detailed numerical studies has been made, and good agreement between predicted results is shown to exist. Systematic measurements of shock wave attenuation in tubes with obstacles made of aluminum rings have been carried out to verify the theory. The data were used to determine momentum losses in shock waves by means of fitting the measured and predicted results. The drag coefficients thus obtained are shown to be larger by a factor of about four than the corresponding values measured in steady flow conditions within the whole experimental range of roughness configurations. Finally, we apply the results for calculating quasidetonation velocity as a function of ring height and spacing.

Introduction

In principle, one should consider any real detonation wave as nonideal since it is well known that the Zel'dovich-Neumann-Doring model of detonation in its one-dimensional form is not realized in practice. However, in this study the term "nonideal detonation" implies a detonation wave characterized by momentum and energy losses due to interaction of the gas with wall roughness. Taking into account this interaction is of great practical importance because it allows for prediction of the behavior of detonating systems under various conditions and explains why their characteristics are different in laboratory-scale and real, large-scale conditions. Extensive experience in studying gaseous mixtures indicates that momentum and energy losses in detonation waves should be taken into account along with the nonsteady nature of the shock wave-reaction zone complex and the cellular structure of detonation. Clearly, even such highly detonable mixtures as $C_2H_2 + O_2$ and $2H_2 + O_2$ have, in tubes of sufficiently small diameter, detonation velocities somewhat lower than those calculated without allowance for losses.

The losses markedly modify the interrelation of such factors as multidimensionality, the nonsteady nature of the flowfield behind the detonation wave front, and the total losses responsible for detonation limits; they also decrease the detonation velocity. This modification may turn out to be so dramatic that new regimes of detonation propagation come into effect and considerable difference may be observed in the wave structure, compared to the standard CJ case. Roughness elements on the channel wall disturb the flowfield, distort the regular three-dimensional, nonsteady pattern of the detonation cells, and generate additional hot spots that make the reaction proceed in a more stable fashion. This indicates that one-sided approaches, using either stability criteria of the front and the reaction zone without regard for losses or one-dimensional models accounting for the effect of losses, are inadequate for interpreting the detonation process. Unfortunately, a comprehensive analysis of the multidimensionsal flow in detonation waves is not possible at present. A rather detailed qualitative generalization of quasidetonations in tubes with obstacles has been recently suggested[1].

Since a quantitative analysis of the multidimensional flow behind a detonation wave is extremely complex, even for modern computers, it is worth considering possible variations of a one-dimensional theory, compared to the standard CJ theory, when taking momentum losses into account. The one-dimensional approach has already demonstrated its benefit for treating not only stationary waves but also such an essentially multidimensional and nonsteady phenomenon as detonation initiation. Solution of the governing equations for the average quantities behind the detonation front with properly chosen drag and heat coefficients may be expected to give not only qualitatively but also quantitatively correct information on the average nonideal detonation velocity under real conditions. That is, it can serve as a basis for engineering estimates and future multidimensional theories.

When considering gaseous detonation in rough tubes, one introduces into the governing equations the drag and heat transfer coefficients that are usually taken from steady unreacted flow measurements in the form of empirical correlations.[2-4]. It is pointed out[3,4] that this approach underestimates considerably the momentum losses in the reaction zone of a nonideal detonation wave. Some modifications of the empirical correlations are proposed for the drag coefficient. It appears that in the one-dimensional theory the local drag coefficient is infinite just behind the detonation front and drops downstream of the reaction zone to a certain, almost constant value. Depending on the length of the reaction zone, the average drag coefficient in it may become considerably larger than that determined under the steady flow conditions.

The present study deals with further development of the question. We propose a simple approximate theory of shock wave attenuation in a rough tube, including an a priori unknown drag coefficient. In order to determine the latter, we have carried out experiments in a shock tube equipped with properly designed roughness and fitted the experimental and theoretical curves of shock wave attenuation along the tube. The drag coefficient thus obtained is used in the sequel for calculating nonideal detonation velocity.

Mathematical Modeling of Shock Wave Attenuation in a Rough Tube

Consider propagation of a shock wave through a tube with roughness elements distributed uniformly on the internal tube surface. The obstacles interact with the flow behind the shock front, exerting dynamic influence on it and distorting its spatial structure. Laser schlieren photographs and density distribution measurements indicate[6,7] that behind the shock wave front a regular wave structure appears, the shock front becomes curved to a certain degree, and the density is distributed almost uniformly across the tube except for a narrow layer attached to the wall. One may believe that these findings justify the one-dimensional approach to the problem of shock wave attenuation.

Interaction of the flow with roughness elements results both in momentum losses and kinetic energy transformation owing to bow shock formation and turbulence generation and dissipation. Heat transfer to tube walls and roughness elements is relatively insignificant in the total energy balance.[4] Experimental studies of steady stabilized flows in tubes with rough walls reveal the existence of a quadratic drag law for large Reynolds numbers Re.[8] By analogy we assume from now on that the drag coefficient in the nonsteady flow behind a shock wave depends only on obstacle geometry and is independent of the Reynolds and Mach numbers of the flow.

Formulation

Let us consider evolution of a stepwise shock after its entry into a rough section $0 < x < \infty$ of a tube of radius r. The shock wave is assumed to travel initially along the smooth section $\infty < x \leq 0$ of the tube at a constant velocity. On the assumptions formulated previously, the governing equations for perfect gas flow are as follows:

$$\rho_t + u\rho_x + \rho u_x = 0 \tag{1a}$$

$$\rho u_t + \rho u u_x + p_x = -F \tag{1b}$$

$$p_t + up_x - a^2 (\rho_t + u\rho_x) = (\gamma - 1) Fu \tag{1c}$$

$$p = \rho RT \tag{1d}$$

where ρ is the density, u is the velocity, p is the pressure, a is the sound velocity, T is the temperature, R is the gas constant, γ is the ratio of specific heats, and indices t and x denote derivatives in time and coordinates. In the smooth section F=0, whereas in the rough section $F = (C_f/r)\,\rho u|u|$, where C_f is the local drag coefficient referred to henceforth to as the coefficient of momentum losses (in order to distinguish it from the drag coefficient C_f') in a steady stabilized flow.

The initial conditions at the rough section inlet are

$$t = 0, -\infty < x \leq 0, \tag{2a}$$

$$M = M_o \tag{2b}$$

$$\rho = \rho_o \frac{(\gamma + 1) M^2}{(\gamma - 1) M^2 + 2}, \tag{2c}$$

$$u = \frac{2}{\gamma + 1} a_o \left(M - \frac{1}{M}\right), \tag{2d}$$

$$p = p_o \left[\frac{2\gamma M^2}{\gamma+1} - \frac{\gamma-1}{\gamma+1} \right] \tag{2e}$$

and

$$t = 0, \ 0 < x < \infty \tag{3a}$$

$$\rho = \rho_\delta, \tag{3b}$$

$$u = 0, \tag{3c}$$

$$p = p_o \tag{3d}$$

where M_o is the incident shock wave Mach number and index 0 labels the quantities in the unperturbed flowfield. At any instant the conditions given by Eqs. (2a)-(2e) hold for $x \to -\infty$ and those given by Eqs. (3) hold for x ∞.

Results of the Analysis

In spite of a number of simplifications introduced in Eqs. (1a)-(1d), their exact solution as far as we know is not available at present. In most of the cases considered, the problems involving nonlinear interaction of a shock wave with confinement walls are treated only approximately. Quite successful in this sense is the approximate approach suggested by Whitham.[9] This phenomenological approach is based on a simple assumption that has no fundamental substantiation to date. Following Ref. 9 we assume that the perturbed flow behind the shock wave exerts no influence on shock wave propagation; that is, variations in the shock wave velocity are solely due to interaction of the front itself with the obstacles. The method discussed consists in substitution of Eqs. (2a)-(2c) into the equation for C_+-- characteristics derived from Eqs. (1a)-(1d). The solution of the ordinary differential equation thus obtained assumes the form

$$G(m_o) - G(m) = C_f \, x/r \tag{4}$$

In the linear approximation $m = M-1$, $m_o = M_o - 1$, and $G(M) = -(\gamma + 1)/2m$. For the case of arbitrary initial shock wave Mach number $m = M$, $m_o = M_o$ and

$$G(m) = 4\frac{0.4m - 1}{m^2 - 1} + 4\ln(m^2 - 1) + 0.8\ln\frac{m + 1}{m - 1} \qquad (5)$$

Equation (5) is an approximation of the exact function G(M). Deviation of the results given by Eq. (5) from the exact solution of Eq. (4) is less than 5% in the range $1.01 < m < 4$.

In order to determine the limits of the validity of Eq. (4), numerical integration of the set of Eqs. (1a)-(1d) was carried out by the method outlined in Ref. 10. Figure 1 shows the results of numerical calculations in the form of dependence of the shock overpressure Δp_f on the dimensionless complex $C_f\, x/r$ for three different initial shock Mach numbers. Solid lines represent the numerical solution; the dotted and dashed lines represent respectively the linear approximation of Eq. (4) and the approximate solution given by Eqs. (4), and respectively, (5). A comparison of the results indicates that for $M_o <$

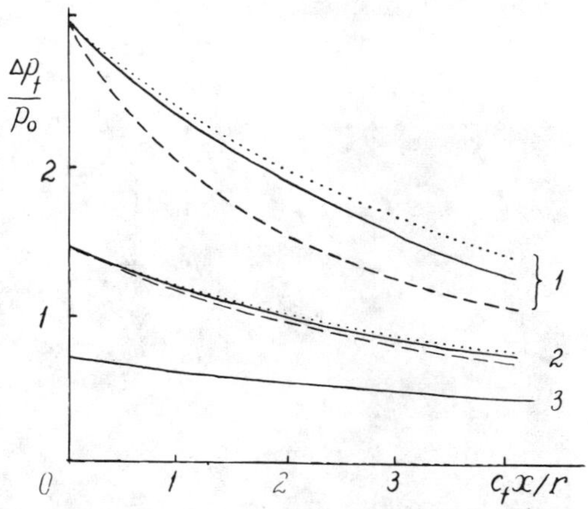

Fig. 1 Predicted shock wave attenuation in a rough tube. 1-M_o = 1.877, 1-1.497, 3-1.272. Solid line is approximate solution given by Eq. (4), dashed line is linear approximation, and dotted line is numerical simulation.

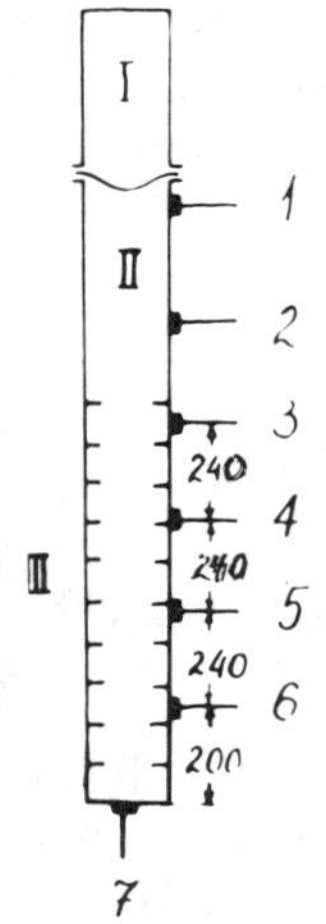

Fig. 2 Experimental setup.

Fig. 3 Obstacle configuration.

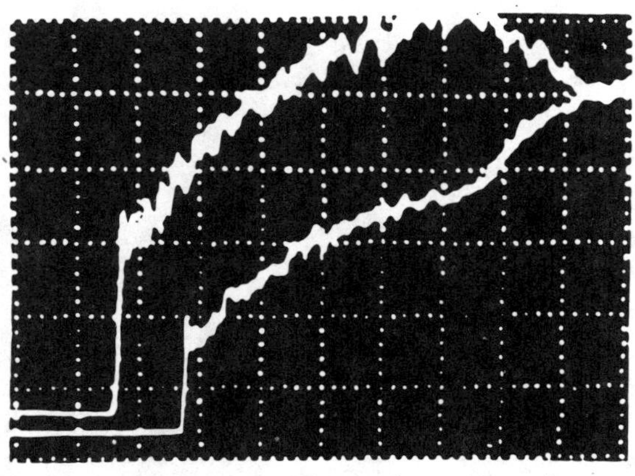

Fig. 4 Typical pressure-time histories in a shock wave propagating through a rough tube.

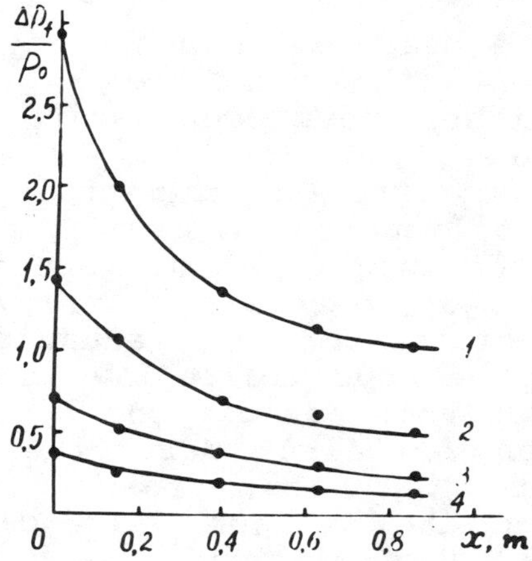

Fig. 5 Measured shock wave attenuation in a rough tube. (1) M_o = 1.877; (2) 1.497; (3) 1.272; (4) 1.154; k = 5 mm, s = 53 mm.

1.5 there exists a good agreement (less than 10% deviation) of linear approximation and the solution based on Eqs. (4), and (5) with the results of numerical calculations.

An increase in M_o deteriorates the consistency of the predictions based on linear approximation; however, the accuracy of approximation given by Eqs. (4), and (5) remains quite satisfactory in a wide range of C_f x/r values. This allows us to put forward a concept that Eqs. (4), and (5) are applicable to approximate calculations of the overpressure in a shock wave propagating through a rough tube. Equation (4) contains a priori the unknown coefficient C_f. In order to determine it, an experimental study of shock wave attenuation in a tube with rough walls was undertaken.

Experimental

A sketch of the experimental setup is shown in Fig. 2. We used a 50-mm-diameter steel vertical shock tube, consisting of a high-pressure chamber (I) 0.5-m long and a low-pressure chamber (II) 1.5-m long. An aluminum ring assembly (see Fig. 3) properly inserted in a low-pressure chamber created obstacles in the rough section III. Six flush-mounted piezoelectric pressure transducers were used to register shock front overpressure and pressure histories (see Fig. 4), as well as to measure the shock wave velocity. In the experiments the quantities M_0, ring spacing s, and ring height k have been varied. Rings with k= 5, 10 mm and s= 8, 13, 28, 53, 78, 103, and 128 mm were used in the experiments. The unknown coefficient of momentum losses C_f, pertaining to a given obstacle configuration (specified by values of k and s), was determined on the basis of a comparison of measured attenuation curves (see Fig. 5) and the approximate law given by Eqs. (4), and (5). For this purpose the optimal value of C_f for each run was first determined by the least-squares method. Then for a given obstacle configuration the average value C_f (k, s) in the range $1.1 \leq M_0 \leq 2$ was derived.

The experimental points thus obtained are shown in Fig. 6. The ordinates of the points correspond to measured values of Δp_f. The abscissa of each point is calculated by the use of the averaged value of C_f (k, s) and the distance x between the inlet of the rough section and the proper pressure transducer. As follows from Fig. 6, the assumption that C_f is independent of M_0 and Re numbers agrees satisfactorily with the experimental results since the discrepancy between the points varies insignificantly with variations

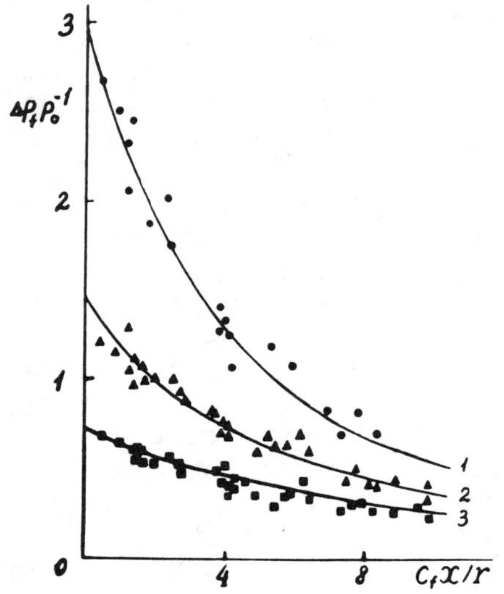

Fig. 6 A comparison of the predicted (solid lines) and measured (points) results on shock wave attenuation in a rough tube. (1) M_0 = 1.877; (2) 1.497; (3) 1.272.

of Δp_f and obstacle configuration. The experimental points corresponding to essentially different levels of losses are grouped close to the theoretical curves. This is evidence in favor of the fact that the parameters M_0 and $C_f\, x/r$ are indeed the quantities governing the process of shock wave attenuation.

Figure 7 shows the dependence of C_f on k (1-k= 5 mm, 2-10 mm) and s ($n = s^{-1}$) (solid lines). Clearly, the dependence is nonmonotonic. If s/k = 5-10, the coefficient C_f attains its maximum value; that is, shock wave attenuation is the most efficient in this case. For a quantitative comparison of the coefficient of momentum losses C_f, with the hydraulic drag coefficient λ measured in a steady stabilized flow, the data by Koch[8] were used (dashed lines). In the quoted study the tube diameter (50 mm) and obstacle configurations (assemblies of thin rings) were the same as those employed in our study. Therefore, a direct comparison of results is justified. It follows from the comparison of solid and dashed lines in Fig. 6 that the coefficient of momentum losses in the nonsteady flow behind a shock wave correlates unexpectedly well with the hydraulic drag coefficient λ measured in conditions of a steady stabilized flow. Note that the one-dimensional

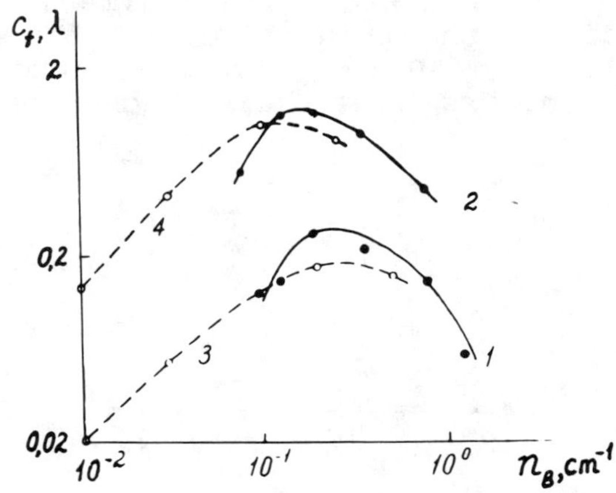

Fig. 7 Dependence of the coefficient of momentum losses C_f and the hydraulic drag coefficient λ on ring spacing and height. Curves 1,3 correspond to k = 5 mm; curves 2,4 to k = 10 mm; tube radius r = 25 mm.

theory of a steady flow operates with the quantity $C_f' = \lambda/4$. Hence the coefficient of momentum losses C_f in the flow behind a shock wave is approximately four times larger than C_f'. This conclusion confirms the theoretical prediction[3,4] dealing with the drag coefficient behind a nonideal detonation wave.

Nonideal Detonation

The dependence $C_f\,(k,s)$ (see Fig. 7) permitted us to suggest a complete one-dimensional theory of gaseous detonation in a rough tube, and it showed a way purposefully to vary the detonation velocity and suppress detonation. Application of the available models[2-4] seems to produce a reasonable explanation of some experimental observations.

Figure 8 shows the calculated dependence of the detonation velocity on ring spacing ($n_B = s^{-1}$), drawn with due regard for the results[4] and the data of Fig. 7 (D_o is the CJ detonation velocity). The curves represent only preliminary results and are based on model chemical kinetics having no relation to any particular detonable mixture.

Curve 1 corresponds to the detonation regime with combustion in a boundary layer[2,4] in a 50-mm-diameter tube with a ring height k of 5 mm. Curve 2 corresponds to the detonation regime with the bulk ignition behind the front[3,4] in a tube of the same diameter but k = 10 mm. In the latter case the kinetic parameters of the model chemical reaction

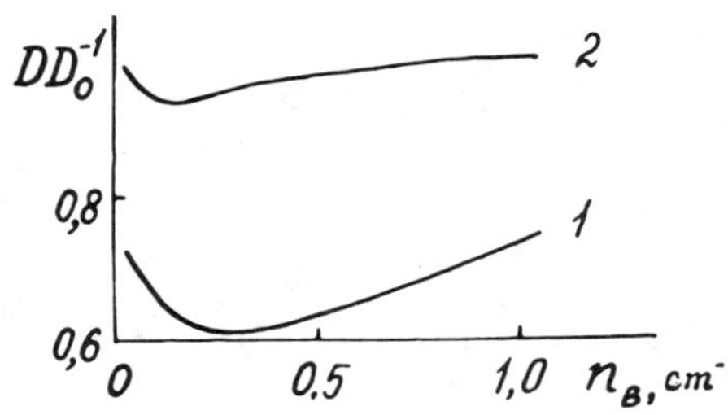

Fig. 8 Predicted dependence of the detonation velocity of a model detonable mixture on ring spacing. Curve 1 corresponds to the detonation regime with boundary layer combustion; curve 2 corresponds to the detonation regime with bulk ignition.

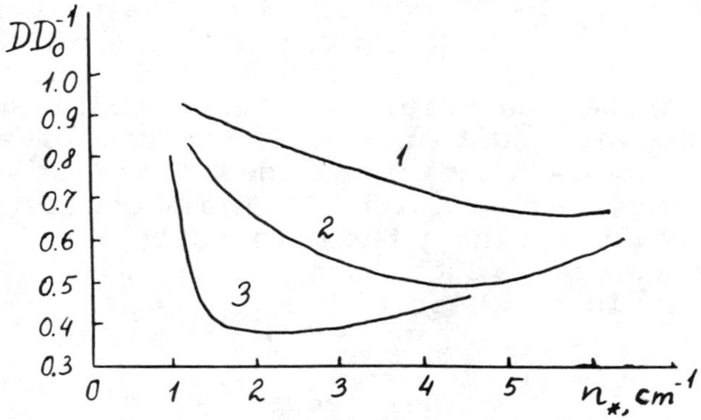

Fig. 9 Measured dependences of the detonation velocity of $2C_2H_2 + O_2$ mixture in a rough tube 11: (1) d=9 mm, k=1 mm: (2) d=7 mm, k=1 mm; (3) d=7 mm, k=1.6 mm.

with a rate constant of the Arrhenius type are as follows: activation energy 2 MJ/kg, heat release 2 MJ/kg, preexponential factor 10^{10} s^{-1}, and reaction order 1. The D/D_0 (n) curves have a strongly pronounced minimum. Qualitatively, the similar dependences of D on n_B are obtained in Ref. 11 (see Fig. 9). Thus, variation of obstacle geometry allows one to control the detonation velocity. The minimum dependence of detonation velocity on obstacle spacing observed experimentally seems to be due to the fact that the hydraulic drag coefficient λ attains its maximum value. It is of importance that under the purposeful variation of the detonation velocity by means of a proper choice of the roughness configuration, one may use available data on hydraulic characteristics of channels with different roughness. This considerably widens the possible engineering solutions for safety measures aimed at mitigating detonations.

Conclusion

Our study of shock wave attenuation in rough tubes reveals the following features. The drag coefficient in nonsteady flow conditions behind a shock wave correlates satisfactorily with C, the hydraulic drag coefficient, measured in steady, stabilized flow conditions. The principal parameters governing shock wave attenuation are the initial Mach number and the dimensionless complex $C_f\, x/r$. The approximate law of shock wave decay given by Eq. (4) agrees satisfactorily with the numerical calculations. The configuration of wall roughness with s/k = 5-10 produces a maximum drag and consequently provides the most efficient shock wave attenuation. The results obtained allow calculating the velocity of nonideal detonation in a rough tube within the framework of one-dimensional theories.

References

[1]Lee, J. H. S., "The Propagation of Turbulent Flames and Detonations in Tubes," Advances in Chemical Reaction Dynamics, D. Reidel, New York, 1986, p. 345.

[2]Zel'dovich, Ya. B., Gelfand, B. E., Borisov, A.A., Frolov, S.M., and Polenov, A.N., "The Reaction Zone in Low-Speed Detonation in Tubes with Rough Walls," Soviet Journal of Chemical Physics, vol 4, 1985, p. 279.

[3]Zel'dovich, Ya. B., Gelfand, B. E., Kazdan, Ya. M., and Frolov, S.M., "Detonation Propagation in a Rough Tube with Due Regard for Deceleration and Heat Transfer," Soviet Journal of Physical Fire/Heat Detonation, vol. 23, 1987, p. 103.

[4]Zel'dovich, Ya. B., Borisov, A.A., Gelfand, B.E., Frolov, S.M., and Mailkov, A.E., "Nonideal Detonation Waves in Rough Tubes," Progress in Astronautics and Aeronautics: Dynamics of Explosions, vol. 114, edited by A. L. Kuhl, AIAA, New York, 1988, p. 211.

[5]Zel'dovich, Ya. B., "On the Theory of Detonation Propagation in Gaseous Systems," Soviet Journal of Experimental Theoretical Physics., vol 10, 1940, p. 542.

[6]Graig, J. E. and Haines, K. A., "Shock Propagation in a Tube with Transverse Ribs," Proceedings of the 12th Symposium (International) on Shock Tubes and Waves, Jerusalem, Israel, 1980, p. 248.

[7]Merzkirch, W., and Erdmann, W., Measurements and Calculation of Shock Attenuation in a Channel with Perforated Walls, AFOSR-TR-82-1081, 1982.

[8]Koch, R., Loss of Pressure and Dissipation of Heat by (verwitzbelter) Electricity, VDI-Forch.-Heft, 1958, p. 469.

[9]Whitham, G. B., Linear and Nonlinear Waves, Wiley, New York, 1974.

[10]Richtmyer, R.D., and Morton, K.W., Difference Methods for Initial-Value Problems, Interscience, New York, 1967.

[11]Guenoche, P. and Manson, N. "Influence of Conditions on the Transversal Limits for Propagation of Shock and Combustion Waves," Review Inst. Fr. Petr., vol. 2, 1949, p.53.

Studies of the Mechanism of Detonation Propagation in Porous Structures

G. O. Thomas,* D. H. Edwards,† and S. H. M. Jones‡
University College of Wales, Aberystwyth, United Kingdom

Abstract

Results are presented of an experimental program to investigate detonation propagation in porous media. Macroscopic properties were determined in tubes tightly packed with perforated, stainless steel foil. To investigate the microscopic behavior, the interfoil spacing was progressively increased and detonation propagation studied using pressure, smoked foil, and spark schlieren techniques. The present results confirm the conclusions of previous work that the macroscopic behavior is controlled by the form drag. On the microscopic scale, the interaction of the flow with the medium results in a complete breakdown of the natural transverse structure of the detonation.

However, the new externally induced frontal structure continues to provide sufficient gasdynamic heating for autoignition of the gas to occur. This mechanism was observed for cases where the interfoil spacing was only a small fraction of the initial cell width. Thus, despite the presence of severe loss mechanisms, successful propagation is possible. As such, the present results indicate that a knowledge of the natural cellular structure is by itself insufficient to determine whether an established detonation would propagate in a relatively densely packed environment.

Introduction

Previous experiments have already demonstrated that not only can gaseous detonations propagate successfully in tubes filled with porous media, but these detonations can also propagate with steady state velocities significantly lower than that possible if the tube were unfilled. It is generally accepted that unsupported detonations that exhibit a velocity deficit greater than 10% will fail -- the Shchelkin criterion.

However, Kauffman et al.[1] observed that detonations in methane- or propane-oxygen mixtures could propagate successfully with velocity deficits as high as 25% in tubes packed with metal or ceramic spheres. In addition to the experimental

---

† Professor, Department of Physics.

* Lecturer, Department of Physics.

‡ Research Assistant, Department of Physics.

observations, they derived a model to account for the measured velocity deficits in terms of the characteristic dimensions of the porous media and heat and momentum loss effects. They did not, however, consider in detail the microscopic mechanisms that must also play a role in this mode of detonation propagation, most specifically, the modification of the transverse wave structure as it interacts with the individual elements that constitute the porous media.

In the present study, further results are presented on the macroscopic propagation of detonation in a porous structure, in this case a roll of perforated metal foil, a form which might be considered suitable for use in an industrial plant. Subsequently, the packing density and orientation of the foil layers were modified to allow a more detailed study of the microscopic behavior of the transverse wave structure as a detonation propagated through the medium. In addition to providing insight into the mechanism of propagation in porous media, the results are also relevant to understanding the mechanism of detonation propagation in general.

Experimental Details

Two basic experimental configurations were used for the present tests. In the first, a 50 mm diam., 18 m long stainless steel pipe was instrumented with four pressure and seven ionization gauges. These gauges were used to determine time-of-flight information, which was recorded using a multichannel event recorder. A twin-channel transient recorder was also used to record the static pressure profile from a Kistler 603 pressure gauge and the output from a microwave interferometer system[2]. The latter gave a sinusoidal output whose frequency was proportional to the velocity of the reaction zone of the wave after it emerged from the foils.

This general experimental arrangement was used to determine the macroscopic behavior of detonations when a section of the tube was packed with various lengths of a perforated and expanded stainless steel foil in the form of a roll. The foil used is shown in Fig. 1. A range of gaseous mixtures were tested, all of which were premixed by the method of partial pressures. The incident detonation was initiated by means of a 1.2 m long oxy-acetylene booster section. The second experimental arrangement consisted of a 5.5 m long duralumin tube with an internal cross-section of 38 x 76 mm. This could be fitted with special sections to allow smoked foil and schlieren studies to be undertaken to investigate the microscopic propagation behavior in the foils. In this case sections of the tube were packed in one of two ways: First, with up to 45 uniformly spaced layers of the perforated foil (38 x 300 mm in size), or second, with solid metal or perforated sheets aligned axially for both orientations relative to the tube walls. In these cases the interfoil spacing was 6 or 12 mm depending on the orientation.

Experimental Results

Macroscopic Behavior

The initial tests investigated the velocities and stability of detonations propagating in the foil for a range of mixtures and initial pressures. The length of foil in this particular study was 1.86 m. The results obtained are summarized in Table 1.

For undiluted oxy-acetylene, a stable wave was observed to propagate along the entire length of the foil for initial pressures as low as 25 Torr, for which case a 15%

Fig. 1 Perforated foil used to pack the 50 mm diameter detonation tube.

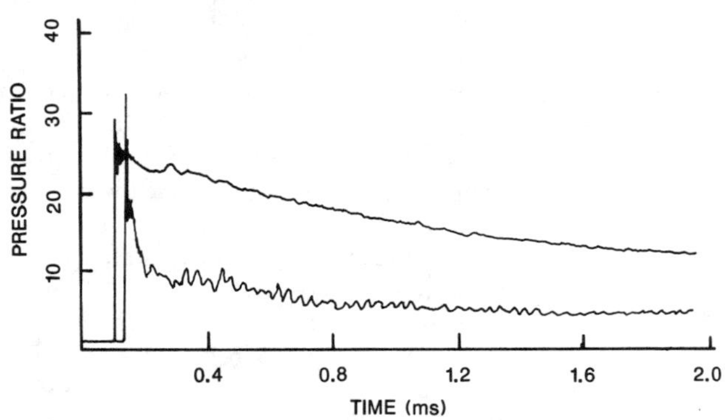

Fig. 2 Static pressure records obtained in the 50 mm diameter tube for tests with and without foil packing.

velocity deficit was recorded. In all cases the microwave records indicated a rapid re-establishment of the wave as it left the foil. Static pressure measurements obtained with and without the foil in place are shown in Fig. 2. A similar reduction in impulse was observed when reflection pressure measurements were attempted. With undiluted oxy-hydrogen the detonations in the foil were also stable, even though velocity deficits of up to 39% were recorded.

For these tests the re-establishment of detonation after the wave emerged from the foil was increasingly delayed as the initial pressure decreased and the observed velocity deficit increased. These measured velocities are compared with the predicted

Table 1 Summary of measured velocity deficits in the foil obtained in the 50 mm diameter tube.

System	Initial Pressure, Torr	Cell Spacing, mm	Velocity mm/μs Measured	CJ	% of CJ
C_2H_2+2.5O_2					
	25	7	1.96	2.25	85
	50	4	2.11	2.28	91
	100	2	2.22	2.32	94
C_2H_2+2.5O_2+40%Ar					
	25		1.60	1.94	82
	50		1.80	1.97	91
	100		1.95	2.00	98
C_2H_2+2.5O_2+75%Ar					
	200	3	1.28	1.73	74
	250	2.5	1.40	1.74	80
	380	2	1.50	1.75	86
2H_2O_2					
	75	15	1.67	2.72	61
	150	5.5	2.15	2.76	78
	250	2.5	2.40	2.79	86
	380	1.5	2.56	2.81	91
	620		2.68	2.84	94
	767		2.84	2.85	99

theoretical CJ velocities in Fig. 3. The corresponding distance-time plots are shown in Fig. 4 for a range of initial pressures. Waves at lower initial pressures were rapidly attenuated as the shock and reaction zone decoupled. In the tests where a stable self-sustaining wave propagated in the foil, there was no observable separation of the ionization front and the leading shock front.

Tests were also conducted in which the fuel-oxygen mixtures were diluted with an inert gas. The microwave records shown in Fig. 5 were obtained with 75% argon dilution in oxy-acetylene. The re-establishment time of the wave after the foil is considerably increased as the initial pressure decreases and the resulting measured velocity deficit in the foil increases.

Microscopic Behavior

For these tests the mixture C_2H_2 + 2.5O_2 + 80%Ar was used exclusively with initial pressures of 40 and 75 Torr. This mixture was chosen because of the extreme regularity of its transverse structure, which resulted in detonations with two and seven cells across the width of the tube at the respective initial pressures used.

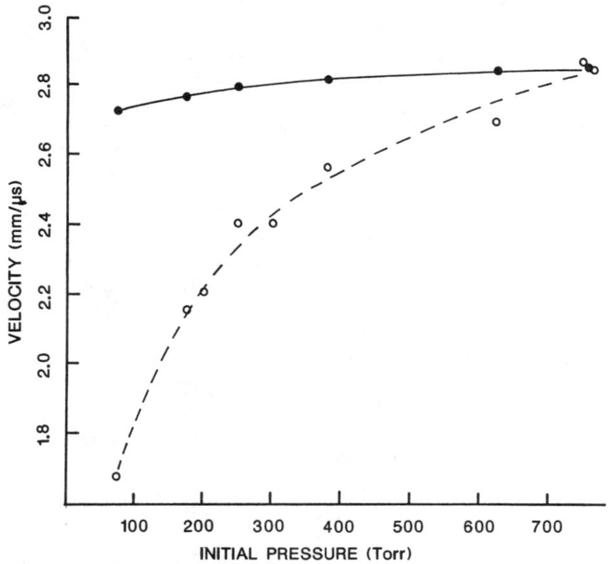

Fig. 3 Comparison of theoretical CJ velocities with measure velocities in oxy-hydrogen detonations propagating in foil for a range of initial pressures (50 mm diameter tube); solid line -- theoretical; dashed line -- experimental.

Smoked foil records taken at 40 and 75 Torr with seven layers of foil are shown in Fig. 6, interfoil spacing in these cases is approximately 10 mm. It can be seen that the natural regular structure is severely disturbed due to the interactions with the foil elements. However, it is interesting to note that for the corresponding pressure measurements (shown in fig. 7), then apart from the attenuation of the transverse waves, the average profile of the detonation does not change significantly until the number of layers of foil used is increased to 30 and above. Beyond this point the wave fails rapidly. Severe disruption of the regular transverse structure is also evident in the spark schlieren photographs shown in Fig. 8. With the foils present the front is still highly nonplanar and evidence of localized explosions are apparent. Other regions show a clear separation of the lead shock and reaction zone.

Discussion

In considering the probable behavior of the detonation as it entered the foil, it would appear reasonable to assume that the failure point should be related in some manner to the detonation cell size and to the separation of the foil layers. The results of Vasil'ev[3] and Knystautus et al.[4] for example, have shown that the critical conditions for propagation in smooth tubes and tubes filled with repeated annular obstacles occur when the controlling dimensions are of the order of one or two cells. However, cell size data for certain of the mixtures tested in the present study are shown in Table 1, and from these it is readily seen that detonations can in fact propagate successfully in the packed tube even when the interfoil spacing (of the order 0.5 mm) is only a fraction of the initial undisturbed cell size.

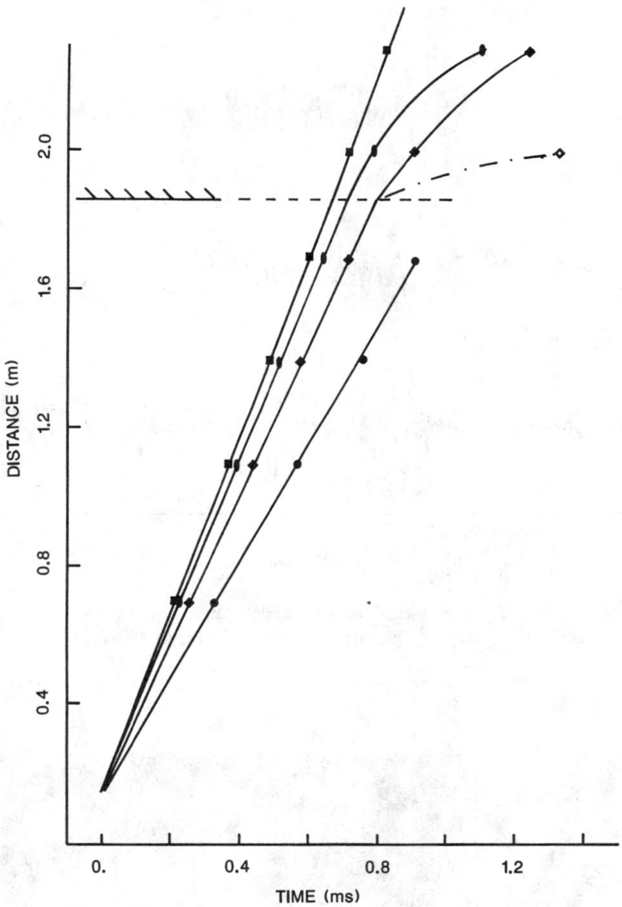

Fig. 4 Distance time plots for oxy-hydrogen detonations propagating through foil for a range of initial pressures in the 50 mm diameter tube. + -- 380 Torr, o -- 250 Torr, + -- 150 Torr, o -- 100 Torr.

Once a detonation wave enters the packing its initial structure is rapidly modified due to interactions with the individual elements that constitute the medium. In the present tests a multitude of possible interacting sources exist, so that even when only a few layers of foils are present the regular structure is rapidly degraded. In these cases there is no visible effect on the macroscopic propagation behavior of the detonation, i.e., on the average velocity and pressure profile. However, as the packing density is increased, then a decrease in the observed velocity becomes apparent, together with a reduction in the observed peak pressure at the detonation front. The most obvious effect caused by the presence of the packing medium is the rapid decrease observed in the downstream pressure after the wavefront has passed. The packing medium, therefore, plays a dual role in the propagation mechanism.

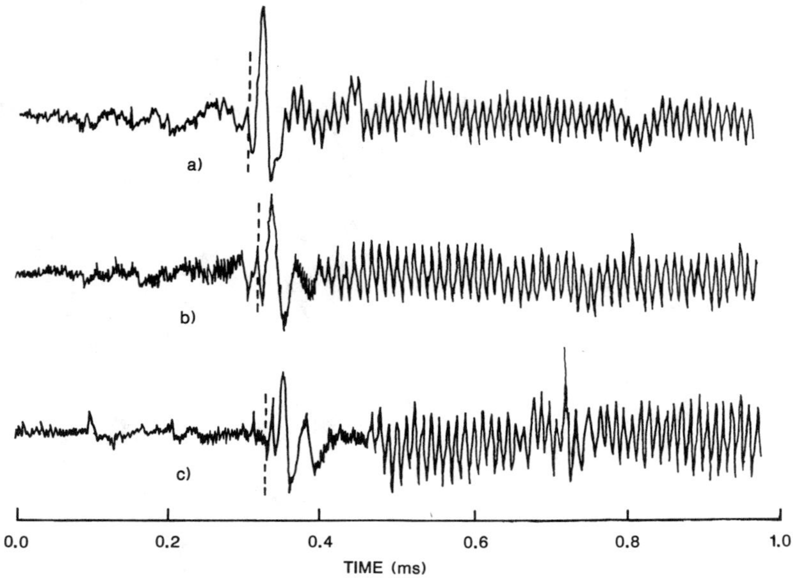

Fig. 5 Microwave records of the re-establishment of detonation in waves that fail as they exit the foils. Mixture is $C_2H_2 + 2.5O_2$ + 75% Ar; initial pressures are a) 380 Torr, b) 250 Torr, and c) 300 Torr.

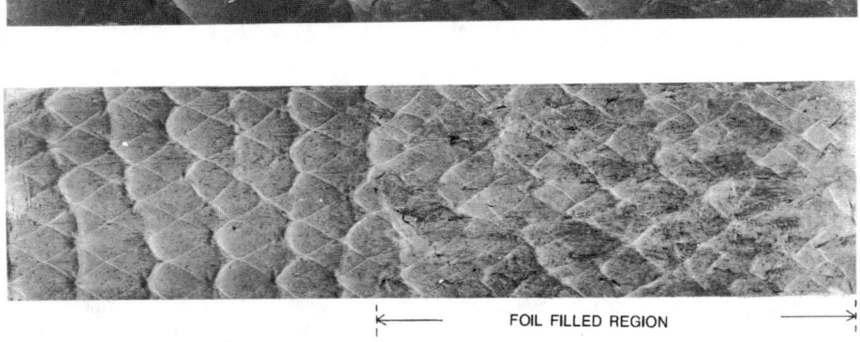

Fig. 6 Smoked foil records obtained with oxy-acetylene diluted with 80% argon at an initial pressure of 40 Torr with a) four and b) seven layers of foil in place (38 x 76 mm tube).

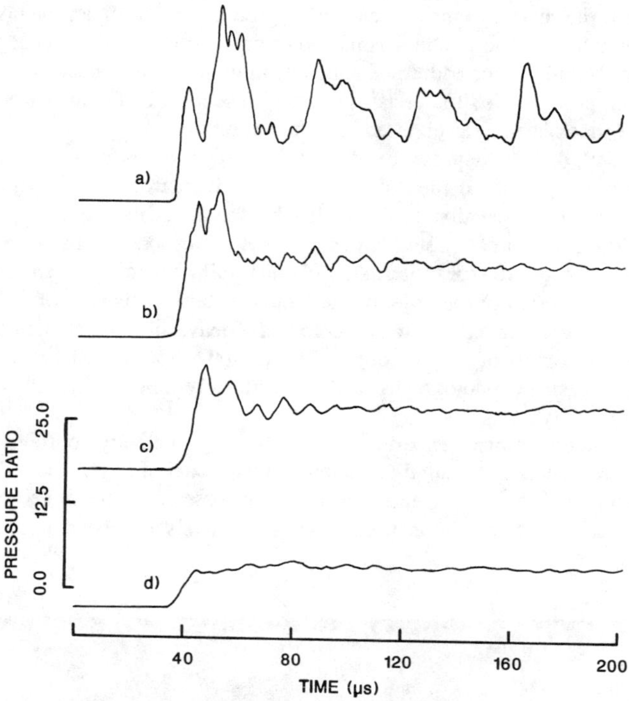

Fig. 7 Measured static pressure histories obtained for an oxy- acetylene detonation diluted with 80% argon propagating in various layers of foil (38 x 76 mm tube) at an initial pressure of 40 Torr; a -- 0 foils, b -- 15 foils, c -- 30 foils. and d -- 45 foils.

Table 2 Comparison of experimental and calculated velocity deficits for two reactive systems in a smooth tube.

Initial pressure, Torr	$2H_2$+O_2+3Ar Exptl. %	$2H_2$+O_2+3Ar Theor. %	C_2H_6+3.5O_2+3.5N_2 Exptl. %	C_2H_6+3.5O_2+3.5N_2 Theor. %
50	5.0	6.0	4.1	6.4
100	3.8	2.6	3.2	3.2
150	3.2	1.7	2.6	2.1
200	2.5	1.2	1.8	1.6

First, the mixture ignition and energy release are sufficiently fast to provide self-sustaining flow. The medium could contribute to this process through either the generation of turbulence or additional shock heating due to the Mach waves generated. The dominant process is still uncertain, but the present limited results would appear to favor simple shock heating as opposed to turbulent mixing.

Second, due to the presence of the foil, energy loss effects become important at later times; as a result, if the induction delay is increased sufficiently to allow the incident front and the reaction to decouple, the wave will be rapidly quenched. The failure condition in this case could still, therefore, correspond to that of a simple one-dimensional Shchelkin type criterion, i.e., that failure occurs when the change in induction time becomes of the order of the induction time. This type of failure criterion has also been recently investigated by Terao and Kobayashi[5] who studied the failure of detonations in channels of varying depth. The frontal velocity measurements provided by the present spark photographs are unfortunately not sufficiently accurate to investigate this fully.

In smooth tubes, experimentally observed velocity deficits have been successfully accounted for using the boundary layer model of Fay[6], where the integrated losses due to boundary layer heat and momentum losses are determined over a length scale characteristic of the detonation. Such an analysis has been used successfully

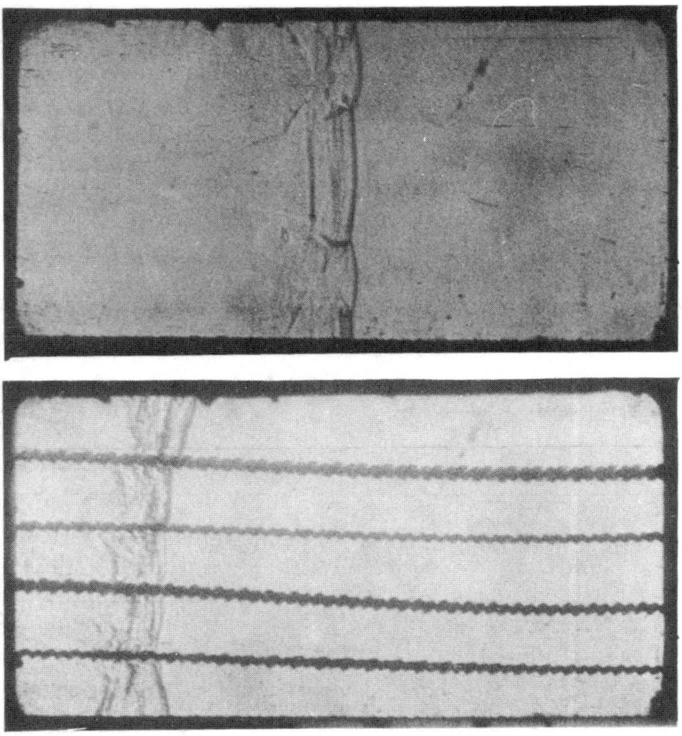

Fig. 8 Spark schlieren records obtained with oxy-acetylene diluted with 80% argon at an initial pressure of 40 Torr with and without four layers of foil packing.

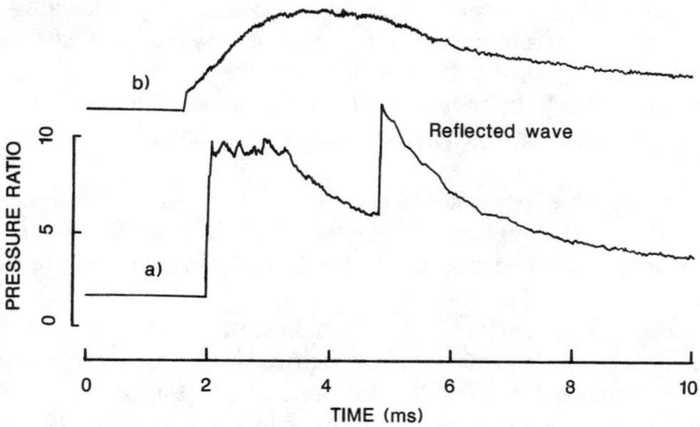

Fig. 9 Comparison of measured static pressure profiles recorded for shock wave transmission and reflection from a section of the material shown in Fig. 1; a) before the foil, and b) within the foil.

Table 3 Comparison of experimental and calculated velocity deficits for two reactive systems in a tube filled with plain and perforated foils, respectively.

Initial pressure, Torr	$2H_2+O_2+3Ar$ Exptl. %	$2H_2+O_2+3Ar$ Theor. %	$C_2H_6+3.5O_2+3.5N_2$ Exptl. %	$C_2H_6+3.5O_2+3.5N_2$ Theor. %
Plain				
50	10.0	6.8	10.0	15.7
100	3.8	7.3	4.0	8.9
150	3.2	5.0	3.0	6.0
200	2.5	3.2	1.5	4.4
Perforated				
50	31.5	11.6	26.7	15.7
100	21.0	7.5	21.0	8.9
150	14.2	4.7	17.0	6.0
200	11.2	3.3	14.0	4.4

during the present study to predict the observed velocity deficits in the unfilled tube. The results are shown in Table 2. The method was not as successful in predicting the deficits when layers of plain foils spaced 6 mm apart were placed in the tube. As would be expected, when the plain foils were replaced by perforated foils, the form drag thus introduced resulted in a much increased deficit as compared to that predicted by boundary layer shear drag effects alone. See Table 3.

The effect of form drag on nonreactive shock propagation in the material used in the macroscopic test (see Fig. 1) can be seen in the pressure records shown in Fig. 9.

A non-reactive decaying shock wave was readily generated by detonating the booster gas with only air in the tests section. The resulting blast was rapidly attenuated after it entered the foil, with a gradual buildup in pressure after the arrival of the lead shock. This profile is significantly different from that observed when a detonation propagates through the foil and where the general detonation pressure profile is maintained. See Fig. 2.

The limited role of heat transfer was further shown in tests where the perforated metal foils were replaced by similar plastic foils whose thermal diffusivity was three times smaller but that only resulted in a small percent change in the measured velocity deficit.

The results presented in Table 3 are interesting for a second reason. The mixtures used for these measurements were specifically chosen for the near equivalence of their dominant cellular structure over the pressure range used. The ethane system, however, displays an extremely irregular substructure as compared to the highly regular structure of the argon diluted hydrogen system. In recent studies Moen et al.[7] have shown that irregular systems are considerably less prone to perturbations and are less affected by wall loss mechanisms. However, for the present tests on propagation through foils there appears to be no significant difference between the two systems, and one is drawn to the conclusion that the disruption caused by the foil must overide the natural structure of the systems.

Thus, if deficits are determined by integrated losses over a characteristic length scale, the length scales associated with the detonation are now determined by the structure that which results from the interaction of the wave with foil, not by the natural structure. The induction and reaction times are the true controlling length scales in detonations, and cell length is only characteristic of the coupling between the gasdynamics and the chemistry in self- sustaining detonation waves. If this is true, it means that knowledge of the cellular structure of self-sustaining detonations alone is insufficient to determine the behavior of waves that are initiated or sustained by external mechanisms or interactions; it is the chemical time scales at the prevailing gasdynamic conditions that are important. These results also have implications for current large-scale tests on flame acceleration and transition to detonation. The failure to achieve transition to detonation does not rule out the possibility that the array could support detonation; it merely indicates a failure to achieve sufficiently high reaction rates, wheras simple cell size arguments cannot be invoked to argue that detonation would not be possible.

Conclusions

The present investigation has shown that the propagation of detonations in porous media possessing a high degree of interconnectivity is possible if sufficiently high reaction rates can be maintained. Such propagation can occur within media whose natural spacing is significantly less than the natural self-sustaining transverse structure of the detonation wave. This is significantly different from the situation that obtains for quasidetonations in discrete repeated obstacles. The medium gives rise to severe momentum losses arising from form drag, but these losses can be tolerated if the chemical heat release rates are sufficiently fast; the wave merely propagates with a lower velocity as required by the governing conservation equations.

References

[1]Kauffman, C.W., Chuanjan, Y., and Nicholls, J.A., "Gaseous Detonations in porous media," 19th Symposium (International on Combustion, The Combustion Institute, Pittsburgh, PA, 1982, pp. 591-597.

[2]Edwards, D.H, and Lawrence, T.R., "Ionization Measurements in Detonation waves," Proceeding of the Royal Society of London, A 286, 1965, pp. 415-439.

[3]Vasil'ev, A.A., Geometric Limits of Gas Detonation Propagation," Fizika Goreniya i Vzryva, Vol 18, No. 2, 1982, pp. 132-136.

[4]Knystautus, R., Lee, J.H., Peraldi, O., and Chan, C.K., "Transmission of a Flame from a Rough to a Smooth-walled Tube," Dynamics of Explosions: AIAA Progress in Astronautics and Aerodynamics (edited by Bowen, J.R., Leyer, J.-C. and Soloukhin, R.I.), Vol 106, AIAA, New York, 1986, pp. 37-52.

[5]Terao, K., and Kobayashi, H., "Experimental Study on Suppression of Detonation Waves," Japanese Journal of Applied Science, Vol. 21, No. 11, 1982, pp. 1577-1579.

[6]Fay, J. A., "Two-dimensional Gaseous Detonation: Velocity Deficit," Physics of Fluids, Vol. 2, No. 3, 1959 p. 283.

[7]Moen, I.O., Sulmistras, A., Thomas, G.O., Bjerketvedt, D., Thibault, P.A., "Influence of Cellular Regularity on the Behavior of Gaseous Detonations," Dynamics of Explosions: AIAA Progress in Astronautics and Aerodynamics (edited by Bowen, J.R., Leyer, J.-C. and Soloukhin, R.I.), Vol 106, AIAA, New York, 1986, pp. 220-234.

Propagation of Gaseous Detonations over Liquid Layers

A. A. Borisov and A. H. Mailkov
USSR Academy of Sciences, Moscow
and
V. V. Kosenkov and V. S. Aksenov
Moscow Institute of Physical Engineering, Moscow

Abstract

The possibility for using thin water layers to suppress waves propagating in tubes of propane/air mixtures at amospheric pressure is demonstrated. Water was spilled on the bottom wall of a tube of round or rectangular cross section. The detonation velocity dropped very rapidly (from about 1800 to about 1300 m/s over a distance of 300 mm) when the wave entered the section of tube holding the water layer. As high-speed schlieren photographs show, the detonation wave decay starts near the wall surface. The heat release zone grows progressively until the chemical reaction and the lead shock wave decouple completely. The detonation waves were only arrested by liquid layers when the number of the cells on their front was less than 3. Reduction of the nitrogen content in a stoichiometric $C_3H_8/O_2/N_2$ mixture decreases the deficit of the reaction front velocity (as compared to the thermodynamic detonation velocity) at the end of the 3.53-m-long section of the tube with the water layer from 1720 m/s at 76% N_2 to 25 m/s at 68.8% N_2 and 0% at 38.8% N_2. The effect of the layer thickness and the nature of a liquid on detonation suppression is investigated. It is shown that transfer of a part of the liquid from the layer into the gas phase occurs much earlier (in a few microseconds after shock wave arrival) than the development of Kelvin-Helmholtz instability. The first rapid rise of the liquid is due to interaction of the pressure waves with the interface.

Introduction

In real conditions gaseous detonation waves often propagate through ducts whose walls are covered with a liquid film (for example, in wet galleries and tubes, in fuel pipelines, etc.). The presence of a liquid may affect both the limiting conditions of detonation propagation and the velocity and structure of detonation waves. This was noticed experimentally in long tubes with hydrocarbon/air mixtures. In tubes that were not dried properly, detonation waves propagated at velocities that were appreciably lower than calculated, especially in those sections of the tube where water from previous experiments had accumulated.

The present study was undertaken to find out at what conditions the effect of a liquid layer on detonation is marked and to elucidate what processes are responsible. The latter question is of special interest since all the estimates demonstrate that the conventional instability considerations (for example, Kelvin-Helmholtz instability) furnish no reasonable explanation of the strong detonation velocity variations observed in wet tubes. The characteristic times of instability development are substantially longer than the duration of the heat evolution process in gaseous detonation waves.

Experimental

Experiments were carried out in a stainless steel detonation tube 70 mm in diameter and 21.5 m long and in a copper tube of rectangular (72x44 mm^2) cross section 4 m long and equipped with windows. Detonation in the two tubes was initiated by exploding a propane/oxygen mixture in tubes of the same cross section attached to the test sections. The sections with the initiating mixture were 1.5 and 1 m long, respectively, in setups with round and rectangular tubes. The initial pressure of the initiating mixture was 1 or 1.5 bar. The mixture was deontated using a long tube of small diameter (where deflagration-to-detonation transition occurred) and a conical transition section. Propane/air or propane/oxygen/nitrogen mixtures were prepared in a stainless vessel and mixed with a fan. The tubes were pumped out before introducing the mixture in them. Water was poured on the bottom wall before pumping out the tube and was unfrozen after mixture admittance. The layer thickness did not exceed 4 mm in the long tube and 13 mm in the shorter one. The lengths of the sections with a water layer were 3.5 m and 0.5 m in the round and rectangular tube, respectively. The sections with a liquid were positioned 15 m and 3.2 m away from the initiating sections and were followed by receiver sections with no water layer.

The detonation (or flame) velocity was measured by ionization probes (a total of nine probes in the long tube and six in the shorter one) mounted in the sections both with and without a water layer. The error in velocity measurements was no higher than ± 10 m/s. Detonation was considered to be stable when the spread in the velocity measurements at different bases was no worse than 1.5%.

Schlieren photographs were taken employing two techniques: 1) a high-speed cine-camera with a rotating mirror prism (a time interval between frames of 2 μs and exposure time of 2 μs) in combination with a long-duration (about 300 μs) pulse light source and schlieren system with

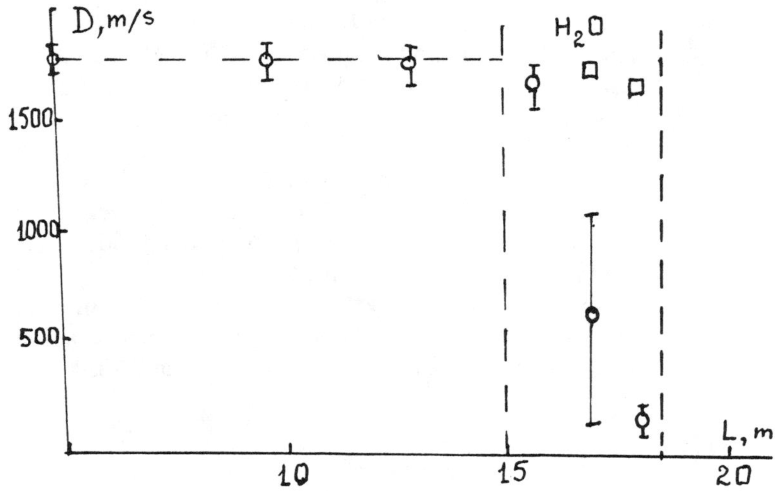

Fig. 1 The reaction-front velocity as a function of distance from the beginning of the water layer in the round tube. Experiments using stoichiometric propane/air mixture, water layers of various thickness. ○ = experiments with continuous layers and decaying detonation; □ = runs where the layer continuity is doubtful.

visualization field 260 mm in diameter; or 2) ruby pulse laser with the light pulse duration of $3x10^{-8}$ s together with the same schlieren system for taking snapshots of better resolution.

Results and Discussion

Detonation in a stoichiometric propane/air mixture propagating steadily through the 15-m section of the tube decayed rapidly upon entering the section with a water layer, provided the layer was sufficiently thick. The latter condition is associated with the great length of the section with the water layer and difficulties in precisely adjusting the section in a horizontal direction. When the layer was thinner than 2 mm, water usually did not cover the entire bottom surface of the tube.

Figure 1 shows the detonation velocity in the tube as a function of distance. The points in Fig. 1 represent many measurements and the vertical bars indicate the real spread of the data in different runs carried out with various layer thicknesses. It is noteworthy that with layers that are confidently considered to be continuous all over the tube section, the detonation wave decays completely to form a flame propagating at velocities of the order of 100 m/s (or even lower). As the schlieren photographs show, the detonation velocity decreases approximately linearly with distance from the beginning of the layer. The rate of the velocity drop at the initial stage of wave propagation over the layer is of the order of $3x10^6$ m/s^2 and grows as the layer becomes thinner. The squares in the figure represent those runs where the layer is suspected to be either frozen or not continuous. After the wave again entered the section without a water layer its velocity very often increased to a value that was appreciably higher than the Chapman-Jouguet (C-J) detonation velocity. It follows that overdriven detonation arises in this section due to self-ignition of the gas between the decoupled shock wave and reaction zone (as in a galloping detonation). The reinitiation usually occurred within the first 1.5 m after the end of the water layer.

Of natural interest is the question of what range of initial conditions the detonation-quenching effect can be observed. From general considerations it follows that detonation may be suppressed when the reaction zone in the wave is large enough for the heat losses to progressively increase it due to the temperature drop in the shocked unreacted gas. This implies that the greater the reaction zone length, the more pronounced should be the quenching effect of the liquid, since the only way the liquid may affect the detonation wave is by decreasing the gas temperature near the wall at the expense of cold mass injection and entrainment into the flow. Bearing this in mind, we performed experiments under identical conditions with stoichiometric

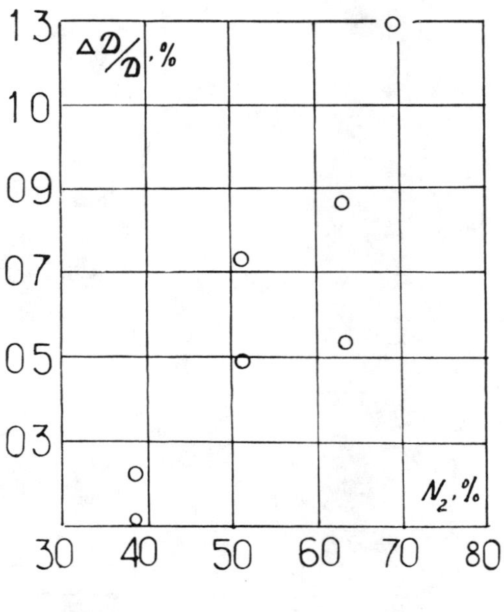

Fig. 2 Deficit of the reaction-zone velocity (as compared to the C-J detonation velocity) in a round tube at the end of the 3.5-m-long tube section with a water layer 4 mm thick as a function of the nitrogen concentration in a propane/oxygen/nitrogen mixture.

propane/oxygen/nitrogen mixtures in which the nitrogen content was varied. The results of the experiments are presented in Fig. 2 where the relative detonation velocity deficit (compared to the detonation velocity in a dry tube) at the end of the tube section with the water layer (2.5 m) is plotted as a function of the nitrogen concentration in the mixture. As is obvious from the figure, the quenching effect becomes significant when the nitrogen concentration is higher than 70% (at 70% N_2 the velocity deficit amounts to 13% and at 75.5% N_2 it is of the order of 90%). Since the reaction zone in a detonation wave is characterized by cell size, we have compared the cell size and tube diameter for the mixture with nitrogen content between 70 and 75.5%. Their ratio turned out to be about 1/3. Thus, detonation suppression may apparently be expected when the cell size is not smaller than 1/3 of the tube diameter. As estimates show, the aforesaid condition indicates that mass transfer at the liquid/gas interface must be appreciable over a time period of the order of 10-30 μs, which is the approximate duration of the reaction zone in the detonation wave.

The second important question is how the layer thickness influences detonation propagation over the liquid surface. The experiments were carried out in a tube of square cross section. It should be noted that the quenching capabilities of the layers in both tubes were nearly the same. Figure 3 presents the measured velocity decrease at a distance of 300 mm

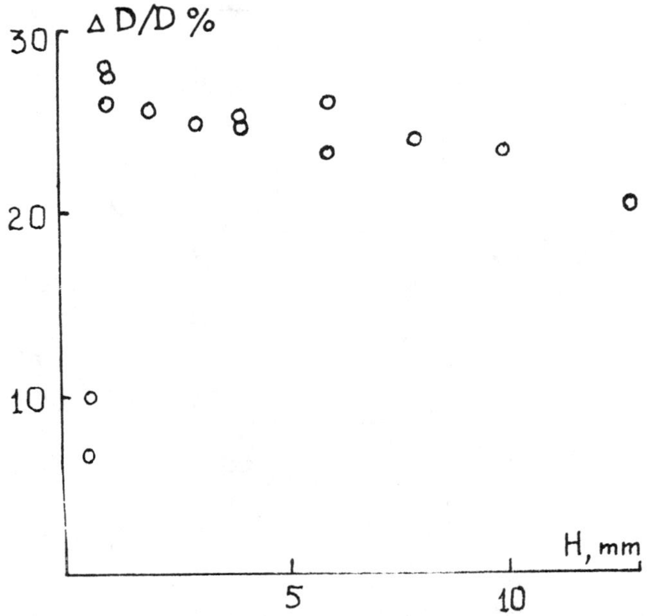

Fig. 3 Detonation velocity deficit at 300 mm from the layer, beginning as a function of the layer thickness. A stoichiometric propane/air mixture in the tube of the 72x44 mm^2 cross section.

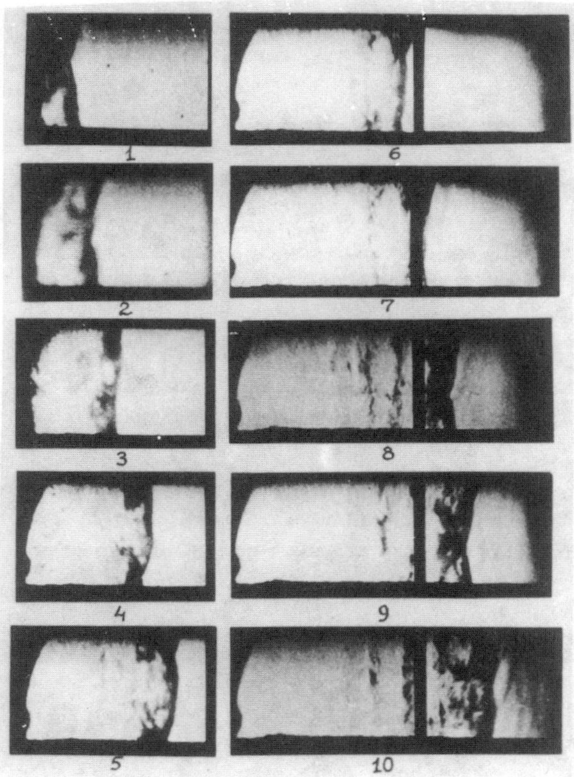

Fig. 4 Detonation wave interaction with a water layer 1 mm thick. Time interval between frames is 13 μs.

from the beginning of the layer as a function of layer thickness. As is seen from the figure, the velocity decrease achieves maximum at H=1 mm. The fall-off of the curve at layer thicknesses below 1 mm is presumably associated with the fact that very thin layers are not continuous because of the poor wettability of the bottom tube wall. The decrease in the quenching capability of layers as their thickness grows is noteworthy.

Experimental measurements and theoretical estimates made by A. A. Borisov et al.[1] have demonstrated that the characteristic times of liquid rise from the wall due to Kelvin-Helmholtz instability behind detonation waves at atmospheric initial pressure are of the order of several hundreds of microseconds; that is, this type of instability cannot be responsible for the effect observed.

In order to elucidate the mechanism of reaction quenching by water ejected from the layer, we used high-speed schlieren photography. The sequence of frames for a thin and thick layer are shown in Figs. 4 and 5. In the case of a thin layer, the zone with density gradients behind the wave

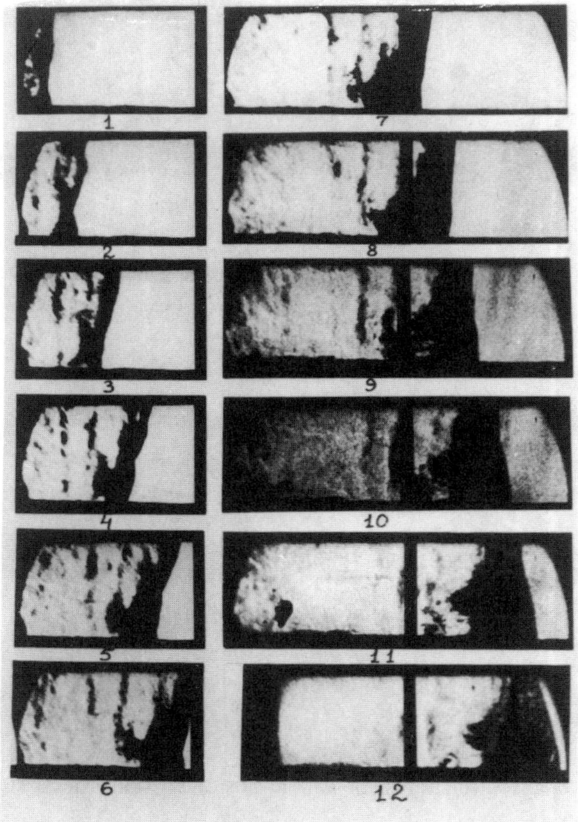

Fig. 5 Detonation wave interaction with a water layer 4 mm thick. Time interval between frames is 13 μs.

front (that presumably includes unreacted mixture volumes) grows almost uniformly all over the tube height as the wave propagates over the layer with a decaying velocity. In contrast to this, in the case of a thick layer the zone with density gradients grows predominantly near the bottom wall covered with water; and only after some time does this zone fill up the entire tube cross section.

The single photographs of better resolution (Fig. 6) make clear the nature of the process responsible for fast mass transfer at the interface. Three features attract attention in these photographs: 1) a series of compression and rarefaction waves in the liquid layer downstream of the detonation front, 2) strong perturbation of the gas flow near the layer in the region where the pressure waves in the layer are most intense, and 3) a clear-cut black triangular strip of shattered liquid rising due to Kelvin-Helmholtz instability far downstream of the reaction zone.

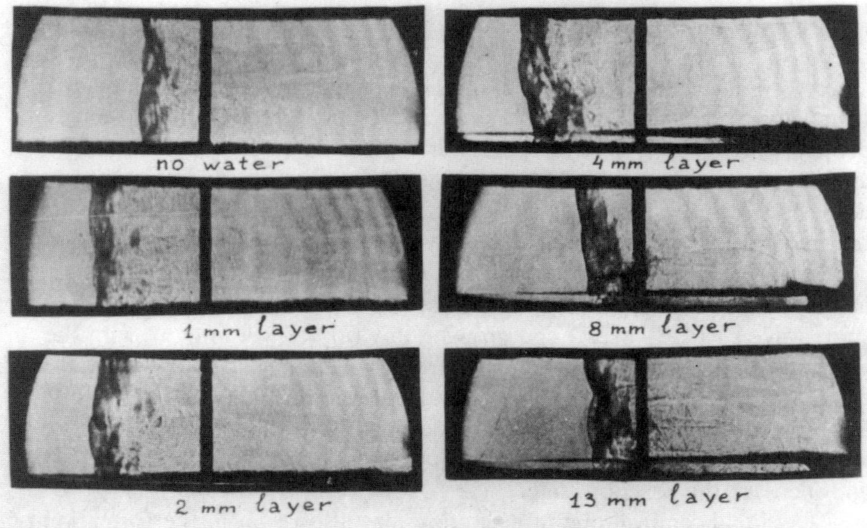

Fig. 6 Photograph of detonation waves propagating over water layers of various thickness.

Several conclusions may be inferred from the photographs. The contribution of Kelvin-Helmholtz instability to the energy and momentum losses in the reaction zone is insignificant because of the low rate of mass transfer at the interface due to this process. Furthermore, multiple reflections of compression and rarefaction waves in the layer are undoubtedly responsible for the growth of the reaction zone. The liquid/gas interface in the region near the detonation wave front seems to be smeared, especially at points where the reflected pressure disturbances in the layer reach the interface. And finally, the rate of detonation velocity decrease is higher, the thinner the layer.

In order to verify the conclusion that the liquid/gas interface seems smeared, we measured the intensity of light scattered by the layer. The measurements were performed as follows. Two photodiodes installed opposite each other recorded the light emitted by the detonation wave and passed, respectively, through the top and bottom windows. The bottom window was covered with the water layer. The sensitivites of the two photodiodes were equalized in the presence of the water layer with the help of a light source introduced in the tube. The photodiodes received the light emitted from a gas volume 3 mm thick.

The signals recorded are illustrated in Fig. 7. It is noteworthy that the signals are practically identical except for the amplitude. The bottom photodiode recorded appreciably lower light intensities, which furnishes evidence of strong scattering of the light by the layer. The scattering angles capable of producing the light attenuation observed in our experiments should be too large to account for the effect by scattering only on the inhomogeneities induced by the acoustic waves in the layer. Therefore, it is believed that the main contribution to the effect comes from surface instability waves. The liquid velocity behind acoustic waves propagating in the layer is $\Delta p/\rho a \approx 1$ m/s (here Δp is the pressure amplitude, ρ is the liquid density, and a is the speed of sound in the liquid). When the rarefaction wave reflected from the tube wall that stops the liquid reaches the gas/liquid interface, the surface layer undergoes acceleration in the right direction for developing the Rayleigh-Taylor instability. Thus, one should expect the appearance and fast growth of short-wave surface waves. The growing tiny ripples will interact with the gas flow behind the detonation wave front, which may strip and entrain them. As the photographs show, the turbulence near the surface is very intense even immediately behind the detonation front; therefore, the stripped liquid will rapidly mix with the gas, decrease the gas temperature, and increase the reaction time.

Liquid removal due to cavitation may also result in the formation of micromist near the surface. Indeed, strong multidimensionality of the detonation wave front may cause rapid variations of the pressure at the liquid surface (with characteristic times of the order of 1 μs). Thus, reflections of the oblique and transverse detonation and shock waves at the interface may generate negative pressure gradients in the layer that are necessary for cavitation and removal of a very thin surface layer. The thickness of this layer depends on the rate of pressure drop behind the acoustic wave front in the liquid layer.

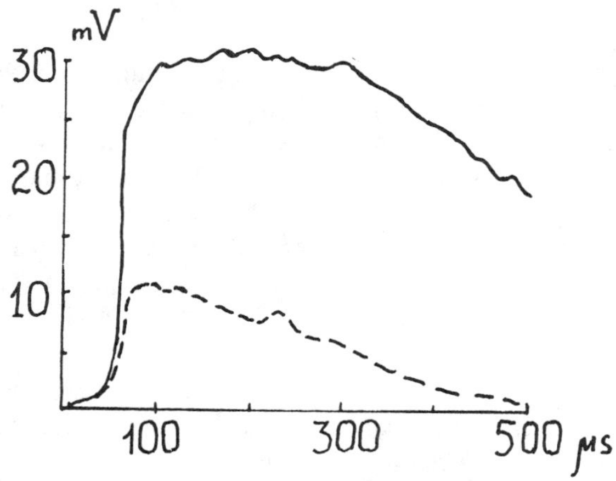

Fig. 7 Intensities of the light emitted by the detonation wave as recorded by photodiodes through the top (solid trace) and bottom (dashed trace) windows.

Let us estimate the amount of the liquid that should be stripped or spalled to produce the effect observed. To reduce the induction period of a propane/air mixture in the conditions existing behind the detonation wave front, it is necessary to lower the temperature by about 200 K, which can be achieved by adding to the mixture about 5% water. Thus, about 1.5×10^{-5} g of water should be ejected from a 1-cm^2 area of the liquid surface in order to lower the temperature in one cm^3 of the gas to the level needed (according to the photographs, this approximately corresponds to the region with quenched reaction at the initial stage of the detonation-wave interaction with the layer). This amount of water corresponds to a 0.15-μm layer of stripped liquid. If we assume that the ripples on the liquid surface grow with the velocity of liquid motion in the acoustic wave (about 1 m/s), their amplitude may reach 0.15 μm in 0.15 μs, that is, almost immediately after the rarefaction wave reflects at the interface.

According to the explanation suggested, the amount of liquid introduced into the gas should be proportional to the number of rarefaction wave reflections at the interface during the reaction time behind the detonation front. Hence, thinner layers should exert a greater effect on the detonation wave. It should be emphasized that thinner layers eject the liquid at the very beginning of the induction period where the reaction is the most sensitive to temperature variations. This is probably why the photographs exhibit more or less uniform growth of the reaction zone and more rapid velocity reduction in the case of very thin layers. It should be noted that the reaction zone growth in the regions outside the boundary layer affected directly by the injected liquid is caused by rarefaction waves generated near the wall (where the reaction is stopped by the liquid mist) that propagate transversally in the subsonic reacting mixture flow.

In order to further verify the idea explaining the behavior of a detonation wave in the presence of a liquid layer, we performed some experiments with another liquid possessing strongly different properties. We anticipated that the use, say, of gasoline, would not change the detonation quenching effect since the wave interaction pattern should not depend appreciably on such liquid properties as viscosity and surface tension. The experiments confirmed that expectation: the effect of gasoline layers was virtually identical to that of water layers.

Conclusion

The experimental studies performed in this work have shown that

1) Detonation waves can be quenched by liquid layers, provided the reaction zone is long enough (the detonation front cell size is no less than about 1/3 of the tube diameter).

2) Thinner layers quench detonation waves better.

3) The effect is due to fast liquid ejection from the layer that occurs within a time interval of the order of 10 μs (which is substantially shorter than the characteristic time of Kelvin-Helmholtz instability development).

4) The effect is most likely associated with the combined action of two types of hydrodynamic instability, Rayleigh-Taylor and Kelvin-Helmholtz, and liquid cavitation, and with fast evaporation and mixing of the stripped liquid with the gas in the reaction zone, which results in decoupling of the reaction zone and the shock wave.

Reference

[1]Borisov, A. A., Kogarko, S. M., and Lyubimov, A. V., "Ignition of Fuel Films behind Shock Waves in Air and Oxygen," Combustion and Flame, Vol. 12, No. 1, Jan. 1986, pp. 465-468.

Surface Detonations and Indirect Ignition Processes

B. Plewinsky,* W. Wegener,* and K.-P. Herrmann†
Bundesanstalt für Materialforschung und -Prüfung, Berlin, Germany

Abstract

The combustion behavior on the surface of a liquid was studied using tetramethyl-dihydrogen-disiloxane. Pure oxygen was used as oxidizer. The combustion processes were observed by means of a rotating prism camera and a rotating drum camera. The investigations showed that detonations or detonation-like phenomena take place on the surface of a compact liquid. This type of heterogeneous detonation is termed surface detonation. By using different experimental setups, it was possible to study one-dimensional and two-dimensional detonation processes on the surface of a liquid. Detonation velocities were observed between 700 ms^{-1}–1100 ms^{-1} (one-dimensional surface detonation).

Ignition of diffusion flames and transition to detonation took place even in the case of a spacial separation by an obstacle placed between the surface of the liquid and the ignition source. Possible mechanisms of this kind of indirect ignition are also discussed in this paper.

All of the effects described were observed at compositions of the gas phase well below the lower flammability limit.

I. Introduction

Detonations in two-phase systems have been known since 1952. In that year Loisson[1] reported on detonations

---

*Research Chemist

†Mechanical Engineer

travelling along liquid films. Two years later Burgoyne and Cohen[2] observed heterogeneous detonations during experiments concerned with the burning behavior of liquid aerosols. In 1968 Strauss[3] proved the existence of detonations in aluminum dust/oxygen mixtures. Furthermore, heterogeneous detonations were also observed in monopropellant liquid aerosols, in single-drop columns,[4] along a wick,[5,6] and in quartz sand when covered with a liquid fuel.[6,7]

This paper presents a new type of two-phase detonation: a detonation or a detonation-like phenomenon on the surface of a compact liquid. Such heterogeneous detonations shall be termed surface detonations. By using different experimental setups it was possible to induce propagation of the detonations in one or two directions. These processes are referred to as one-dimensional or two-dimensional surface detonations.[8] In addition, a special type of ignition is presented, for which the terms indirect ignition or indirect ignition processes are suggested.

All experiments were performed with tetramethyl-dihydrogen-disiloxane (TMDS) as the fuel. This compound was used in the experiments for two reasons. On the one hand, TMDS contains two particularly reactive Si-H-bonds in its molecules. The reactivity of these bonds exceeds by far the reactivity of the analogous carbon/hydrogen bonds, and, on the other hand, it produces extremely luminous flames, which are characteristic of the burning behavior of silico-organic compounds. For all experiments pure oxygen was used as the oxidizer.

Berthold has conducted experiments with different systems that led to similar results.[9]

II. Surface Detonations

All types of heterogeneous detonations investigated so far had one common feature: the surface of the condensed phase has been enlarged in one or the other way. This surface enlargement caused a better mixture of fuel and oxidizer and thus an increase of the burning velocity as well as a greater possibility for a detonative reaction process to occur. Surface detonations, i. e., detonations spreading on the surface of compact liquids, show, however, the greatest possible separation between liquid and gaseous phase. Therefore, contrary to other types of heterogeneous detonations, surface detonations should only be expected to take place in a smaller number of systems.

In the chemical industry pure oxygen is used increasingly as an oxidizer for liquid organic compounds. In order to enable safe design of equipment containing organic liquids and pure oxygen, investigations of one- and two-dimensional surface detonations have been performed.

A Experimental Details

1. One-Dimensional Surface Detonations

The setup used to investigate one-dimensional surface detonations is given in Fig. 1. The central part of this apparatus is a liquid reservoir (length 29.5 cm, width 1.0 cm, and height 0.5 cm). The reservoir, which contained 3 ml TMDS, was placed in a tube made of polymethyl methacrylate (PMMA). The inside diameter of this tube was 3 cm. The experiments were conducted at oxygen pressures of 25-30 bar. Figure 2 shows the applied ignition device in detail. For the experiments described in this subsection only setup 1 is of relevance. The incandescent nickeline wire was installed above the surface of the liquid at a distance of 0.8 cm.

The TMDS combustion processes in the reaction tube were observed by means of a rotating drum camera and a rotating prism camera (HYCAM). The photograph frequency of the rotating prism camera was always 8000 s^{-1}.

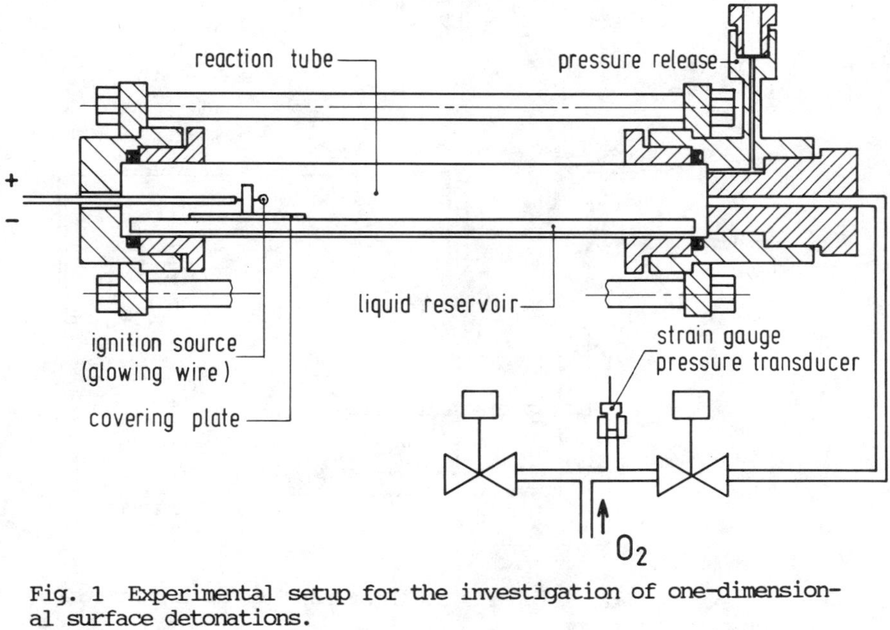

Fig. 1 Experimental setup for the investigation of one-dimensional surface detonations.

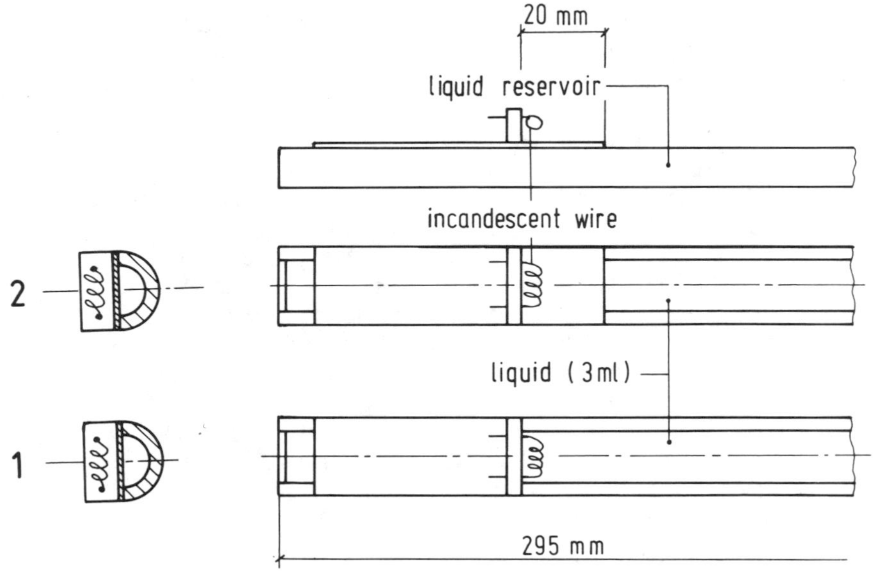

Fig. 2 Ignition device for the investigation of one-dimensional surface detonations (2.1) and indirect ignition processes (2.2).

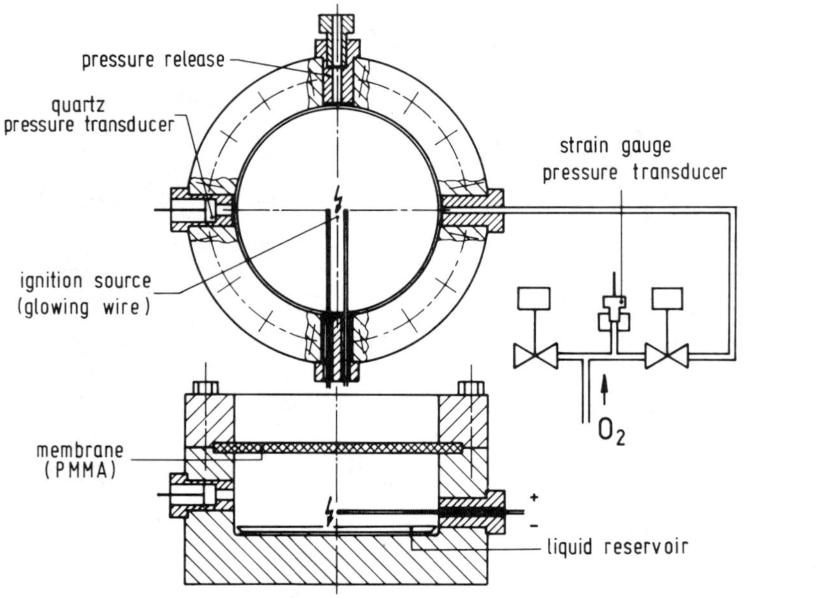

Fig. 3 Experimental setup for the investigation of two-dimensional surface detonations.

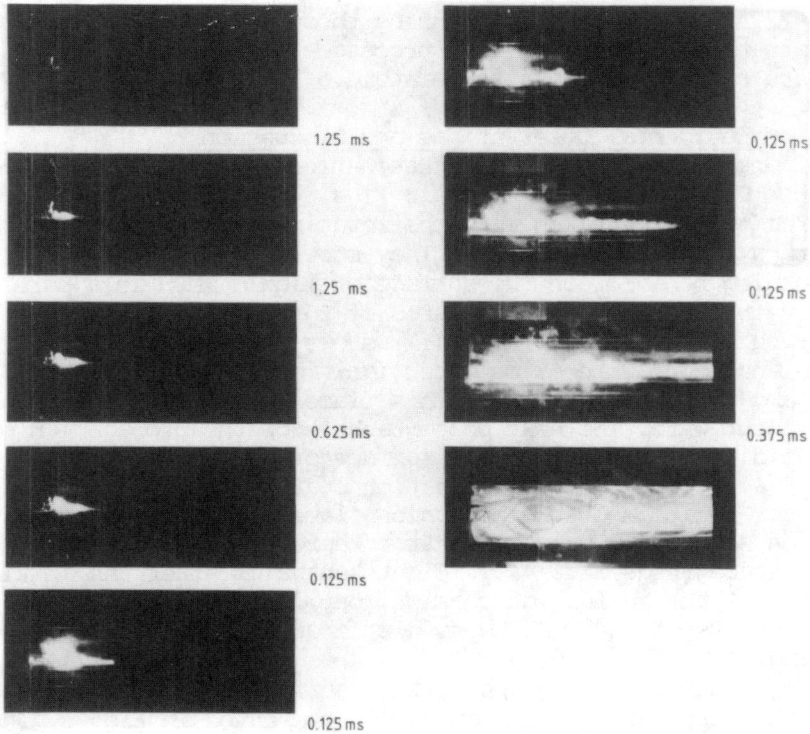

Fig. 4 Rotating prism camera photograph of a one-dimensional surface detonation, $p(O_2)$=25 bar; the time interval between two photographs is marked on the right-hand side.

2. Two-Dimensional Surface Detonations

Figure 3 shows the autoclave that was used to study the two-dimensional surface detonations. The upper part of this figure shows the top view of the autoclave. The lower part shows the sectional view. Membranes made of PMMA (thickness up to a maximum of 3.5 cm) were used as windows. The inside diameter of the autoclave was 34 cm, the height (up to the PMMA membrane) was 13.5 cm. Combustion was initiated in the center of the liquid surface by an incandescent wire made of nickeline, which was placed 2 mm above the surface of the liquid. The volume of the liquid was always 140 ml, which corresponds to a height of 1.5 mm. The liquid fuel used was TMDS, and the oxidizer was pure oxygen. Experiments were carried out at oxygen pressures between 10-25 bar. The combustion processes inside the autoclave were observed by means of a rotating prism camera (HYCAM) with a photograph frequency of 6000 s^{-1} and 8000 s^{-1}.

All experiments, including those performed to investigate the indirect ignition processes (see section IV), were carried out at a temperature of 18±1° C.

B. Experimental Observations and Discussion

1. One-Dimensional Surface Detonations

The photographs shown in Figs. 4 and 6 were selected from two films taken with the rotating prism camera (photograph frequency 8000 s^{-1}). They show an example of the burning behavior on the surface of liquid TMDS in a pure oxygen atmosphere at pressures of 25 bar. The first photograph of the series (Fig. 4.1) shows the glowing of the incandescent wire 43 ms after ignition of a diffusion flame on the liquid TMDS. This flame is close to the incandescent wire. The next photographs show the propagation of the diffusion flame on the surface of the liquid (Figs. 4.2-4.4). The surface detonation starts 3 ms (Fig. 4.5) after ignition of the diffusion flame close to the ignition source and propagates from there towards the end of the tube (Figs. 4.6-4.8). Eight-tenths ms after the initiation of the surface detonation, one can clearly recognize fragments of the PMMA being distributed in the vicinity (Fig. 4.9).

A streak photograph of this surface detonation, taken with the rotating drum camera, is presented in Fig. 5. In this case, the velocity of the process was 710 ms^{-1}. For the system investigated here, the velocity varies between

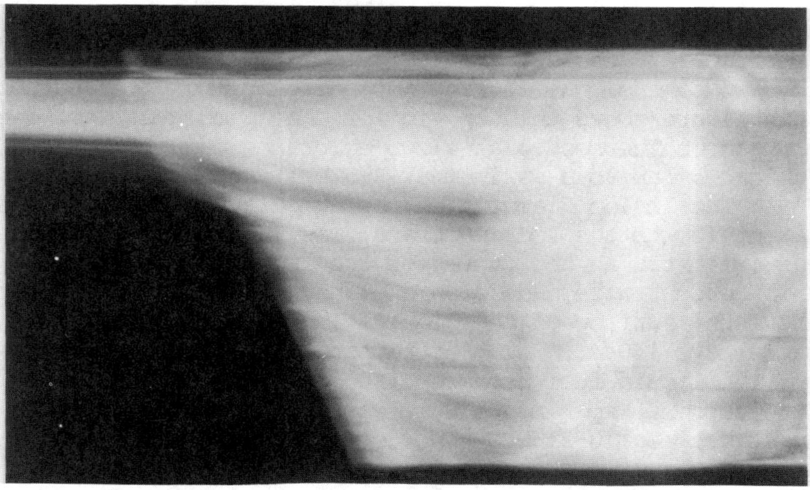

Fig. 5 Streak photograph of a one-dimensional surface detonation, $p(O_2)$=25 bar.

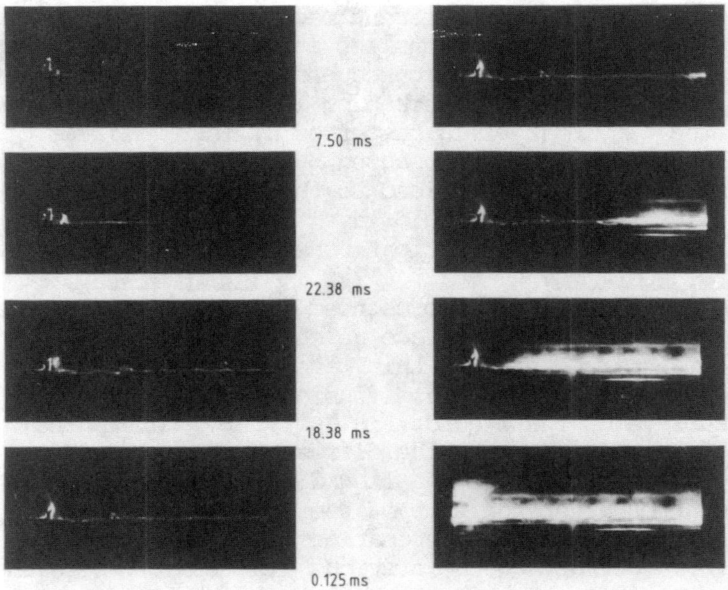

Fig. 6 Rotating prism camera photograph of a one-dimensional surface detonation, $p(O_2)=25$ bar; the time interval between two photographs is marked on the right-hand side.

about 700 ms^{-1}-1100 ms^{-1}. This statement, however, only refers to oxygen pressures between 25-30 bar. From the results of the experiments performed, it cannot be established clearly whether the described burning processes are stationary detonations. This would require experiments with considerably longer equipment.

In liquid TMDS the velocity of sound has been measured at 25° C and amounted to 857 ms^{-1}. This means that the values observed for the velocity of the surface detonation were above as well as below the velocity of sound.

The velocities of surface detonations are even smaller than those of wick detonations. At the same oxygen pressure the velocities of wick detonations are about 1300 ms^{-1}.[6]

Photographs taken with the rotating drum camera show that detonations or detonation-like processes can occur on surfaces of compact liquids, too. During surface detonations, as well as during film and liquid aerosol detonations, the mixing of the liquid fuel with the gaseous oxidizers has to be regarded as the slowest process. It is, therefore, the speed-controlling step in the reaction process that finally leads to the burning of the fuel. For this mixing process two different mechanisms are discussed. In one of these mechanisms it is assumed that the liquid fuel vaporizes as a consequence of "turbulent heating processes." The other mechanism is based on the assumption that a liquid layer is dispersed by instabilities of the phase boundary that are caused by a leading shock wave.[10,11] Both mechanisms lead to the conclusion that only a relatively small liquid layer is transported into the gas phase and can participate there in the burning process.

The photograph series presented in Fig. 6 shows, once again, the process of a surface detonation with the same experimental parameters. In this case, 115 ms after the first glowing of the wire, a diffusion flame was ignited on the surface of the liquid (Fig. 6.1). The next photographs of this series show the diffusion flame spreading over the whole surface of the liquid (Figs. 6.2 -6.4). Forty-eight ms after ignition of the diffusion flame, the transition to the surface detonation could be observed (Fig. 6.5). Contrary to the previous experiment, where the surface detonation started close to the ignition source, it started in this case at the opposite end of the setup. The next photographs show the progress of this heterogeneous detonation (Figs. 6.6-6.8).

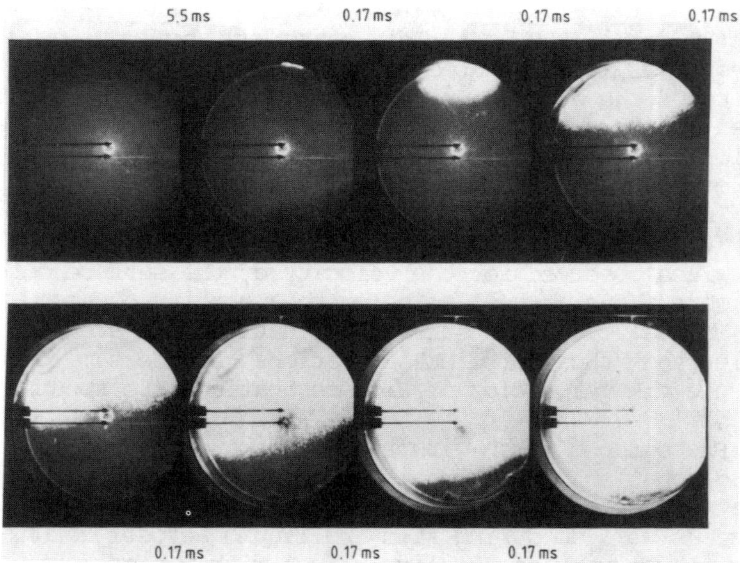

Fig. 7 Rotating prism camera photograph of a two-dimensional surface detonation, $p(O_2)$=25 bar; the time interval between two photographs is marked on the top or on the bottom of the photographs.

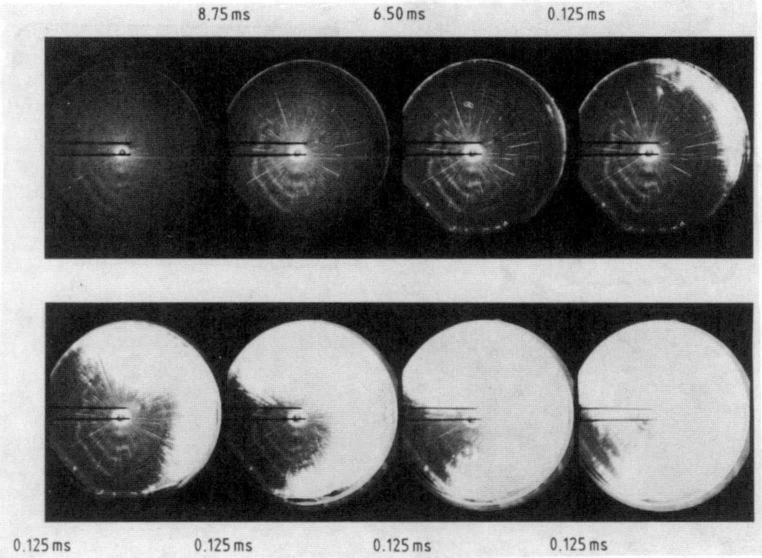

Fig. 8 Rotating prism camera photograph of a two-dimensional surface detonation, $p(O_2)=20$ bar; the time interval between two photographs is marked on the top or on the bottom of the photographs.

2. Two-Dimensional Surface Detonations

The following figures show three characteristic examples of two-dimensional surface detonations.

For the first experiment an oxygen pressure of 25 bar was used. The photograph frequency of the rotating prism camera was 6000 s^{-1}. The results of this experiment are presented in Fig. 7. Eighteen ms after ignition a diffusion flame spread over the whole of the liquid surface (visible as a weak glow in Fig. 7.1). This flame showed no structure. Twenty-three ms after ignition it was possible to observe the start of the surface detonation (Fig. 7.2). It took another 0.8 ms for the surface detonation to spread over the whole liquid surface (Figs. 7.3-7.8). Five and one-half ms after the initiation of the surface detonation, the PMMA membrane was destroyed.

The second experiment was carried out at a 5 bar lower oxygen pressure, i.e., at 20 bar. The photograph frequency of the rotating prism camera was 8000 s^{-1}.

The results of this experiment are given in Fig. 8. Nineteen ms after ignition a diffusion flame spread over the whole surface of the liquid. Twenty-one ms after ignition the diffusion flame showed a dissipative structure that consists of concentric circles (Fig. 8.1). This structure had so far only been observed during experiments

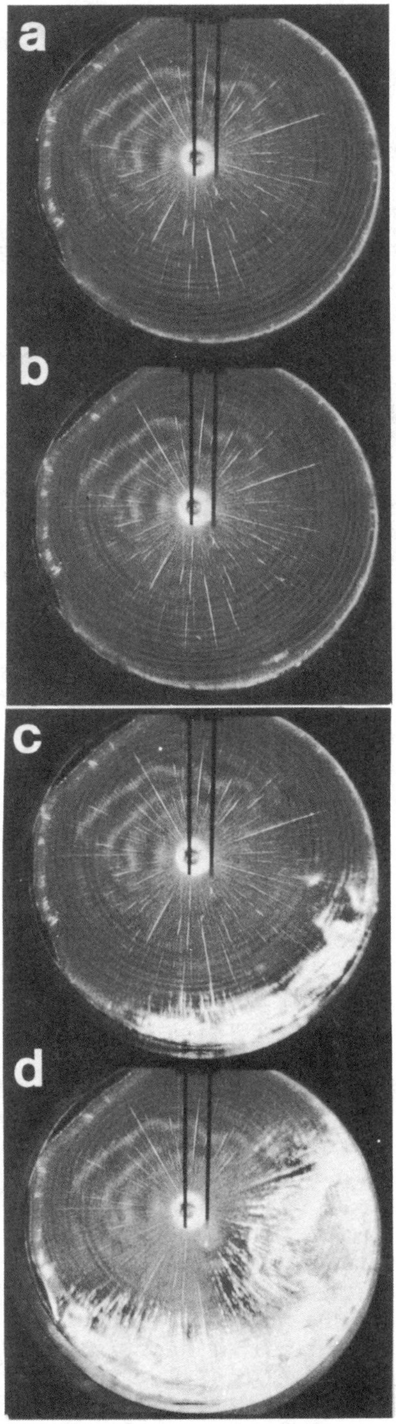

Fig. 9 Rotating prism camera photograph of a two-dimensional surface detonation; detail of Fig. 8; time interval between two photographs 0.125 ms.

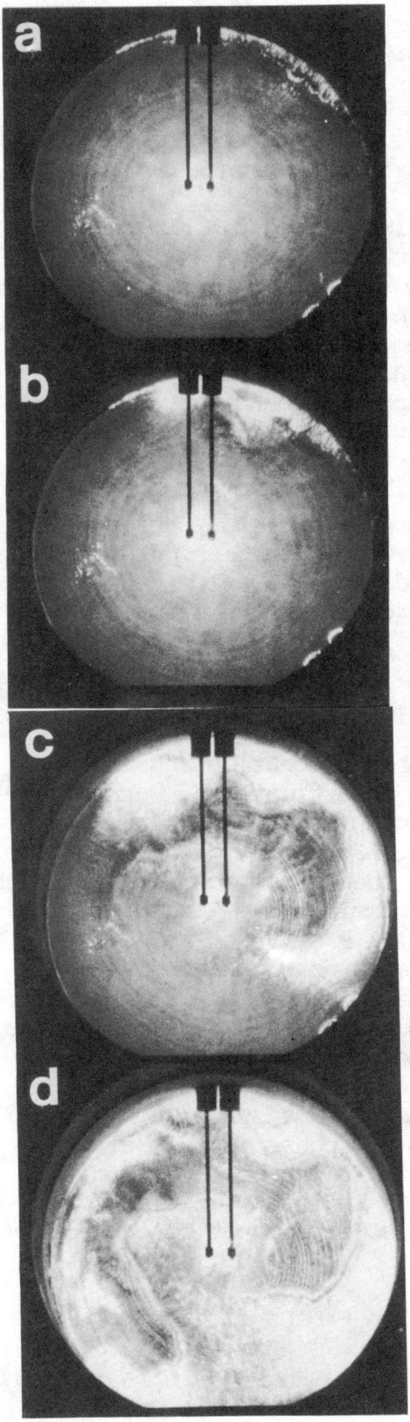

Fig. 10 Rotating prism camera photograph of a two-dimensional surface detonation, $p(O_2)$ = 18 bar; time interval between two photographs 0.125 ms.

with oxygen pressures under or at 20 bar. The surface detonation started 38 ms after ignition of the diffusion flame in the maximum of a dissipative structure, which indicated that the diffusion flame had its maximum intensity there (Fig. 8.3). The dissipative structure was partly maintained during the process of the surface detonation (Fig. 8.4).

The surface detonation also needed 0.8 ms to spread over the whole of the liquid surface.

The following four photographs (see Fig. 9) show more details of this experiment. These pictures were taken at intervals of 0.125 ms. The sequence started 0.125 ms before the ignition of the surface detonation.

In the third cinematographic sequence, as shown in Fig. 10, another demonstration of the dissipative structure is given. For this experiment an oxygen pressure of 18 bar was used. The four photographs of the sequence were taken at intervals of 0.125 ms. Before the ignition and shortly afterwards the dissipative structures are not yet well defined (Figs. 10a-10b). During the first stage of the surface detonation, the structure becomes very sharp in some parts of the surface (Figs. 10c-10d).

III. Gas Phase

In order to discuss surface detonation and indirect ignition, it was important to investigate the burning behavior of saturated TMDS vapor in oxygen. At a temperature of 18° C such gas mixtures detonate at oxygen pressures of up to 4 bar. At this pressure the transition to detonation distance is already of quite considerable length. In the 200-cm-long tube (inside diam. 1.9 cm) used here, distances of about 140 cm were measured. Depending slightly on the oxygen pressure, the velocity of the gas detonation amounts to 2185 m/s at an oxygen pressure of 4 bar and to 2002 m/s at an oxygen pressure of 1 bar.

At an oxygen pressure of over 7 bar, deflagrations were no longer observed in the gaseous system mentioned above. The composition of the gas phase is now below the lower flammability limit. Thus, it can be concluded that the surface detonations and the indirect ignitions observed at oxygen pressures up 30 bar cannot be caused by reactions of the gaseous phase.

IV. Indirect Ignition Processes

It is well known that a gaseous mixture of a fuel and an oxidizer cannot be ignited if its composition is outside the flammability region. It will be shown in the following paragraphs that this statement has to be modified if the gaseous phase is in contact with a liquid phase. An ignition source will always influence a gaseous atmosphere surrounding it, even if this atmosphere is not explosible. If a combustible liquid is present and certain conditions concerning the geometry of the experiment and the chemical reactivity of the liquid are fulfilled, it is possible that a diffusion flame or even a surface detonation may be initiated by some sort of "information" transported from the ignition source through the nonexplosible atmosphere to the surface of the liquid.

This effect can be of considerable importance for the safety of chemical plants, especially for oxidation and chlorination processes.

The experiments, described as follows, were performed with TMDS as the fuel and pure oxygen as the oxidizer. In all cases the composition of the gas phase was below the lower flammability limit.

A. Experimental Details

The setup for the investigation of indirect ignition processes corresponded to the one used for the observation of one-dimensional surface detonations, except for a different ignition device. Figure 2 gives one of the two ignition devices used (marked 2). Here a covering plate made of polyvinyl chloride or of aluminum was used as obstacle and installed below the incandescent wire. This obstacle was 0.1 cm thick and 5.0 cm long and projected 2.0 cm over the incandescent wire.

A further ignition device used during these experiments is given in the next photograph (Fig. 11). Here the incandescent wire is covered by a glass tube. Experiments with a length of x=1.2 cm and x=2.0 cm have been carried out.

B. Experimental Results and Discussion

First of all, the results of the experiments with covering plate are discussed (see Fig. 2.2). The unexpected result: a diffusion flame or a surface detonation developed whether a covering plate was present or not. The kind of covering material also had no influence. This is demonstrated by the next series of photographs (see Fig. 12).

Figure 12.1 shows the condition of the incandescent wire 85 ms after it first started to glow. The ignition of a diffusion flame on the liquid TMDS is illustrated in Fig. 12.2. The two photographs were taken at a time inter-

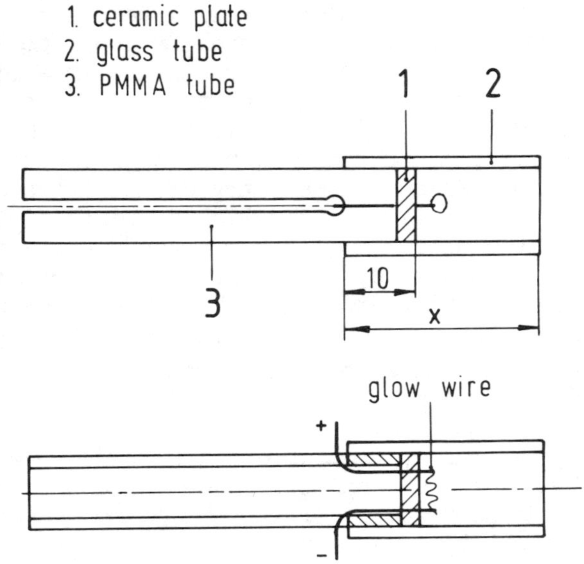

x mm
12
20
40
80
160

Fig. 11 Ignition device for the investigation of indirect ignition processes.

val of only 0.125 ms. Thirteen ms after the flame had spread over the entire surface of the liquid (Fig. 12.4), the surface detonation occurred (Fig. 12.5). In this case the surface detonation started at the end of the tube, close to the ignition source.

In the following discussion, results of experiments are presented during which the nickeline wire was surrounded by a ring-shaped covering tube. If the end of the tube was only 1.2 cm beyond the end of the wire, no ignition took place. For the ignition device with a covering plate as an obstacle, the corresponding length was 2.0 cm, i.e., nearly twice as long. Rotating prism camera photographs of experiments with the ring-shaped covering tube as an obstacle that did not lead to an ignition showed a considerable weaker glowing of the incandescent wire. Because the ignition device was not destroyed during these experiments, circular deposits of silicon dioxide that developed during the burning process of the TMDS were found above the wire.

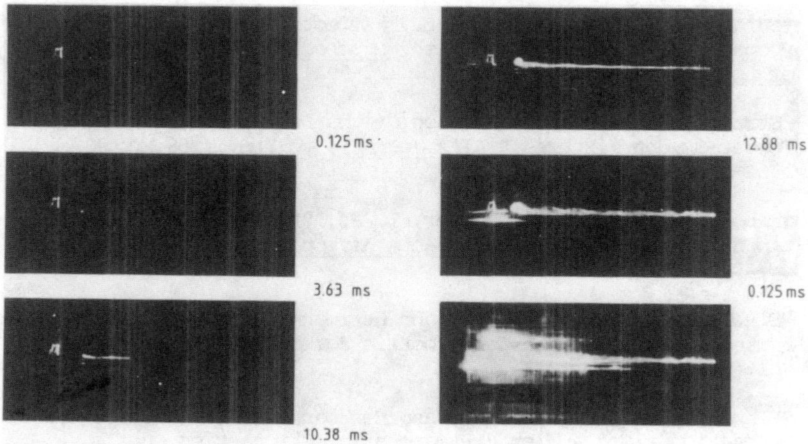

Fig. 12 Rotating prism camera photograph of a one-dimensional surface detonation, $p(O_2)$=25 bar, showing the effect of indirect ignition; the time interval between two photographs is marked on the right-hand side.

How can the results of the experiments be interpreted?

Along the incandescent wire an enforced burning reaction of the TMDS occurred, which was caused by the high temperatures prevailing there. This burning reaction could not spread into the gas phase because its mixture composition was below its lower flammability limit due to the high oxygen pressures. The hot reaction products formed at the incandescent wire were transported to the surface of the liquid and initiated the ignition process there.

The ring-shaped obstacles have influenced the transport process in such a way that an indirect ignition was no longer possible.

Acknowledgments

The authors wish to express their gratitude to Reinhard Kowall, Hansjörg Seeger, Peter Pietsch and Arne Björner for their assistance in the performance of the experiments.

References

[1]Loison, R., "The Propagation of Deflagration in a Tube Covered with an Oil Film," Comptes Rendus de l'Académie des Sciences, Vol. 234, 1952, pp. 512-513.

[2]Burgoyne, J. H., and Cohen, L., "Effect of Drop Size on Flame Propagation in Liquid Aerosols," Proceedungs of the Royal Society of London, Vol. A225, 1954, pp. 375-392.

[3]Strauss, W. A., "Investigation of the Detonation of Aluminium Powder-Oxygen Mixtures," AIAA Journal, Vol. 6, 1968, pp. 1753-1756.

[4]Dabora, E. K., and Weinberger, L. P., "Present Status of Detonations in Two-Phase Systems," Acta Astronautica, Vol. 1, 1974, pp. 361-372.

[5]Plewinsky, B., Wegener, W., and Herrmann, K.-P., "Measurements of Explosion Pressures in Calorimetric Bombs," Thermochimca Acta, Vol. 94, 1985, pp. 33-42.

[6]Plewinsky, B., Wegener, W., and Herrmann, K.-P., "Heterogeneous Detonation along a Wick," Progress in Astronautics and Aeronautics, Vol. 114, 1988, pp. 362-371.

[7]Lyamin, G. A., "Heterogeneous Detonation in a Rigid Porous Medium," Fizika Goreniya i Vzryva, Vol. 20, No. 6, 1984, pp. 134-138.

[8]Rao, C. S. R., Sichel, M., and Nicholls, J. A., "A Two Dimensional Theory for Two Phase Detonation of Liquid Films," Combustion Science and Technology, Vol. 4, 1972, pp. 209-220.

[9]Berthold, W., private communication, 1988.

[10]Borisov, A. A., et al., "Mechanism for Mixture Formation Behind a Shock Sliding over a Fluid Surface," Fizika Goreniya i Vzryva, Vol. 17, No. 5, 1981, pp. 86-93.

[11]Ragland, K. W., and Nicholls, J. A., "Two-Phase Detonation of a Liquid Layer," AIAA Journal, Vol. 7, No. 5, 1969, pp. 859-863.

Chapter IV. Multiphase Detonations

Detonation of Unconfined Large-Scale Fuel Spray-Air Clouds

W. B. Benedick* and S. R. Tieszen†
Sandia National Laboratories, Albuquerque, New Mexico
and
R. Knystautas‡ and J. H. S. Lee‡
McGill University, Montreal, Quebec, Canada

Abstract

Unconfined heterogeneous two-phase detonations in liquid droplet air mixtures have been investigated. The fuel-air cloud is formed by the explosive dissemination of liquid fuel into ambient air surroundings. This is done by pouring the liquid fuel into a V-shaped trough and by detonating an explosive line cord laid along the bottom vertex of the trough. An aerosol-air cloud typically 7 m high by about 1.5 average width can be generated in this way with a fuel mass/explosive charge ratio of about 200. In the present study, the length of the trough is 9 m, giving a detonable fuel-air cloud of about 100 m^3. Initiation of detonation in the cloud is by an end-mounted explosive charge, typically 850 g of Detasheet explosive. For less sensitive fuels, the 1.5 initial section of the trough is filled with propylene oxide to serve as the driver fuel. The propylene oxide driver and the test fuel are disseminated simultaneously. Detonation in the propylene oxide section is initiated by the sheet explosive charge, and the detonation then transmits from the driver section into the rest of the cloud formed from the test fuel. For insensitive fuels requiring a larger cloud dimension, two parallel fuel troughs 1.2 m apart are used. It has been found that detonability of fuel spray-air clouds depends not only on fuel sensitivity but also on the amount of fuel in the vapor phase just before detonation of the cloud. Very sensitive fuels, such as propylene oxide and low-vapor-pressure

---

*Senior Member of the Technical Staff, Shock Waves and Structural Physics Division (Deceased).

†Senior Member of the Technical Staff, Innovative Technology Applications Division.

‡Professor, Department of Mechanical Engineering.

nitrated hydrocarbons, can be detonated quite readily even in aerosol form. For more insensitive fuels, such as hexane, detonation is achieved only when vaporization of the fuel spray in the cloud is allowed to occur. Insensitive fuels with a low vapor pressure, such as decane. could not be detonated at all in the present tests.

Introduction

Detonations of mixtures of liquid fuel sprays (droplets or aerosols) and air have been studied extensively in the past two decades.[1–6] Perhaps the most conclusive demonstration of the existence of self-sustained heterogeneous detonations in low-vapor-pressure liquid fuels (e.g., decane) is the work of Bowen et al.[7] They used very small particles ($\sim 2\ \mu m$) in pure oxygen at atmospheric initial pressure and were able to observe steadily propagating waves with velocity fluctuations of $\pm 1\%$ over many tube diameters. They used a relatively weak ignition source, and the detonations were formed via transition from deflagration to detonation. Thus, there was no question of an excessively powerful initiator overdriving the wave in the region of observation. Furthermore, a multiheaded spinning wave was also observed, indicating the true universal characteristic of the cellular detonation structure. However, for the much less sensitive fuel-air mixtures, attainment of self-sustained heterogeneous detonations of low-vapor-pressure fuels is less convincing. Samirant et al.[6] used perhaps the largest diameter detonation tube (i.e., 0.5 m diameter by 4.5 m long). He reported successful heterogeneous detonations in sprays (mean droplet diameter on the order of 350 μm) of low-vapor-pressure fuels such as decane and hexanol in air. However, a very powerful ignition source ($\sim$ 1 kg of high explosive) was used, and this renders the results obtained questionable. It is known that the strong transverse perturbations generated by a strong initiator decay very slowly, if at all, in a confined tube. These igniter-generated transverse perturbations play the role of true transverse waves in a cellular detonation and, thus, maintain artificially a steadily propagating front. If these strong transverse perturbations are removed (e.g., damping by porous walls), or weakened by relaxing the confinement (e.g., diverging cross-sectional tube area), the "detonation" will fail; it is incapable of generating its own transverse waves.

Perhaps the best way to demonstrate the detonability of an explosive mixture is to study unconfined detonations in it. Bull et al.[5] reported experiments on unconfined detonations in fuel aerosols and air. Fuel droplets of mean diameter of about 15 μm were produced in a 5 m^3 volume by sonic air-blast atomizers. Spherical detonations were initiated by a high explosive charge (typically 25-500 g) and had a propagation length of 1.83 m for observation. Bull et al. found that, for a high-vapor-pressure fuel such as hexane, unconfined detonations can be initiated readily with a charge weight of about 25 g of tetryl, which is typical of that for a homogeneous gas-phase detonation of hydrocarbon-air mixtures. However, for low-vapor-pressure fuels such as decane and dodecane, detonations were not observed even with 500 g of tetryl initiator. On the basis of further experiments with a mixture of hexane and dodecane as fuels, Bull et al. concluded that self-sustained detonations may require a certain quantity of fuel vapor to be present prior to ignition.

From all of these previous studies, one may conclude that vaporization and turbulent mixing of the fuel vapor with air is the controlling mechanism for heterogeneous two-phase detonations. Thus, two-phase detonations may be said to be, in fact, hybrid detonations where the fuel appears in both the vapor and liquid state. Since the scale of Bull's experiments is insufficient to detonate dodecane without the presence of hexane vapor, it seems important to carry out further tests on a larger scale to establish the detonability of low-vapor-pressure fuels in air. The present paper reports recent results of some large-scale tests of unconfined detonations in hexane and decane to elucidate further the controlling mechanisms of heterogeneous detonations.

Experimental Details

The task of producing a large uniform volume of monodisperse liquid droplets in air is extremely difficult, if at all possible. Even if such a monodispersed droplet generator can be designed, gravity settling, droplet coalescence, and induced flow by the spray would still tend to render the mixture nonuniform. Lacking a knowledge of the length scale of heterogeneous detonations (as a function of particle size, vapor pressure, type of fuel, etc.), at present, it is difficult to estimate the requirement as to the uniformity of the mixture itself. Thus, in the present study, it was decided, first, to emphasize the global features of heterogeneous detonations.

The liquid droplet-air mixture is generated by explosive dissemination of the liquid fuel. A horizontal linear V-shaped metal trough (standard structural 90-deg angle) is filled with the liquid fuel, which is then dispersed into the atmosphere by detonating primacord (or mild

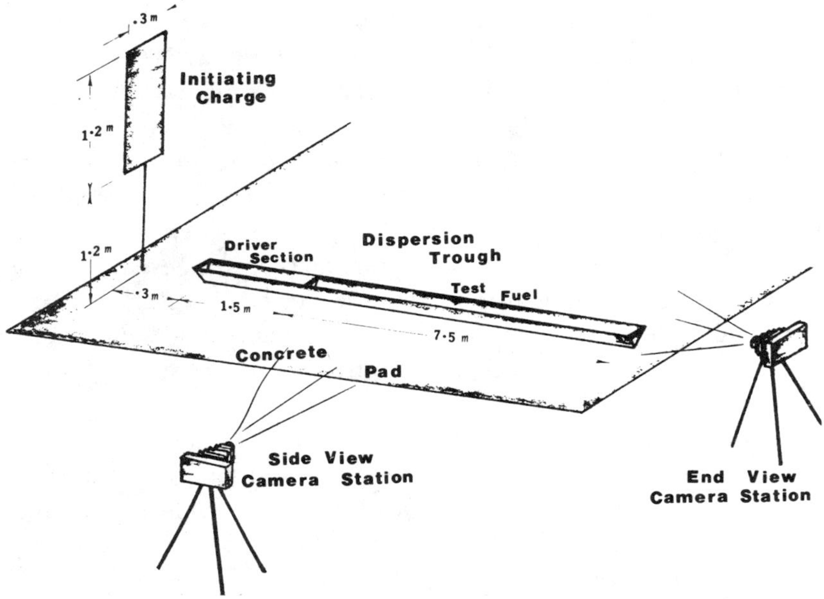

Fig. 1 Schematic diagram of fuel dispersion trough, intiating charge, and camera stations.

detonating fuse) laid along the bottom vertex of the V-channel. Figure 1 shows a schematic diagram of the trough, the plane sheet explosive initiating charge, and the high-speed camera diagnostics. For a given volume (or mass) of the liquid fuel per unit length along the channel, various strengths of the explosive cord have been tried to obtain the optimum cloud volume. Typically, the amount of explosive (PETN) used is 20 grains/ft (4.32 g/m) for a volume of about 1 liter/m of liquid fuel. The typical aerosol-air cloud generated above the trough is about 2.1 m at its maximum width and 7 m high. The total cross-sectional area is on the order of 10 m^2. Figure 2 shows selected frames of a high-speed movie of the cloud from camera stations viewing the time evolution of the length and width of the cloud. Typically, the length of the trough is about 8–10 m, giving a total cloud volume of about 80 m^3 as compared with the 5-m^3 volume of Bull's earlier experiments. Initiation of detonation is at one end of the cloud, giving a propagation length of about the same dimension as the channel itself (i.e., $\sim$ 10 m). Compared with the 1.83-m propagation length of Bull's experiment, the present study permits the detonation to be observed over a much longer distance to assess its self-sustenance effectively uninfluenced by the initiation source. For certain fuels in which detonation could not be obtained, it

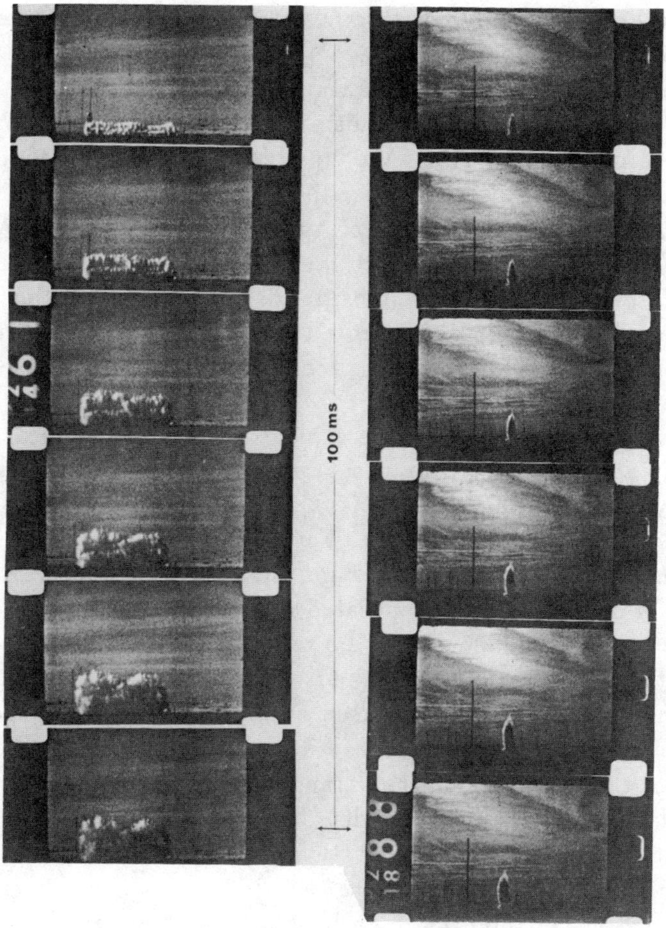

Fig. 2 Sequence of selected high-speed movie frames illustrating side and end views of fuel dispersion to form an Fuel-Air-Explosive cloud (80 m^3).

was thought that the width of the cloud is below the critical dimension to sustain an unconfined detonation. The work on the critical channel width[8,9] and the influence of yielding confinement studied by Renard et al.[10] and Murray and Lee[11] indicate that the minimum dimension of the cloud must be at least greater than the characteristic length scale of the detonation. In the case of homogeneous gas detonations, the characteristic length scale is the detonation cell size λ. However, for two-phase detonations, no clear indication of the existence of a regular cell pattern has been observed to date, even though the work of Bowen et al.[7] on two-phase fuel-oxygen mixtures does indicate the presence of a multiheaded spinning structure.

To provide a means of generating a wider aerosol-air cloud, two parallel troughs separated by a distance of 1.2 m are used. With two parallel troughs, the width of the cloud is approximately doubled. No direct measurement of droplet size in the cloud has been made. However, from simple analysis based on the work of Mayer[12] and Andersen and Wolfe,[13] it is estimated that the particle size ranges from about 20 to 50 μm. The mixture composition is not determined, but an estimate is made on a global basis from the quantity of fuel dispersed and the volume of the cloud formed, as indicated by the high-speed movie records.

Initiation of the detonation is achieved by a sheet explosive charge (typically about 850 g of Detasheet) mounted on a 0.3- $\times$ 1.2-m piece of plywood. The explosive sheet is raised on a post so that its bottom is about 1.2 m above the ground. For the less sensitive fuels, a driver section is used rather than increasing the solid explosive charge. The first 1.5 m of the trough is separated from the rest so that a different (more sensitive) driver fuel can be used. Propylene-oxide is usually used in the present study for this driver section. The propylene-oxide driver and the test fuel are dispersed simultaneously in the experiment. The solid explosive sheet initiates a detonation in the propylene-oxide portion of the cloud, which then transmits into the remaining portion of the cloud containing the droplets of the test fuel. In this manner, the degree of initiation overdriving is minimized.

A number of liquid fuels have been tested (e.g., propylene-oxide, hexane, decane, nitrated hydrocarbons, etc.). However, the emphasis of the experiment has been on hexane and decane in an effort to extend and clarify the previous study of Bull et al.[5] and, in particular, to assess the role of the vapor phase in the detonability of such fuels.

Results and Discussion

A large number of trials have been carried out initially to optimize the proportion of mass of liquid fuel to be dispersed to the dispersion charge weight per unit length of channel. These tests were carried out with water and observed with high-speed movie diagnostics. Figure 2 shows typical results of such dispersion tests. It was found that the best results are obtained for a mass ratio of 200 (i.e., mass of liquid fuel to mass of explosive per meter). From the maximum cloud volume estimated from the movie film records, the amount of liquid dispersed was tailored to yield an average fuel-air cloud composition that was approximately stoichiometric for all of the fuels tried.

For the detonation tests, it was decided to study propylene-oxide first, since it is a very sensitive fuel with a high vapor pressure at ambient temperatures (442 Torr at 293 K). The propagation of detonation through the cloud then can yield information on the dispersion and uniformity in the cloud with the present method of explosive dissemination. Because of the high vapor pressure of propylene-oxide, it is necessary to prechill the metal trough with liquid nitrogen before the liquid fuel is introduced, since significant evaporation can occur prior to detonation if the channel is warm. High-speed movie records show that shortly after dispersal of the propylene-oxide, the cloud becomes practically transparent, indicating that the aerosol has evaporated. Thus, for a high-vapor-pressure fuel such as propylene-oxide, the detonation occurs practically in the gas phase rather than as a two-phase heterogeneous detonation. This explains the relative ease in detonating propylene-oxide-air mixtures. The detonation is observed to propagate along the entire length of the cloud, and the local inhomogeneities in the cloud can be deduced from the shape of the advancing front. The local inhomogeneities, however, are insufficient to cause the detonation to be quenched as it propagates. Because of the unknown detailed stoichiometry of the cloud and its nonuniformity, no attempt is made to measure the local detonation velocity and to compare it with the theoretical Chapman Jouguet values. However, the average velocity of the detonation over the length of the cloud is on the order of 1600 m/s, typical of that for fuel-air mixtures.

It has been found[14,15] that additives such as propyl nitrate and butyl nitrate to kerosene result in the sensitization of the fuel. Some tests in the present series were carried out using nitrated hydrocarbons as fuel. Detonations are easily obtained with results similar to that for propylene-oxide even though the vapor pressure for these fuels is very much lower (typically 0.5 Torr at 293 K). No significant evaporation of the aerosols can be observed (as indicated by changes in the transparency of the cloud after dispersion) prior to initiation. Thus, in these cases, the detonation is truly one in a two-phase heterogeneous unconfined environment.

The main objective of the present test series was very specific: to assess the detonability of hexane-air and decane-air aerosol mixtures. In particular, it was intended to determine the role of the vapor phase

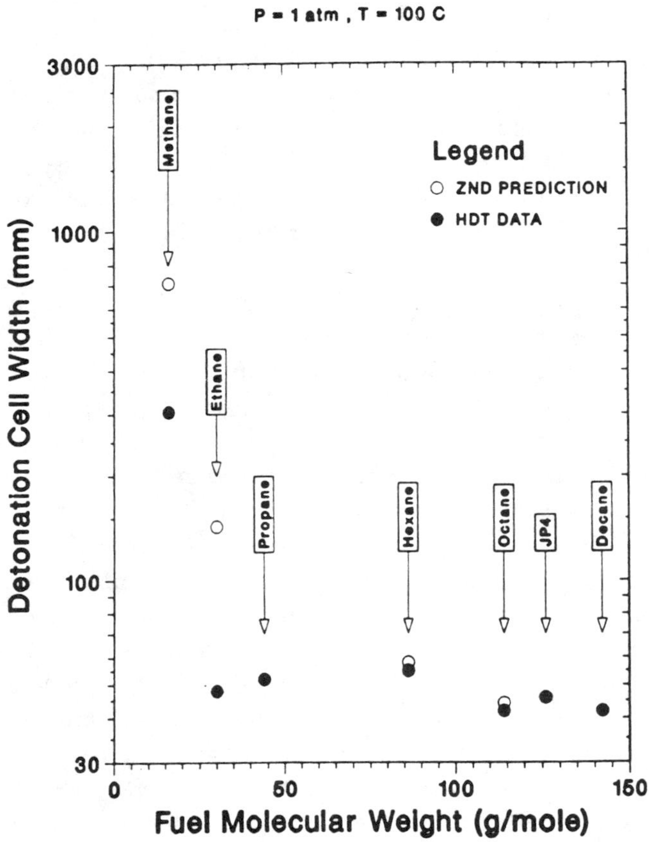

Fig. 3 Detonation sensitivity of gaseous hydrocarbon-air mixtures at stoichiometric composition as a function of molecular weight of n-alkanes.

in the detonability of such mixtures. The choice of the two fuels was intentional in view of their similarities and differences in the context of the above-stated objective. They are similar because, in the gas phase, as a result of studies carried out in the hot detonation tube (HDT) at Sandia,[16] they were shown to have virtually identical detonation sensitivity at stoichiometric composition. This is illustrated in Fig. 3, which shows the detonation cell size for a range of liquid hydrocarbon fuels plotted against molecular weight. The cell size for stoichiomet-

Fig. 4 Selected frames of a high-speed film taken at 500 frames/s of detonation initiation in a hexane-air cloud 150 ms after fuel dispersion.

ric hexane-air and decane-air is around 5-6 cm at 100°C. Hexane and decane are different in terms of their vapor pressure. At 20°C, hexane has a vapor pressure of ~ 127 Torr, whereas decane has ~ 1.5 Torr. Thus, by having two fuels of comparable nominal detonation sensitivity while having drastically different vapor pressure, it is possible to assess the role of the vapor phase in the detonation sensitivity by judiciously exploiting these properties in the present test series.

The approach in the present tests has been to create conditions in the cloud where vaporization can proceed progressively and can be observed directly. This is done by dispersing the fuel as an aerosol and then allowing it to sit or incubate in this suspended form while vaporization proceeds in the ambient surroundings. The typical dispersal time to achieve maximum cloud dimensions for a nominal stoichiometric fuel-air concentration was determined as $\sim$ 100 ms. At this time, the liquid hydrocarbon fuels that have been tried (e.g., hexane, decane) remain predominantly in aerosol form, as observed by high-speed photographic observations of the dispersal process. None of the hydrocarbon fuels tried, including hexane, can be detonated in this state. Hexane represents the highest vapor pressure fuel studied.

In the present test series, hexane fuel and a nitrated hydrocarbon driver were tried in the double-trough configuration. The high-speed film records displayed in Figs. 4 and 5 clearly show the sequence of

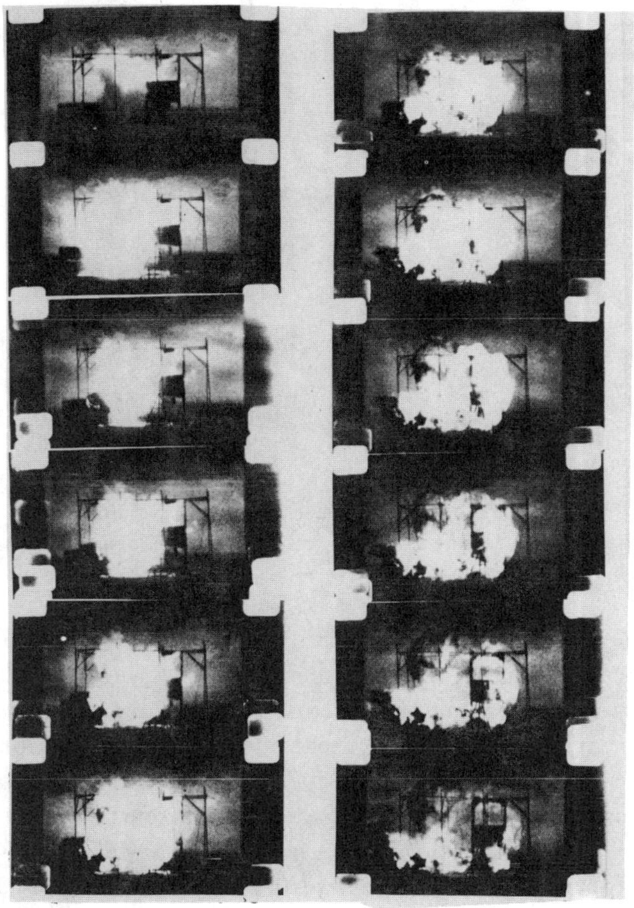

Fig. 5 Selected frames of a high-speed film taken at 6000 frames/s of detonation initiation in a hexane-air cloud 150 ms after fuel dispersion.

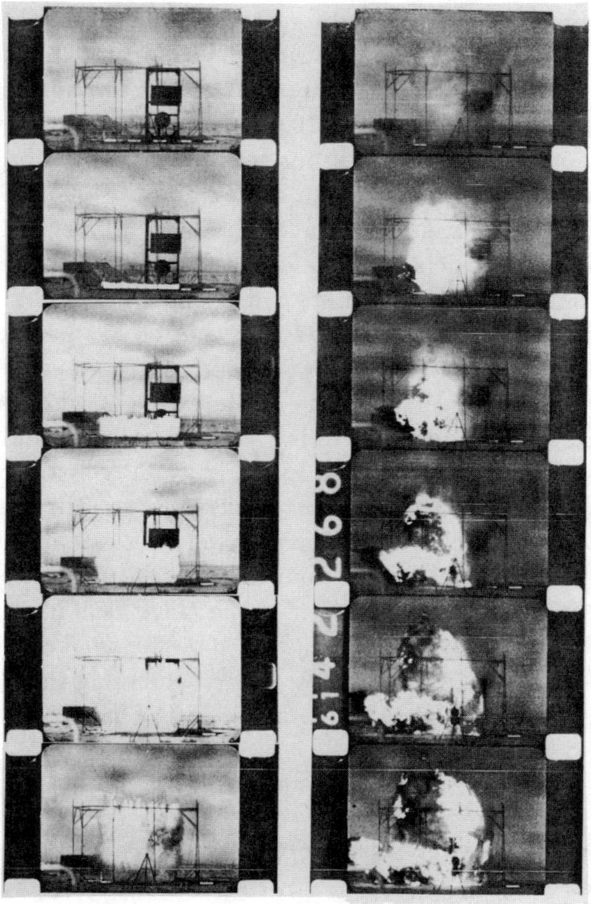

Fig. 6 Selected frames of a high-speed film taken at 500 frames/s of deflagration in a decane-air cloud 200 ms after fuel dispersion.

events. The photographs in Fig. 4 were taken at 500 frames/s and show the dispersion, vaporization, and detonation. The photographs in Fig. 5 were taken at 6000 frames/s and show cloud detonation. In this case, corresponding to Figs. 4 and 5, the delay time to initiation was increased to 150 ms. The film records clearly show vaporization of the hexane-air cloud. This is the segment of the cloud on the right, with the left being the low-vapor-pressure driver aerosol that has a low vapor pressure and, hence, remains in aerosol form. It must be remembered that the cloud is about 4 m (12 ft thick) as seen from the side, and it clearly becomes transparent, indicating total vaporization. When the detonation wave is initiated, it transmits and continues to propagate in the hexane (vapor)-air cloud for the total length of the cloud. The detonation of the hexane-air mixture was confirmed independently for this trial by the cellular structure written on the smoked foil plate. This shot was repeated with a 100-ms delay. Once again, failure to initiate detonation for $t_d \leq 100$ ms was the result. Subsequent trials with increasing delay times up to a maximum of 500 ms all indicate detonation of the hexane-air cloud. The only difference from shot to shot with increasing delay time is that the cloud begins to lift off progressively higher from the concrete pad. Hence, the essential role of the vapor phase for hydrocarbon fuel-air detonation has been convincingly confirmed.

For comparable sensitivity fuel—namely decane, which has a low vapor pressure compared with hexane—no detonation could be obtained, even with delay times up to 400 ms. Figure 6, which corresponds to a 200-ms delay to initiation after the beginning of dispersion, clearly illustrates this. The film sequence is taken at a rate of 500 frames/s. It is clear that the fuel in the cloud remains in aerosol form and the detonation from the driver portion of the cloud fails in the decane-air mixture. No cellular structure is observed on the smoked foil for any of the decane-air shots. Again, it is worth remembering that in the HDT, hexane and decane are equally detonable in the vapor phase (Fig. 3).

Conclusions

The present series of tests using explosive dispersion of liquid fuels to form an aerosol-air cloud clearly shows that sensitive liquid fuels such as propylene-oxide and nitrated hydrocarbons can be detonated easily in aerosol form. Because of its high vapor pressure ($\sim$ 450 Torr at 20°C), it is most likely that propylene-oxide is in the vapor phase at the time the cloud is detonated. However, there is no doubt that at detonation the nitrated hydrocarbon fuels tested are in aerosol form because of their very low vapor pressure and because of direct experimental evidence. The precise mechanisms underlying the detonation of aerosols is not understood currently, and it was beyond the scope of the present tests to attempt to shed light on such mechanisms. With regard to insensitive liquid hydrocarbon (transportation) fuels such as hexane, decane, etc., the present results conclusively indicate that detonation can occur only in the fuel vapor phase. A high-vapor-pressure fuel such as hexane could not be detonated within the first 100 ms after explosive dispersion and while it is still clearly in aerosol form. For delay times of 150 ms and higher, where the fuel aerosol is seen to vaporize, detonation is observed clearly throughout the hexane (vapor)-air cloud. For a fuel such as decane with a low vapor pressure, which remains in aerosol form for long periods of time after dispersion, detonation could not be initiated under any circumstances in the present trials. Tests in the heated detonation tube show that hexane and decane have comparable detonation sensitivities in the vapor phase.

Acknowledgment

The authors are grateful for the support of Marshall Berman who made these tests possible. The authors also gratefully acknowledge assistance in obtaining the data from Aris Makris, Charlie Daniel, Greg Sittler, Rod Oliver, David Weigand and John Thomas. The cinematrography is by Bill Bishop.

References

[1]Dabora, E. K., Ragland, K. W., and Nicholls, J. A., "A Study of Heterogeneous Detonations," Astronautical Acta, Vol. 12, No. 9, 1966, pp. 9-16.

[2]Dabora, E. K., Ragland, K. W., Ranger, A., and Nicholls, J. A., "Two Phase Detonations and Drop Shattering Studies," NASA CR 72225, 1967.

[3]Dabora, E. K., Ragland, K. W., and Nicholls, J. A., "Drop-Size Effects in Spray Detonation," Proceedings of 12th Symposium (International) on Combustion, The Combustion Institute, Pittsburgh, PA, 1969, pp. 19-26.

[4]Ragland, K. W., Dabora, E. K., and Nicholls, J. A., "Observed Structure of Spray Detonations," Physics of Fluids, Vol. 11, 1968, pp. 2377-2388.

[5]Bull, D. C., McLeod, M. A., and Mizner, G. A., "Detonation of Unconfined Fuel Aerosols," Progress in Astronautics and Aeronautics: Gasdynamics of Detonations and Explosions, edited by J. R. Bowen et al., Vol. 75, AIAA, New York, 1981, pp. 48-60.

[6]Samirant, M., Smeets, G., Baras, Ch., Royer, H., and Oudin, L. R., "Dynamic Measurements in Combustible and Detonable Aerosols," Propellants, Explosives, Pyrotechnics, Vol. 14, 1989, pp. 47-56.

[7]Bowen, J. R., Ragland, K. W., Steffes, F. J., and Loflin, T. G., "Heterogeneous Detonation Supported by Fuel Fogs or Films," Proceedings of 13th Symposium (International) on Combustion, The Combustion Institute, Pittsburgh, PA, 1971, pp. 1131-1139.

[8]Liu, Y. K., Lee, J. H., and Knystautas, R., "Effect of Geometry on the Transmission of Detonation through an Orifice," Combustion and Flame, Vol. 56, 1984, pp. 215-225.

[9]Benedick, W. B., Knystautas, R., and Lee, J. H., "Large-Scale Experiments on the Transmission of Fuel-Air Detonations from Two-Dimensional Channels," Progress in Astronautics and Aeronautics: Dynamics of Shock Waves, Explosions and Detonations, edited by J. R. Bowen et al., Vol. 94, AIAA, New York, 1985, pp. 546-555.

[10]Renard, J., Brossard, J., and Leulmi, R., "Propagation des Detonations Gazeuses dans les Tubes Souples," Sciences et Techniques de l'Armement, Memorial de l'Artillerie Francaise, Vol. 60, 1986, pp. 109-133.

[11]Murray, S. B., and Lee, J. H., "Another Look at the Stream Tube Concept in Light of the Cellular Structure of Detonation," Sciences et Techniques de l'Armement, Memorial de l'Artillerie Francaise, Vol. 60, 1986, pp. 521-541.

[12]Mayer, E., "Theory of Liquid Atomization in High-Velocity Air-stream," ARS Journal, Vol. 31, 1961, pp. 1783-1785.

[13]Andersen, W. H., and Wolfe, H. E., "Aerodynamic Breakup of Liquid Drops," Proceedings of 5th Shock Tube Symposium, Silver Spring, MD, 1965, pp. 1145-1182.

[14]Lu, P. L., Slagg, N., Fishburn, B., and Ostrowski, P., "Relation of Chemical and Physical Processes in Two-Phase Detonations," Acta Astronautica, Vol. 6, 1979, pp. 815-826.

[15]Dabora, E. K., "Effect of Additives in the Lean Detonation Limit of Kerosene Sprays in Air," University of Connecticut, Report 0507-129-F, 1980.

[16]Tieszen, S. R., Stamps, D. W., Westbrook, C. K., and Pitz, W. J., "Gaseous Hydrocarbon-Air Detonations," Sandia National Labs., Rept. SAND88-1925J, 1988.

Lean Detonation Limit of Sensitized Kerosene Sprays in Air

Eli K. Dabora*
University of Connecticut, Storrs, Connecticut

Abstract

A shock-tube technique was used to determine the lean detonability limit of kerosene sprays in air when the kerosene is sensitized by either propyl nitrate (PN) or butyl nitrite (BN). The technique accounts for the initiation energy and its rate, or power. Monodisperse sprays having droplet diameter of 780 μm are tested in a 5-cm-square tube, 2.7 m in length. An initiation energy of 50 J/cm^2 at a rate of 160 Kw/cm^2 is used, and three equivalence ratios (ϕ) namely, 0.59, 0.44, and 0.3, are tested. The kerosene is mixed with 10% or 20% of either PN or BN. The results indicate that the addition of PN reduces the limiting ϕ so that, for 10% PN, it is between 0.44 and 0.3 and below 0.3 for 20% PN. Erratic behavior was detected when 10% BN was used. However, when 20% BN is mixed with the kerosene, the limiting ϕ is found to be below 0.3.

Nomenclature

A = driven section area
a_1 = speed of sound of gas in driven section
d = drop diameter
E = energy
E* = critical initiation energy
f = droplet shedding frequent
l_4 = driver length
M = Mach number
M_1 = initial Mach number

*Professor, Department of Mechanical Engineering.

m = mass
P = power
p_1 = original pressure in driven section
u_l = terminal velocity of droplet
X = distance from diaphragm
X_c = distance from diaphragm at which the rarefraction wave interacts with shock front
γ = ratio of specific heats
ϕ = equivalence ratio = fuel-air ratio/(fuel-air ratio) stoichiometric.
ρ = density
ρ_l = liquid density

Introduction

Two-phase detonations have been of interest for the past two or three decades, and the state of our understanding of them has been reported by Dabora and Weinberger[1] and Nettleton[2]. One particular type of two phase detonations, namely spray detonations, have been investigated in some depth for some time. For example, the work of Dabora et al.[3] and Ragland et al.[4] dealt with the properties and structure of spray detonations as manifested in a constant-area tube, whereas Nicholls et al.[5] and Bar-Or et al.[6] were concerned with cylindrical detonations. Lu et al.[7] were concerned with the effects of chemical and physical processes on the detonation, finding that certain additives to the fuel can affect the establishment of detonations. More recent reviews of spray detonations can be found in articles by Dabora[8] and Sichel[9].

Despite the rather extensive work performed so far on spray detonations, the experimental difficulties involved have precluded exhaustive treatment of some aspects of the phenomenon. For example, very few systematic studies have been conducted to determine the detonability limits. This situation is to be contrasted with the detonability limits of gaseous mixtures that are reasonably well known for many different combinations.

Knowledge of detonability limits is of importance for at least two considerations: The fuel-air explosion (FAE) concept, and hazards elimination. In the former, one is interested in conditions that would insure the occurrence of detonation and, in the latter, the conditions that would eliminate the probability of detonations. So far the determination of limits seems to elude theoretical considerations alone and, therefore, almost all limits are based on either direct or indirect experiments.

The most important parameter in the determination of limits is the initiation energy. For detonable mixtures, there is a minimum critical energy that is usually necessary to initiate a detonation and, for a given combination of fuel and oxidizer, the minimum critical energy depends on the mixture ratio. Near-stoichiometric mixtures require the least critical energy, but as the mixture becomes either fuel-rich or fuel-lean the critical energy increases. The composition limits are then reasoned to be those compositions that require inordinate amounts of energy before a detonation can occur. In principle, however, true limits are those that would not support detonation, no matter how large the initiating energy is. Limits according to this definition would, however, be hard, if not impossible, to determine. A more practical approach is to define limits as those compositions that would not detonate when a given initiation energy is used. The choice of the magnitude of the energy can be arbitrary, say, an order of magnitude higher than the stoichiometric critical energy.

Although the initiation energy seems to be the most important parameter, there is reason to believe that the rate of energy deposition is of importance as well.[10, 11]. Thus, it can be argued that there is a minimum initiation energy below which no detonation can occur, no matter how high the power is, and that there is a minimum power below which no detonation can occur, no matter how high the total energy is. This idea is incorporated in the work of Dabora[12] on spray detonations, where critical energies and powers can be estimated.

The main purpose of this paper, however, is to determine in a practical way the lean limit of kerosene spray in air, when two additives are mixed with kerosene, namely, propyl-nitrate and butyl-nitrite. Since one would expect that the limit is affected by drop size, the limit should have been determined for each drop size. Unfortunately, time limitation precluded this, so that only sprays with drop size of 780 μm were considered.

Experimental Method

The shock-tube technique was used to determine the lean limit. In this method, a shock wave of finite strength is induced into the driven section of the shock tube, which is prefilled by the desired mixture. The decay or the acceleration of the wave is observed. If no acceleration is observed, the mixture is deemed nondetonable. If, however, acceleration is observed that is indicative of a detonation, the fuel-air mixture is then leaned in subsequent runs until no detonation takes place and, in this way, the lean limit is identified.

The shock tube that was used in this study is described in detail in Ref. 13 and is shown schematically in Fig. 1. Briefly, the driven section of the tube is 2.75 m long with a 5.08 x 5.08 cm-square cross section. The tube is topped by a tee section to accommodate a dump tank on the side. Another section on top of the tee accommodates a spray generator similar to the one described by Dabora.[14] To insure a reasonable spray along the tube, the tube is oriented in the vertical direction. The driver section is circular in cross section, providing a driver to driven area ratio = 1.65. The driver length is variable except that, for our tests, a 15-cm driver length was used. The tube operates on the double diaphragm principle to insure diaphragm break up at a desired pressure ratio.

The tube is provided with pressure switches at 30.5-cm intervals, starting at 15 cm from the diaphragm. In addition, two piezoelectric pressure transducers are placed at selected locations. At location 6, in addition to the pressure measurement, optical radiation is monitored with an EGG Lite Mike. Photographic observation of the phenomenon is made near that location with a spark source schlieren system.

The wave velocity is measured from time intervals as recorded in a raster display of the pressure switches as well as from two electronic time-interval units. Pressure and radiation measurements are recorded with oscilloscopes as indicated in Fig. 1.

The spray runs are conducted as follows. The tube is purged with air, and fresh air is established in the tube. The driver and the diaphragm sections are filled with helium to the desired pressure. Through a mechanical timer, the drop generator is started and stopped, and the diaphragm section is dumped to the atmosphere via a solenoid valve. The shock is thus initiated, and the measurements are automatically recorded. It should be pointed out that, for the usual use of a shock tube, the drive length is chosen large enough to insure a constant wave velocity

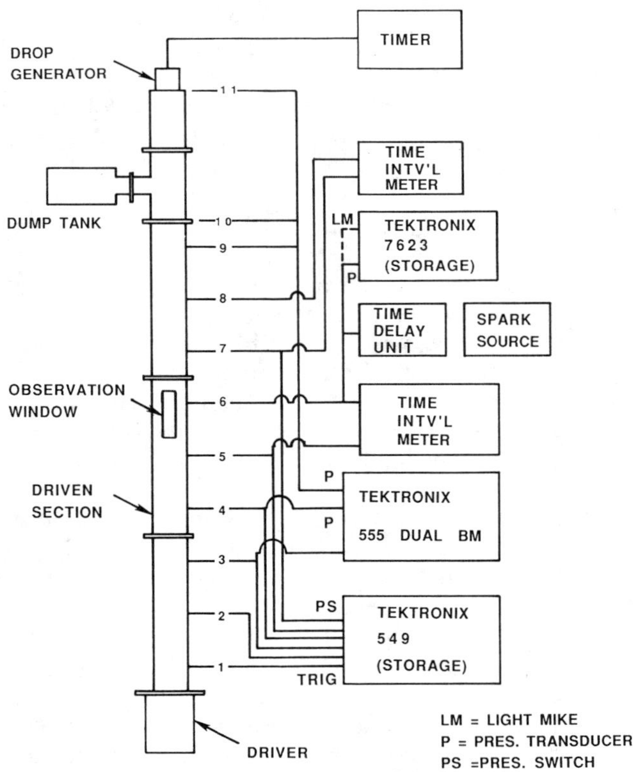

Fig. 1 Schematic of tube and instrumentation.

throughout the driven section. However, for our purposes, since we desired to observe whether a detonation does or does not occur by noting the wave velocity, the short driver used was appropriate. With a short driver, the rarefaction wave in the driver, after reflection, catches up with the wave and reduces its velocity if it is nonreactive. If the shock wave is reactive, as in a detonation, it is expected that the wave will continue to propagate unhampered by the rarefaction wave.

Analytical evaluations of the rate of energy input in a shock tube, the energy provided by the driver gas, and the decay in the Mach number when a non reactive gas is in the driven section have been given by Dabora[15,16] as follows:

$$P = \left[\frac{2\gamma}{\gamma+1} (M_1^2 - 1) + 1 \right] \left[\frac{2(M_1^2 - 1)}{(\gamma+1) M_1} \right] P_1 a_1 \tag{1}$$

$$E = P (X_o/l_4)\, l_4/ M_1 a_1 \tag{2}$$

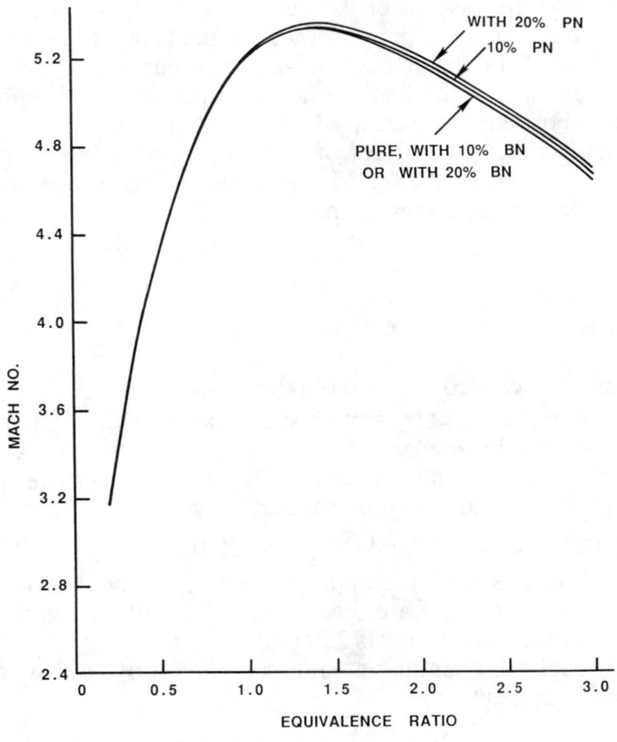

Fig. 2 Detonation Mach number of Kerosene with additives and air.

$$X / X_c = (M_1^2 - 1) / (M^2 - 1) \tag{3}$$

The above-described facility was used to determine the lean limit of H_2-air and it was found that this limit is between the 10% - 14% H_2 mixtures,[15,16] a value in agreement with the currently acceptable one rather than that found in earlier investigations (e.g., Refs. 17 and 18).

Kerosene-Air Experiments

Theoretical Calculations

The Chapman Jouguet detonation properties of mixtures of kerosene and air and of kerosene + additives and air were calculated using the computer code of Gordon and McBride[19]. Kerosene was assumed to have an average formula of $C_{11.6}\ H_{23.2}$ with a heat of formation of -5.6 Kcal/mole. The additives used were n-propyl nitrate ($C_3H_7NO_3$) and n-butyl nitrite ($C_4H_9NO_2$). Their heats of formation were taken as -41.6 and -50.0 Kcal/mole, respectively. In the calculations, the fuel was assumed to be in the gaseous phase, an assumption that has practically no effect on the detonation velocity. However, the fact that the fuel is liquid does have an effect on the detonation Mach number and the pressure ratio. The effect depends on the loading factor or the mass ratio of the liquid component. For lean mixtures, which are of interest in this work, the effect is negligible. The effect of additives for lean mixtures is also negligible, as can be seen from Fig. 2, where the variation of the Mach number with equivalence ratio is shown.

Fuel-Air Mixture Ratio Determination

The tests are carried out in monodisperse sprays. As was indicated before the spray is formed by a method similar to that of Dabora[14] where capillary jets are broken up into equal-sized droplets by inducing deliberate disturbances into the liquid jets at an appropriate frequency. When gauge 22 capillary needles with an internal diameter of 0.413 mm were used, the droplet size produced was 780 μm.

According to Ref.[14] the appropriate frequency and the shedding velocity of a kerosene jet are 390 Hz and 74 cm/s respectively. The terminal velocity of the droplet is 250 cm/s.

The air-fuel ratio per single column of fuel drops can be calculated from the following equation:

$$m_{air}/m_{fuel} = 6A\, u_l\, \rho_{air} / \pi\, d^3 f\, \rho_l \tag{4}$$

a)

b)

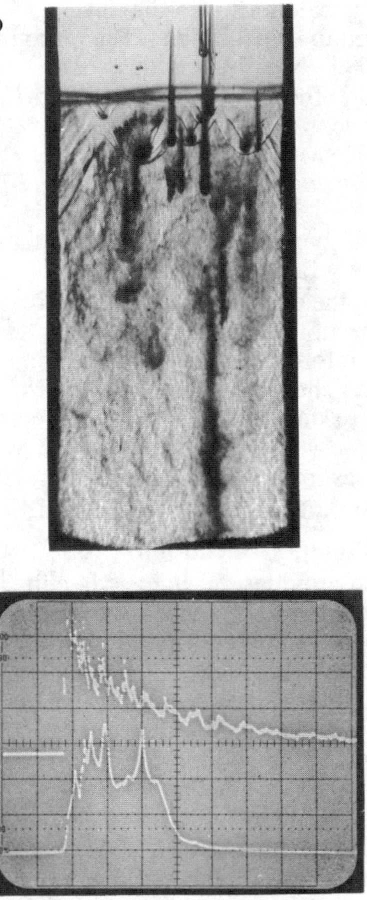

Fig. 3 Schlieren photograph (a), emission and pressure record (b), for kerosene & 10% PN (ϕ = 0.44) at location 6. (a) tube width = 5.08 cm; (b) 200 µs/div, top trace: 100 psi/div, bot. trace: 5 mV/div.

Using the appropriate values (A = 25.8 cm^2), the air-fuel ratio is found to be 99.6 per column of drops. The stoichiometric air-fuel ratio, as determined from the stoichiometric equation is found to be 14.76, so that the equivalence ratio per column is 0.148. Because of the manner in which the spray is generated, ϕ can be varied in a discrete manner only. Thus, ϕ's of 0.3, 0.44, 0.59, etc., are obtainable.

It should be mentioned that, because the shedding velocity is smaller than the terminal velocity, the fuel-air ratio is larger near the top than the bottom of the tube. It is possible to estimate the distance for the drops to travel before they reach a velocity within 95% of the terminal velocity. For our conditions, this turns out to be about 0.9 m.

In the preceding calculations, the effect of propyl-nitrate and butyl-nitrate is assumed to be negligible. The reason is that the maximum amounts used are 20% by weight, and their densities are close to that of kerosene. Furthermore, the carbon-hydrogen ratio is approximately the same as that of kerosene.

Energy Requirement for Initiation

The minimum energy requirement for the initiation of detonations in sprays depends on many factors, such as the drop size, the mixture ratio, the ignition delay, and the energy deposition profile after ignition[12]. The last two properties are not known for either kerosene or kerosene with additives, and it is therefore necessary to make some assumptions before the minimum initiation energy can be estimated. If the ignition delay is assumed to be equal to one-half the drop breakup time and if the heat release from each drop is assumed to be instantaneous, then the method of Ref.[12] can provide an estimate of the minimum initiation energy required for a stoichiometric mixture. From Table 1 and Fig. 2 of Ref.[12], it is found that E^*/d = 100 J/cm^3, which for d = 780 μ, gives E^* = 7.8 J/cm^2. The 15-cm driver used provides 50 J/cm^2 when the initial Mach number is 3.5. It is also indicated in Ref.[12] that if the breakup time is controlling, then the minimum initiation energy is minimally affected by the equivalence ratio. Thus, the 50 J/cm^2 provided by the driver (which is 6-7 times the estimated requirement) used in our experiments can be considered adequate for the determination of limits, only if the drop breakup time is controlling the ignition delay.

Experimental Results

Previous work Ref.[20] has indicated that kerosene-air mixtures are nondetonable. Limited runs conducted here with pure kerosene at ϕ = 0.44 produced no detonation either. Thus, our work concentrated on the effect of the two additives, propyl nitrate and butyl nitrite.

Specifically, tests at ϕ = 0.59, 0.44 and 0.3 were conducted for 10% and 20% of the additives in the fuel. For each test, a schlieren photograph was taken to insure that the spray is present and to obtain a visual feel for the phenomenon. In addition, pressure data and emission records were taken. Also, the progress of the wave is monitored throughout the tube. The criteria used to indicate detonability were light emission and the attainment of a wave speed equal to, or higher than, the CJ wave speed at some point in the detonation tube.

Representative data are reported here. For the 10% propyl nitrate, Figs. 3 and 4 are schlieren photographs, pressure and emission records at ϕ = 0.44 and 0.3. The schlieren photograph in Fig. 3 shows a typical detonation wave in which the front appears as a nonplanar front because of some interactions from waves behind it. The pressure record shows

a)

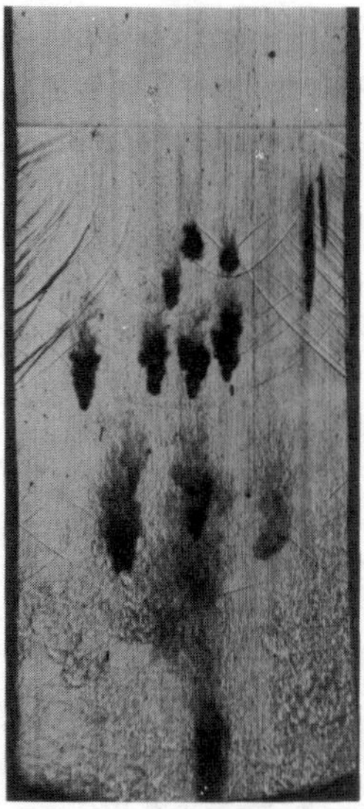

b)

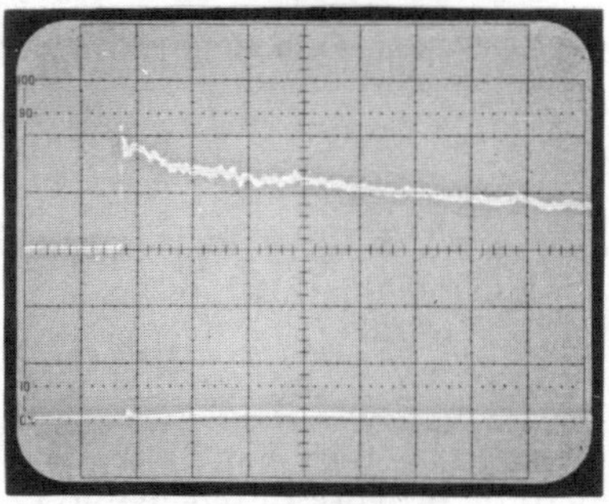

Fig. 4 Schlieren photograph (a), emission and pressure record (b), for kerosene + 10% PN ($\phi = 0.3$) at location 6. (a) tube width = 5.08 cm; (b) 200 μs/div, top trace: 100 psi/div, bot. trace: 5 mV/div.

some oscillations and pronounced peaks that could be due to blast waves from each drop although such blast waves do not seem to be detectable in the schlieren photograph. The emission record seems to indicate the presence of the peaks also. One interesting feature is the very short ignition delay.

The schlieren photograph in Fig. 4 shows a typical nonreactive front. Usually, in such a situation, the front remains planar. The pressure record shows no pronounced peaks, and the emission trace shows no radiation.

The wave velocity variation is shown in Fig. 5. It can be seen that, for $\phi = 0.59$ and 0.44, the propagation velocity increases at some point in the tube and reaches a velocity comparable to the respective CJ velocity.

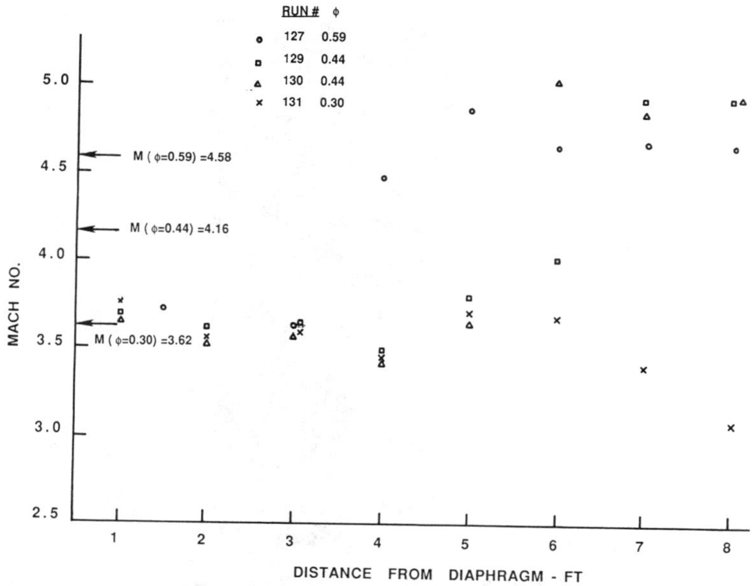

Fig. 5 Variation of propagation Mach number with distance for kerosene + 10% propyl nitrate.

On the other hand, for $\phi = 0.3$, the velocity does not veer too much from the initiation velocity and eventually drops down to below the CJ velocity. This situation, together with the emission records, is interpreted as absence of a detonation. Thus, the conclusion is that the lean limit for kerosene with 10% propyl-nitrate is somewhere between $\phi = 0.44$ and 0.3. As indicated before, because of our method of generating the spray, it is not possible to give a more precise limit.

Analysis of the results on runs with 20% propyl nitrate showed that detonation is possible at $\phi = 0.44$ and 0.3. This can easily be seen from the velocity data shown in Fig. 6 and, therefore, it must be concluded that the lean limit is below $\phi = 0.3$. Thus, it appears that propyl nitrate does affect the limit in a qualitative way: the higher its mass fraction in the fuel, the lower the lean limit.

The results with 10% butyl nitrate were rather erratic. There seemed to be cases of no detonation, even at $\phi = 0.59$, whereas sometimes detonations at $\phi = 0.3$ could be observed. As a result of this situation, the limit for this mixture could not be determined with any certainty. For the 20% butyl nitrite mixture, the data were found to be comparable to the 20% propyl nitrate. Again, detonation appears possible at $\phi = 0.3$, as corroborated by the wave velocity data shown in Fig. 7. Thus, the limit for this mixture is below $\phi = 0.3$.

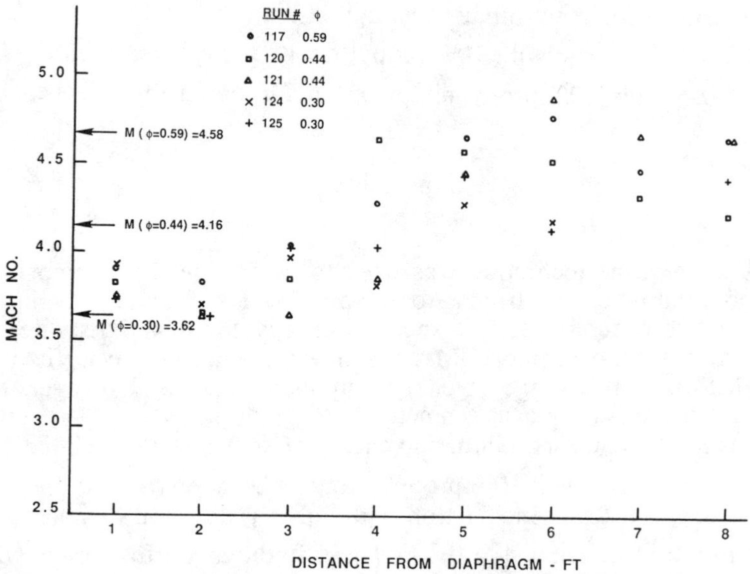

Fig. 6 Variation of propagation Mach number with distance for kerosene + 20% propyl nitrate.

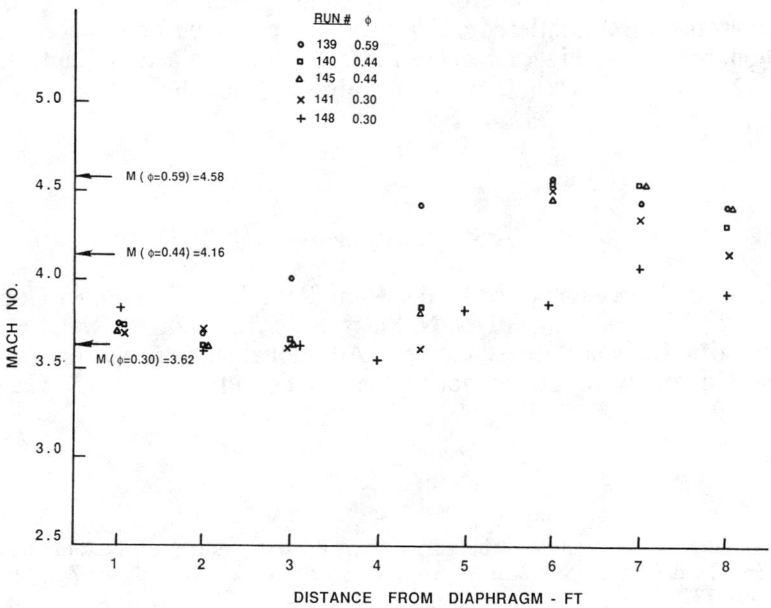

Fig. 7 Variation of propagation Mach number with distance for kerosene + 20% butyl nitrite.

In brief then, at an initiation energy of approximately 50 J/cm^2 the lean limit of kerosene with 10% propyl nitrate in air is between $\phi = 0.3$ and $\phi = 0.44$. For 20% propyl nitrate or butyl nitrite, the limit is below $\phi = 0.3$.

Conclusions

A shock-tube technique was used to determine the composition detonability limit of sensitized kerosene-air mixtures. Although the limits obtained were determined at specific energy input and its rate, as determined by the initiation Mach number, the technique is flexible enough to permit a range of energy inputs and rates. However, the technique is limited to planar geometry.

It is found that, for an initiation energy of 50 J/cm^2, the lean limit of kerosene sensitized with 10% propyl nitrate is between $\phi = 0.3$ and $\phi = 0.44$ hereas, for 20% sensitization with either propyl nitrate and butyl nitrite the limit is below $\phi = 0.30$. These findings are for sprays with droplet diameter of 780 µm.

The initiation energy used is 6-7 times the estimated energy needed if the droplet breakup time is controlling. However, the fact that the detonability of kerosene is affected by the additives indicates that chemical effects are important. Therefore, to get a deeper understanding of our findings, tests on smaller droplet sprays need to be conducted. In addition, breakup and ignition data similar to those obtained by Kauffman and Nicholls[21] on fuel droplets subjected to shock waves are recommended.

Acknowledgment

This work was supported by the Army Research Office under Grant DAAG 29-78-G-0074 with Drs. N. Slagg and P.L. Lu of ARRADCOM as Scientific Liaison Representatives. Additional support was provided by the University of Connecticut Research Foundation through Grant 0507-139.

References

1. Dabora, E.K., and Weinberger, L.P., "Present Status of Detonations in Two-Phase Systems," *Acta Astonautica,* Vol. 1, 1974, pp. 361-372.

2. Nettleton, M.A. "Shock-Wave Chemistry in Dusty Gases and Fogs: A Review," *Combustion and Flame,* Vol. 28, 1977, pp. 3-16.

3. Dabora, E.K., Ragland, K.W., and Nicholls, J.A., "Drop Size Effects in Spray Detonations," *12th Symposium (International) on Combustion*, The Combustion Institute, Pittsburgh, 1969, pp. 19-26.

4. Ragland, K.W., Dabora, E.K., and Nicholls, J.A., "Observed Structures of spray Detonations," *Physics of Fluids*, Vol. 11, 1968, pp.2377-2388.

5. Nicholls, J.A., Sichel, M., Fry, R. and Glass, D.R., "Theoretical and Experimental Study of Cylindrical Shock and Heterogeneous Detonation Waves," *Acta Astronautica*, Vol. 1, 1974, pp. 385-404.

6. Bar-Or, R., "Experimental Study of Cylindrical, Two-Phase Detonations in Monodisperse Sprays," Ph.D. dissertation, Dept. of Aerospace Engineering, Univ. of Michigan, Ann Arbor, MI, 1979.

7. Lu, P.L., Slagg, N., Fisburn, B., and Ostrowski, P., "Relation of Chemical and Physical Processes in Two-Phase Detonation," *Acta Astronautica,* Vol. 6, 1979, pp. 815-826.

8. Dabora, E.K., "Fundamental Mechanisms of Liquid Spray Detonations," in *Fuel-Air Explosions*, Univ. of Waterloo Press, Waterloo, Ontario, 1982, pp. 245-264.

9. Sichel, M., "The Detonation of Sprays: Recent Results," in *Fuel-Air Explosions,* Univ. of Waterloo Press, Waterloo, Ontario, 1982, pp. 265-300.

10. Lee, J.H., Knystautas, R., and Guirao, C.M, "Critical Power Density for Direct Initiation of Unconfined Gaseous Detonations," *15th Symposium (International) on Combustion*, The Combustion Institute, Pittsburgh, 1974, pp. 53-66.

11. Knytautas, R., and Lee, J.H. "On the Effective Energy for Direct Initiation of Gaseous Detonations," *Combustion and Flame,* Vol. 27, 1976, pp. 221-228.

12. Dabora, E.K., "Energy and Power Requirements for Direct Initiation of Spray Detonations," *Proceedings of the 12th International Shock Tube Symposium on Shock Tubes and Waves*, Jerusalem, 1980, pp. 685-694.

13. Fox, G., "An Experimental Study of the Relaxation Zone Behind Shock Waves in Water-Air Mixtures with Emphasis on Drop Breakup," Ph.D. dissertation, Dept. of Mechanical Engineering, Univ. of Connecticut, Storrs, CT. 1976.

14. Dabora, E.K., "Production of Monodisperse Sprays," *Review of Scientific Instruments*, Vol. 38, 1967, pp. 502-506.

15. Dabora, E.K., "Effect of Additives on the Lean Detonation Limit of Kerosene Sprays in Air," Univ. of Connecticut, Storrs, CT., Rept. 0507-129-F, 1980.

16. Dabora, E.K., "The Relation between Energy and Power for Direct Initiation of Hydrogen-Air Detonations," *Proceedings of The Second International Conference on the Impact of Hydrogen on Water Reactor Safety*, Sandia National Laboratories, Albuquerque, NM, 1982, pp. 931-952.

17. Laffitte, P.F., "Flames of High Speed Detonation," *Science of Petroleum*, Vol. IV, Oxford Univ. Press, London, 1938, pp. 2995-3003.

18. Belles, F.E., "Detonability and Chemical Kinetics: Prediction of Limits of Detonability of Hydrogen," *7th Symposium (International) on Combustion*, Butterworth, London, 1959, pp. 745-751.

19. Gordon, S., and McBride, B., "Computer Program for Calculation of Complex Equilibrium Compositions, Rocket Performance, Incident and Reflected Shocks and Chapman-Jouguet Detonations," NASA SP-273, 1971.

20. Lu, P.L, ARRADCOM, Dover, NJ, private communication, 1979.

21. Kauffman, C.W., and Nicholls, J.A., "Shock-Wave Ignition of Liquid Fuel Drops," *AIAA Journal*, Vol. 9, 1971, pp. 880-885.

Explosive Decomposition of Hydrazine by Rapid Compression of a Gas Volume

R. L. Bunker* and D. L. Baker†
NASA White Sands Test Facility, Las Cruces, New Mexico
and
J. H. S. Lee‡
McGill University, Montreal, Quebec, Canada

Abstract

The initiation mechanism and the explosion mode of hydrazine decomposition under rapid compression conditions were investigated. Experiments were performed in which a column of liquid hydrazine (20 cm in length) was accelerated into a column of gaseous nitrogen (38-102 cm in length) in a close-ended cylindrical chamber (1.3 cm in diam). The liquid hydrazine column was physically separated from the nitrogen by a thin Teflon diaphragm (0.05 mm thick) and was accelerated into the nitrogen column by activating a rapid opening solenoid valve, which connected the cylindrical chamber to an accumulator (14 liters) of high pressure. A pressure transducer was placed at the end of the bottom chamber to register the pressure-time history of the event. High-speed video data were taken during some experiments conducted using an apparatus constructed from glass tubes. The video data revealed the formation of a froth produced by the acceleration of hydrazine into nitrogen at the liquid hydrazine-gaseous nitrogen interface. The initiation mechanism of the explosive hydrazine decomposition was shown to be dependent on the formation of a froth at a critical temperature and experimentally dependent on the nitrogen volume and pressure in the bottom chamber and the peak hydrodynamic surge pressure. High-speed video data together with data obtained from nitro-

---

* Aerospace Engineer, NASA
† Senior Scientist, Lockheed-ESC
‡ Professor of Mechanical Engineering

methane detonation studies indicate that under rapid compression conditions, the explosion mode of hydrazine decomposition is a confined thermal runaway reaction. Experiments conducted with water confirmed that the explosive events observed in hydrazine experiments were caused by chemical reactions of liquid hydrazine and not by water hammer effects.

Introduction

The use of hydrazine in propulsion and power systems is widespread in the aerospace industry. Catastrophic rupture of liquid hydrazine lines has been observed when the liquid hydrazine is accelerated into empty lines or lines containing gas or vapor bubbles. Experiments previously have been conducted in which a small quantity of liquid hydrazine was accelerated into a gas bubble in a U-tube arrangement.[1,2,3,4] Tube failures were observed in some of these experiments; however, the initiation mechanism or explosion mode of the hydrazine decomposition reaction was not elucidated. Liquid hydrazine is known to undergo exothermic decomposition when the liquid is heated under quasistatic conditions in a reaction chamber.[5,6] In the vapor state, hydrazine was shown not only to support flame propagation,[7,8] but detonated quite easily as well.[9,10,11,12] In the liquid phase, previous studies failed to establish conclusively the detonability of hydrazine.[1,13,14] In experiments reported by Benz and Pippen,[15] both vapor and liquid hydrazine underwent thermal runaway reactions (rapid exothermic chemical reactions that occur uniformly throughout the liquid or vapor).

Although explosive events have been observed in dynamic hydrazine systems, the mechanism of the violent reactions that occurred are not known at present. The objective of the present study is to determine the initiation mechanism and the explosion mode associated with hydrazine under rapid compression conditions.

Experimental Details

In the present investigations, a simple, straight geometry system is used (Fig. 1) instead of the U-tube arrangements used in the previous studies. The test system consisted of a 14-liter

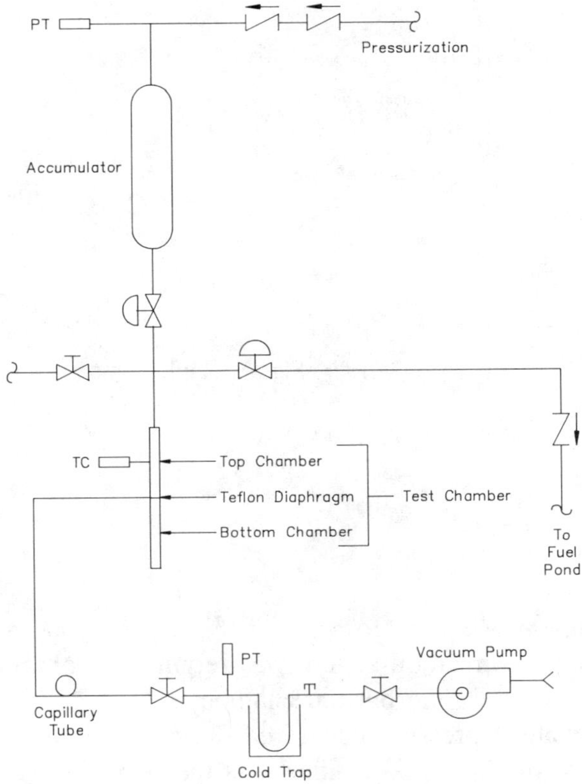

Fig 1 Rapid compression test system

accumulator that could be pressurized to 13.8 MPa, an inert-gas vent-and-purge system, and a high-speed solenoid valve to introduce the pressurization gas into the test section. The high-speed solenoid had a 2.54-cm orifice diameter with an 8-ms actuation time.

A schematic of the test section is shown in Fig. 2. The top chamber of the test section contained the test liquid, and the bottom chamber contained gaseous nitrogen normally at a pressure of 85.5 kPa. The test section was fabricated from two 1.3-cm-diam 304 stainless steel tubes (wall thickness of 0.09 cm). The top chamber was kept at a constant length of 20.3 cm, and the bottom chamber was kept at a length of 38.1 cm for most experiments. A 304 stainless steel union connected the two chambers. The interior volume of the chambers was separated by a 0.05-mm Teflon diaphragm contained within the union. A

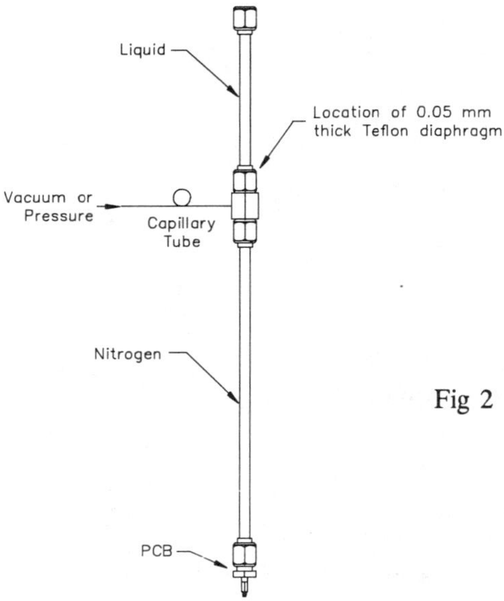

Fig 2 Rapid compression test chamber

modified union was used to connect the test section when a vacuum or pressurization system was required to obtain the various pressures in the bottom chamber.

The dynamic pressure generated during the compression of the nitrogen by the liquid column was measured using a piezo-electric dynamic pressure-transducer placed at the bottom of the test section (Fig. 2). In addition to the pressure measurements, high-speed video was used in some of the experiments. The test liquids used in these investigations were monopropellant-grade hydrazine (per Military Specification MIL-P-26536D), distilled water, diluted hydrazine (containing 10-20% distilled water), and nitromethane.

Prior to testing, each chamber was cleaned with soapy water, rinsed with isopropyl alcohol, and blown dry with nitrogen. The test section was then assembled, loaded with the test liquid in a nitrogen-purged glove box, and installed into the test system. The nitrogen purge on the test system was activated to eliminate air that may have entered the test system during chamber installation. The accumulator was pressurized to a driving pressure of 2.1 MPa for most experiments. The nitrogen purge was then deactivated, and the isolation valve was opened. As a result, the test liquid in the top chamber was pressurized rapidly and accelerated through the diaphragm into the bottom chamber.

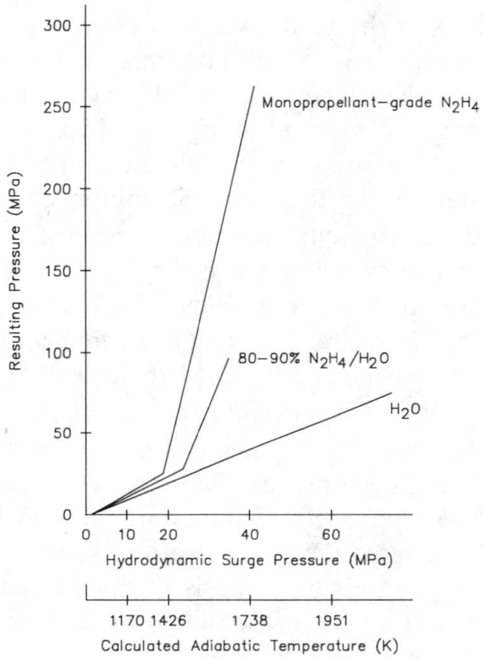

Fig 3 Resulting pressures and calculated adiabatic temperatures at various peak hydrodynamic surge pressures

Three series of experiments were performed during these investigations. In the first series, the effect of the peak hydrodynamic surge pressure on the explosive hydrazine decomposition reaction was studied. The next series examined how the nitrogen volume and pressure in the bottom chamber affected the peak hydrodynamic surge pressure and, thus, the explosive event. The third series of experiments was conducted to determine the explosion mode, which produced the explosive events observed in dynamic hydrazine systems. All experiments conducted were initially at ambient temperatures of approximately 300 K.

Results and Discussion

The effect of peak hydrodynamic surge pressure on the explosive hydrazine decomposition reaction was studied using hydrazine, diluted hydrazine, and water. Experiments were conducted with water to determine impact pressure generated solely from hydrodynamic effects. By varying the driving pressure, significant increases in the peak hydrodynamic surge pressure measured at the closed end of the bottom chamber were obtained (Table 1). When hydrazine and diluted hydrazine were used as test liquids, pressures resulting from hydrodynamic effects plus hydrazine decomposition were obtained. In Fig. 3, resulting pressures generated during the adiabatic compression of nitrogen by the liquid hydrazine column are plotted against the peak hydrodynamic surge pressure generated by water. When the peak hydrodynamic surge pressure was below 17 MPa, resulting pressures generated by hydrazine and diluted hydrazine appear to be caused only by hydrodynamic effects. When peak hydrodynamic surge pressures were above 17 MPa, the resulting pressure produced by hydrazine could be as much as 6 times larger than the peak hydrodynamic surge pressure. For the diluted hydrazine mixtures, peak hydrodynamic surge pressures had to be larger than 34 MPa before large deviations were observed.

Assuming the compression to be isentropic, the adiabatic compression temperatures corresponding to the peak hydrodynamic surge pressures can be calculated as:

$$T2/T1 = (P2/P1)^{\gamma-1/\gamma}$$

These temperatures are plotted on the abscissa with the peak hydrodynamic surge pressures in Fig. 3. The slight increase in

Table 1 Effects of driving pressure on the peak hydrodynamic surge pressure of water[a]

Driving pressure, MPa	Surge pressure, MPa
1.5	14
1.7	16
2.1	23
3.5	28
6.9	37

[a]Data were obtained with a pneumatically operated ball valve (0.64 cm in diameter) instead of the rapid opening solenoid valve. The actuation speed of the ball valve was not determined, although the speed was presumed to be much slower than the solenoid valve.

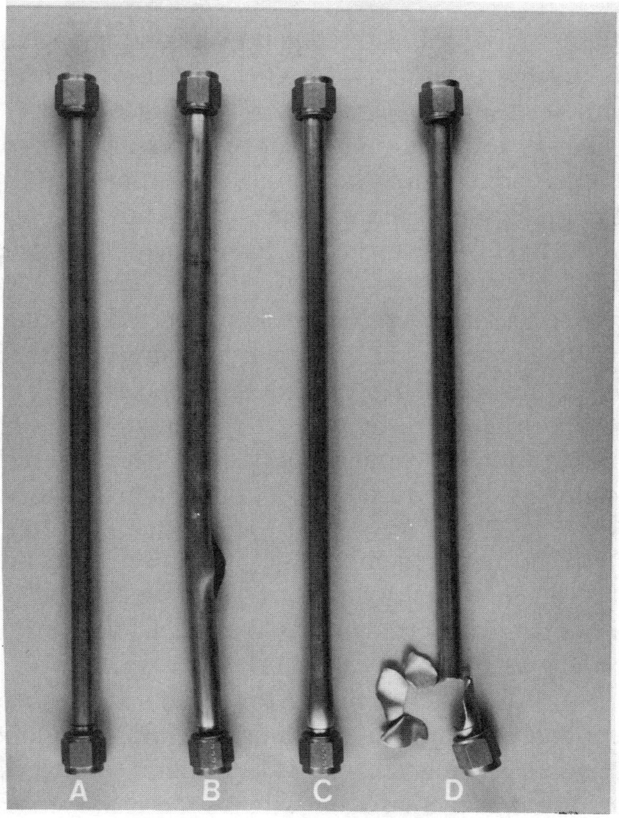

Fig 4 (a) Pretest chamber; (b) Hydrostatic burst chamber; (c & d) Typical chambers after hydrazine compression

the resulting pressure due to the onset of hydrazine decomposition occurs at about 17 MPa (corresponding to a temperature of about 1360 K). These results indicate that there is a critical temperature corresponding to a sufficiently short induction time such that heat losses are minimized and the hydrazine decomposition reaction is accelerated rapidly.

In Fig. 4, two reference chambers and two chambers damaged during this series of experiments are presented. Fig. 4a, a pretest chamber, and 4b, a chamber subjected to a hydrostatic burst test, are included as references. In the hydrodynamic burst test, a pressure of approximately 103 MPa was measured, and the rupture occurred at the weakest portion of the chamber, near the center. This damage is unlike the typical damage observed during the rapid compression experiments shown in Figs. 4c and 4d. In Fig. 4c, the chamber did not rupture but expanded significantly at the point where the liquid hydrazine impacted the closed end. In this experiment, resulting pressures exceeded 69 MPa. Figure 4d displays the most typical result observed. In this case, the chamber exhibited only local ruptures and expansion near the closed end, which is the strongest portion of the chamber. Resulting pressures measured in this type of failure were in excess of 276 MPa. No evidence of flames or burning was observed.

The next series of experiments was conducted to study the effects of nitrogen volume and nitrogen pressure on peak hydrodynamic surge pressure and hydrazine decomposition. The nitrogen volume was varied by varying the length of the bottom chamber from 13 to 102 cm at a constant initial pressure of 85.5 kPa. The nitrogen pressure in the bottom chamber was varied from 0.005 to 141 kPa at a constant initial volume of 36 cm^3. In general, the peak hydrodynamic surge pressure of water decreased as nitrogen volume or pressure increased. The resulting pressures obtained with hydrazine as the test liquid were found to increase initially as the nitrogen volume increased. Chamber deformation was observed in the 12-cm^3 chamber, whereas chamber ruptures and pressures in excess of 276 MPa were obtained in chamber volumes ranging from 24 to 72 cm^3. When the initial chamber volume was increased to 84 cm^3, the chamber was deformed but did not rupture. When experiments were performed with the largest volume chamber (96 cm^3), no deforming of the chamber was observed, and a pressure of 38 MPa was measured. Experiments with hydrazine at nitrogen pressures less than 0.005 kPa were similar to those with water. As the nitrogen pressure was increased between 7 and 55 kPa, no definite trends were observed. The resulting pressures, however, were much greater than those observed for water and indicate hydrazine decomposition. Further increases in nitrogen pressures between 69 and 127 kPa caused chamber rupture and resulting pressures in excess of 276 MPa. When nitrogen pressures were increased above 134 kPa, the chambers did not rupture and the resulting pressures decreased. At a nitrogen pressure of 141 kPa, no evidence of chamber damage was observed, and the resulting pressures, which were approximately 48 MPa, approached the peak hydrodynamic surge pressures obtained with water.

From the experimental data, the final volume and the adiabatic temperature of the nitrogen gas after compression were

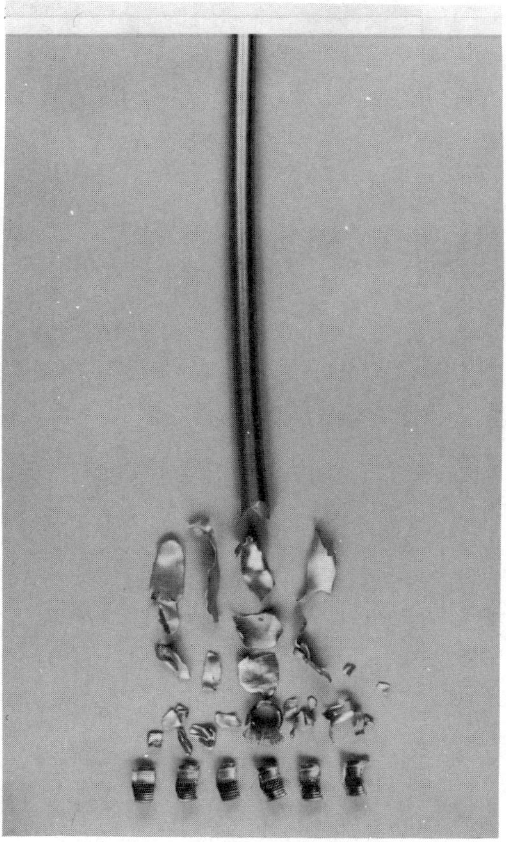

Fig 5 72-cm^3 chamber with extreme damage

calculated and are shown in Table 2 with the experimental conditions and results. Table 2 has been arranged so that the final volume of the nitrogen is increasing from top to bottom. At a final nitrogen volume of 0.0002 cm^3, the actual temperature of the nitrogen in the chamber is much lower than the calculated adiabatic temperature due to significant heat losses. Thus the temperature required to accelerate hydrazine decomposition rapidly is not reached, and the resulting pressure agrees with the peak hydrodynamic surge pressure of water. As the final nitrogen volume increases, heat losses become less significant and more heat is transferred to the hydrazine, which results in hydrazine decomposition and higher pressures than the peak hydrodynamic surge pressures. When the final nitrogen volume is between 0.22 and 0.82 cm^3, the critical temperature required to initiate significant hydrazine decomposition is reached and the

Table 2 Initial conditions, calculated conditions, and results

Chamber length, cm	Nitrogen pressure, kPa	Initial volume, cm^3	Calculated final volume, cm^3	Calculated adiabatic temperature, K	Hydrodynamic surge pressure water, MPa	Resulting pressure hydrazine, MPa
38	0.005	36	0.0002	39,500	131	131
38	7	36	0.04	4,478	90[a]	203[a]
38	14	36	0.07	3,742	96[a]	182[a]
38	21	36	0.10	3,230	86	141[a]
13	85	12	0.11	1,970	62	230[a]
38	28	36	0.12	2,903	79[a]	196[a]
38	41	36	0.17	2,575	76	210[a]
38	55	36	0.22	2,303	69	169[a]
25	85	24	0.22	1,970	62	[b]
38	69	36	0.26	2,159	69	[b]
38	85	36	0.30	2,034	69	[b]
38	99	36	0.36	1,889	62	[b]
38	113	36	0.43	1,757	55	[b]
51	85	48	0.47	1,903	55	[b]
38	127	36	0.54	1,605	45	[b]
64	85	60	0.65	1,830	48	[b]
38	134	36	0.79	1,380	28	179[a]
76	85	72	0.82	1,797	45	[b]
38	141	36	0.92	1,302	24	48
89	85	84	1.24	1,616	31	96[a]
102	85	96	1.53	1,569	28	38

[a]Chamber deformed, no rupture. [b]Chamber ruptured, pressure increase was greater than 276 MPa.

chamber fails. As the final volume increases, the peak hydrodynamic surge pressure and, thus, the calculated adiabatic temperature decrease. As a result, hydrazine decomposition is not significant, and the resulting pressure differs only slightly from the peak hydrodynamic surge pressure obtained with water.

The chambers used in the nitrogen volume and nitrogen pressure experiments showed damage similar to that shown in Figs. 4c and 4d. In one case, however, a 72-cm^3 chamber experienced extreme damage (Fig. 5). The resulting pressure was not measured, but was large enough to rupture the bottom fitting. As in Figs. 4c and 4d, no evidence of flames or burning was observed.

To aid in determining the explosion mode, the last series of experiments was conducted using nitromethane, a fluid known to detonate in the liquid phase, for comparison and using a high-speed video camera and a Pyrex glass tube (0.64 cm in diameter) as the bottom chamber to record the hydrazine decomposition reaction. In Fig. 5, the fragmentation of the chamber suggests that a detonation could have been initiated when the liquid column impacted at the closed end of the bottom chamber, but it subsequently failed as the detonation propagated. Since the critical diameter for a hydrazine detonation was unknown, nitromethane was used in some experiments to determine if a detonation can, indeed, be initiated under rapid compression conditions. The results shown in Fig. 6 indicate that similar damage patterns were obtained for both hydrazine and nitromethane during rapid compression experiments. Recently, modified cardgap experiments have shown liquid nitromethane to propagate a detonation in a 1.3-cm-diam 304 stainless steel tube.[16] Subsequent experiments to detonate hydrazine liquid indicated that the critical diameter for liquid hydrazine was in excess of 10.2 cm.[14] Thus if a detonation was initiated in either liquid under rapid compression conditions, the detonation was a transient decaying wave that quickly failed; however, in the nitromethane case, the detonation should have persisted. Additional data are required to resolve conclusively if detonation is indeed initiated by the impact.

Some researchers[3,4] have indicated that the damage observed during U-tube experiments may have resulted from the initiation of a hydrazine vapor detonation, which appeared to fail subsequently. Pure hydrazine vapor is detonated easily at pressures as

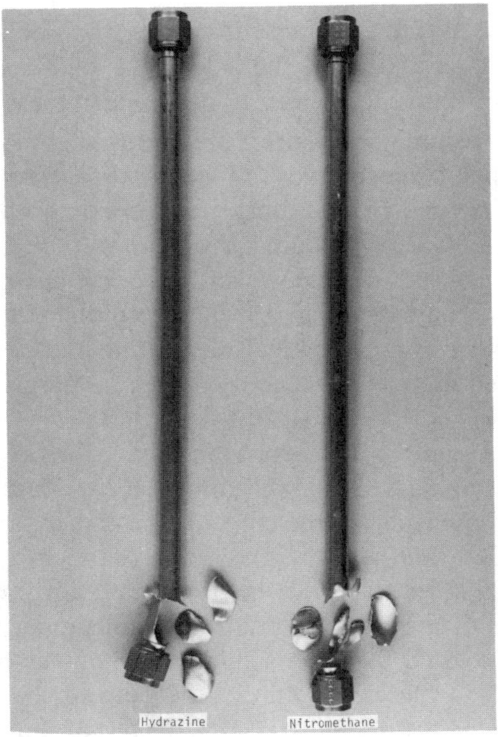

Fig 6 Comparison of chamber damage from hydrazine and nitromethane

low as 0.5 kPa;[12] however, the detonability of pure nitromethane vapor was unknown. To determine the possibility of a vapor detonation producing the test-section ruptures observed in both hydrazine and nitromethane experiments, detonation studies using pure nitromethane vapor were conducted.[17] Preliminary experimental results indicate that pure nitromethane vapor is not detonable in pressures up to 50.7 kPa. From these results, a vapor detonation appears to be unlikely; however, additional studies should be performed.

High-speed video data were taken during experiments in which hydrazine or water was accelerated by a 3.5-MPa driving pressure into a 38-cm-long, 0.64-cm-diam Pyrex glass chamber pressurized with nitrogen at 85.5 kPa. A 6000-frames/s video camera was used to record the events. The peak hydrodynamic surge pressure was not measured in these experiments.

Photographs of selected frames from the high-speed video data are shown in Fig. 7. These photographs display the lower

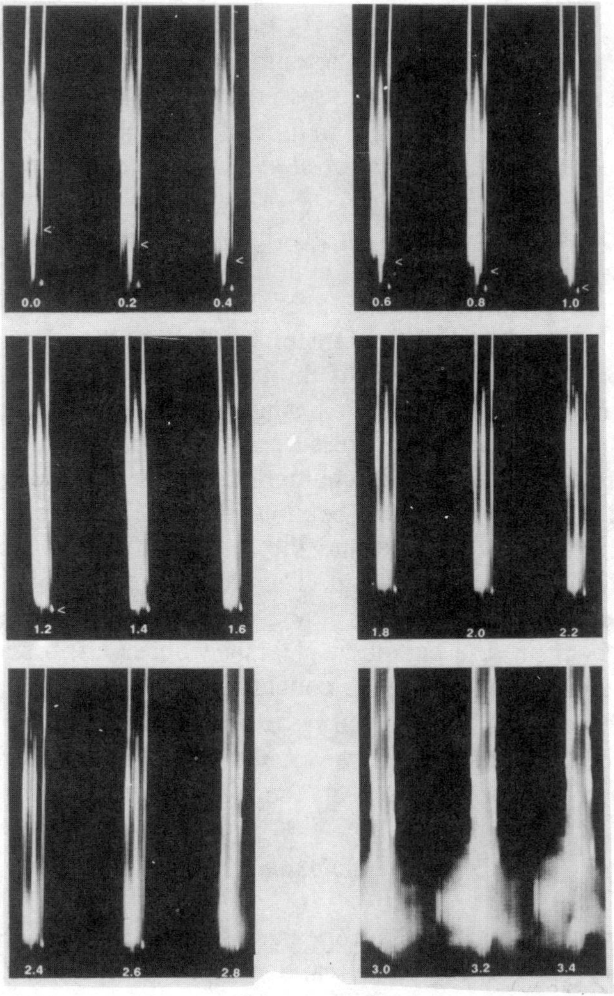

Fig 7 High-speed video during hydrazine compression, seen through glass lower chamber

section of the bottom chamber, which failed during the hydrazine experiments. The time sequence (which begins at zero) was set arbitrarily at a point when liquid hydrazine had already entered the bottom chamber. The elapsed time between frames was 0.2 ms.

The location of the leading edge of the hydrazine in the chamber is shown by the pointers and is observed to be cloudy. This region of cloudiness is believed to be a froth created by Taylor instabilities at the liquid-gas interface resulting in Helmholtz mixing of gas and liquid.[18] As the liquid hydrazine column (which is transparent in these photographs) was accelerated down the chamber, the froth was compressed at the bottom fitting. The size of the froth reached a minimum at approximately 2.4 ms, and the glass chamber ruptured somewhere between 2.6 and 2.8 ms.

The froth creates a medium in which gas bubbles are surrounded by a thin film of liquid hydrazine, maximizing the surface area between the nitrogen gas and the liquid hydrazine. When the froth is compressed, heat transfer to the liquid hydrazine is large compared with heat loss to the system. Hydrazine in the froth is heated instantaneously, causing rapid thermal runaway reactions that generate large pressures.

Local rupture of the glass chamber is evident in these photographs and is believed to have been caused by rapid decomposition of hydrazine in the froth and the local confinement provided by the initial inertia of the liquid. The experiments also show that the chamber yielded before the liquid column of hydrazine could be pushed back up the chamber. Damage observed in stainless steel test sections support these conclusions.

A froth was also formed during experiments conducted with water. The glass chamber, however, did not fail, and the water was observed to rebound back up the glass chamber after impact.

Conclusions

The experimental results appear to indicate that explosive events observed in these studies and possibly in dynamic hydrazine systems are thermal explosions initiated by a thermal runaway reaction.

Experiments performed using nitromethane as the test liquid indicate, although not conclusively, that a liquid- or vapor-phase detonation did not occur in these experiments. High-speed video data show the formation of a froth at the nitrogen gas-liquid hydrazine interface. The froth, which is thought to be nitrogen bubbles surrounded by thin layers of liquid hydrazine, provided a means for efficiently transferring heat generated in the nitrogen bubbles to the thin layers of liquid hydrazine. When the froth was compressed by the accelerating column of liquid hydrazine, the nitrogen bubbles were heated adiabatically, the generated heat was transferred to the thin liquid hydrazine layers, and a thermal runaway reaction was initiated in these hydrazine layers.

Experimentally, the conditions required to initiate the explosive decomposition of hydrazine are set up by the nitrogen volume, nitrogen pressure, and the peak hydrodynamic surge pressure. From the high-speed video and calculations, a froth must be formed and a critical temperature must be reached before the explosive decomposition of hydrazine can be initiated. An indication that the critical temperature was reached is obtained by evaluating the final nitrogen volume.

Acknowledgments

The authors would like to acknowledge F. J. Benz for his technical support, and M. T. Armendariz and D. Fernandez for their support in conducting experiments that generated data used in this paper. Further acknowledgment is made to Col. Ben Lucas, Air Force Space Division, and Walter Karakulko, Johnson Space Center Propulsion and Power Division, for their funding and technical support.

REFERENCES

[1]Vander Wall, E. M., Suder, J. K., Beegle, R. L. Jr., and Cabeal, J. A., "Propellant-Material Compatibility Study," Aerojet Liquid Rocket, Sacramento, CA, AFRPL-TR-71-41, Dec. 1971.

[2]Briles, O. M. and Hollenbaugh, R. P., "Adiabatic Compression Testing of Hydrazine," presented at the Fourteenth Joint Propulsion Conference of the American Institute of Aeronautics and Astronautics and the Society of Automotive Engineers, Las Vegas, NV, 1978.

[3]Benz, F. J., Long, T. L., and Weary, D. P., "Explosive Hydrazine Decomposition Due to Rapid Gas Compression (Adiabatic Heating)," JANNAF Safety and Environmental Protection Subcommittee Meeting, Johns Hopkins Univ., Laurel, MD, May 1984, p. 21.

[4]Briles, O. M., Hagemann, D., Benz, F. J., and Farkas, T., "Explosive Decomposition of Hydrazine Due to Rapid Gas Compression," 1985 JANNAF Propulsion Meeting, edited by K. L. Strange, Johns Hopkins Univ., Laurel, MD, April 1985, pp. 519-525.

[5]Benz, F. J., "Thermal Hazard Determination of Various Materials with Hydrazine at Elevated Temperatures," JANNAF Safety Environmental

Protection Subcommittee Meeting, Johns Hopkins Univ., Laurel, MD, 1981, pp. 105-113.

[6]Wedlich, R., Davis, D. D., and Peters, T., "Evaluation of Thermal Hazards of Hydrazine Decomposition by Accelerating Rate Calorimetry," JANNAF Safety and Environmental Protections Subcommittee Meeting, Johns Hopkins Univ., Laurel, MD, 1988, pp. 33-41.

[7]Adams, G. K. and Stocks, G. W., "The Combustion of Hydrazine," Fourth Symposium (International) on Combustion, Williams and Wilkins, Baltimore, MD, pp. 239-248.

[8]Benz, F. J., Bishop, C. V., and Pedley, M. D., "Ignition and Thermal Hazards of Selected Aerospace Fluids," NASA White Sands Test Facility, Las Cruces, NM, Rept. RD-WSTF-0001, 1988.

[9]Jost, W., "Investigation of Gaseous Detonations and Shock Wave Experiments with Hydrazine," Aeronautical Research Laboratories, Office of Aerospace Research, Wright-Patterson Air Force Base, OH, Rept. ARL 62-330, April 1962.

[10]Jost, A., Michel, K. W., Troe, J., and Wagner, H. Gg., "Detonation and Shock-Tube Studies of Hydrazine and Nitrous Oxide," Aerospace Research Laboratories, Office of Aerospace Research, Wright-Patterson Air Force Base, OH, Rept. ARL 63-157, September 1963.

[11]Heinrich, H. J., "On the Propagation of Detonations in Hydrazine Vapor," Zeitschrift für Physikalische Chemie, Vol. 42, (1974), translated by Eckart W. Schmidt, Rocket Research Company, Redmond, WA, September 1986, Rept. RRC-86-TT-28.

[12]Pedley, M. D. et al., "Hydrazine Vapor Detonations," Paper presented at Eleventh International Colloquium on Dynamics of Explosions and Reactive Systems, Warsaw, Poland, Aug. 1987, pp. 45-63.

[13]Scott, F. E., Burns, J. J., and Lewis, B., "Explosive Properties of Hydrazine," Bureau of Mines, Pittsburgh, PA, Rept. BM-RI-4460, 1949.

[14]Beeson, H. D., Bunker, R. L., and Plaster, M., "Investigation of Detonations in Liquid Hydrazine," Paper presented at JANNAF Safety and Environmental Protection Subcommittee Meeting, Johns Hopkins Univ., Laurel, MD, 1988, pp. 49-55.

[15]Benz, F. J. and Pippen, D. L., "Autoignition, Flammability and Explosion Properties of Hydrazine and Monomethylhydrazine," JANNAF Safety and Environmental Protection Specialist Session, edited by J. A. E. Hannum, Johns Hopkins Univ., Laurel, MD, 1980,
pp. 603-616.

[16]Unpublished, NASA White Sands Test Facility, 1989.

[17]Unpublished, NASA White Sands Test Facility, 1989.

[18]Comer, R. H., "Ignition and Combustion of Liquid Monopropellants at High Pressures," Sixteenth Symposium (International) on Combustion, The Combustion Inst., Pittsburgh, PA, 1977, pp. 1211-1219.

Detonation Structure of Corn Starch Particles-Oxygen Mixtures

F. Zhang* and H. Groenig†
Shock Wave Laboratory, RWTH Aachen, Germany

Abstract

The detonation structure of corn starch particles suspended in oxygen was investigated in a horizontal tube having an inside diameter of 141 mm. The stable self-sustained detonation exhibits spinning behavior in a wide particle concentration range. The combined results permit the deduction of a fairly detailed description of the stable regime of the single spin detonation which possesses two constant parameters at the wavefront, i.e., propagation velocity D and angular velocity Ω.

Introduction

It is well known that the transverse wave plays a dominant role in propagation of gaseous detonations. The interest is whether such a transverse wave exists also in a two-phase system composed of solid combustible particles suspended in an oxidizing gas, when a stable self-sustained detonation is achieved.

Previous experimental studies provided definitive proof that such a heterogeneous system is able to display a stable self-sustained detonation.[1,2] The present investigation is devoted to the structure problem. Spin detonations have been observed in a corn starch particles-oxygen mixture in a wide particle concentration range (equivalence ratio $\varphi = 0.27 - 8.11$). The detailed structure near the tube wall is strikingly similar to that in a homogeneous gaseous mixture for a single spin detonation.[3-6] In the present study, however, it appears not only in the marginal regions, i.e., near the composition limits or ini-

*Assistant Professor, on leave from East China Institute of Technology, Nanjing, China.

†Professor, Department of Mechanical Engineering.

tial pressure limit. Study of the interior structure shows that spin structure exists also in the inner region and that the angular velocity Ω at the front is constant. Thus, for a stable single spin detonation there are two constant parameters at the wave front, that is, axial propagation velocity D and rotation velocity Ω.

Experimental Setup

Experiments were conducted in a horizontal tube (Fig. 1) with a 141-mm inside diameter. It consists of a 4-m initiation section, a test section 17.4-m long and a 2.36-m^3 expansion tank. The test section contains 44 gauge stations and an optical section. The initiation pulse is produced by detonating a stoichiometric mixture of hydrogen and oxygen of 1 bar in the initiation section.

The dust can be fed onto the bottom of the test section by a small car, which is calibrated showing a deviation within 5 % of particle concentration per unit tube length. The dust dispersion in the cross section is achieved by an oxidant flow through the dispersion tubes of 12-mm outside diameter, which have a row of nozzles, according to Laval's pressure principle, opposite to the bottom of the test section. The whole dispersion process is detected by a camera in an auxiliary lucite

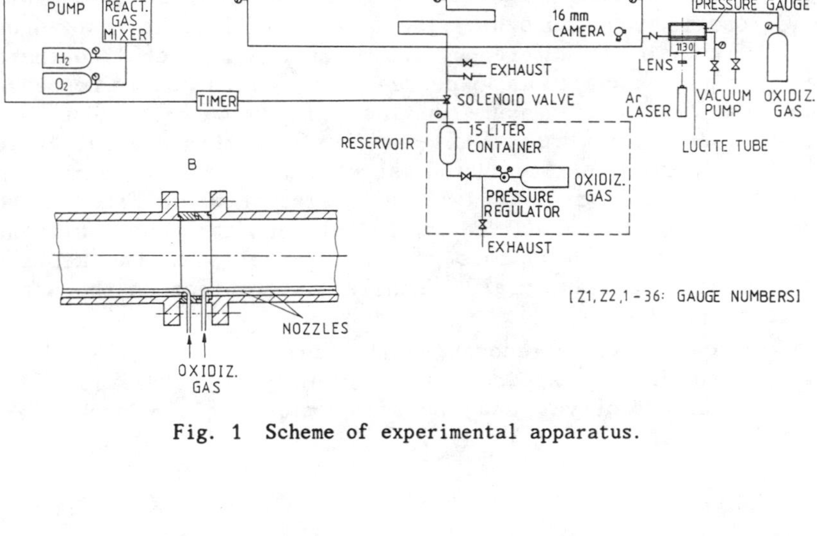

Fig. 1 Scheme of experimental apparatus.

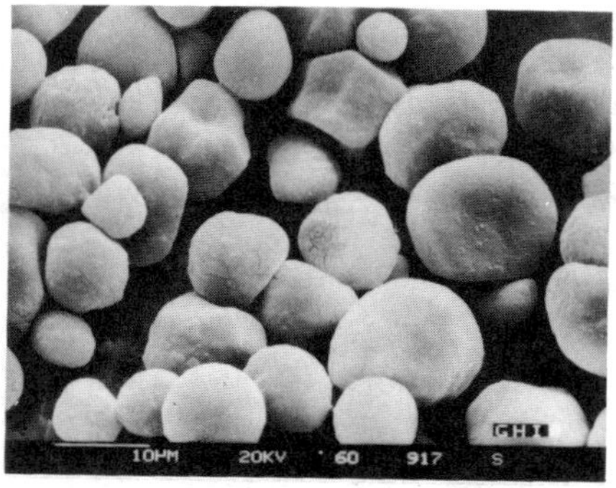

Fig. 2 Picture of corn starch particles obtained by electron microscope.

tube that has the same dispersion arrangement and thus the same constant pressure rate as in the test section. This method provides a fairly homogeneous particles-gas suspension both along the tube length and within the cross section.[1]

The measurements of the pressure and wave velocity are accomplished using Kistler 603 B pressure transducers. The flame is detected by fast pin silicon photodiodes and ion gauges. A streak camera is used to record the self-luminous parts of the wave pattern. The initial temperature is $T_0 = 293 \pm 2$ K.

A corn starch particles-oxygen mixture is used in the present study. A picture of the particles obtained by electron microscope is shown in Fig. 2. The particle size varies and is essentially under 15 μm with an average size of 10 μm.

Spin Structure near the Tube Wall

A ring mounted in the test section is used to investigate the detonation structure near the tube wall. It contains gauges arranged on two circumferences 166-mm apart. In each line there are eight gauges spaced at 45-degree intervals. A double response probe is developed to measure the locations of both the shock and the reaction front (Fig. 3). It consists of a thin piezoelectric film 2 mm in diameter and an ion gauge in a distance of 3 mm from the film. A record from this gauge is also shown in Fig. 3. The shock front location is given by the film and the reaction front by the ion gauge. Thus, this gauge, together with the Kistler 603 B pressure gauges, yields detailed information on the outer circle of the waves.

Typical pressure records on the outer circle are shown in Fig. 4 and display apparently a single spin detonation. The propagation velocity D and the spin track angle α are deter-

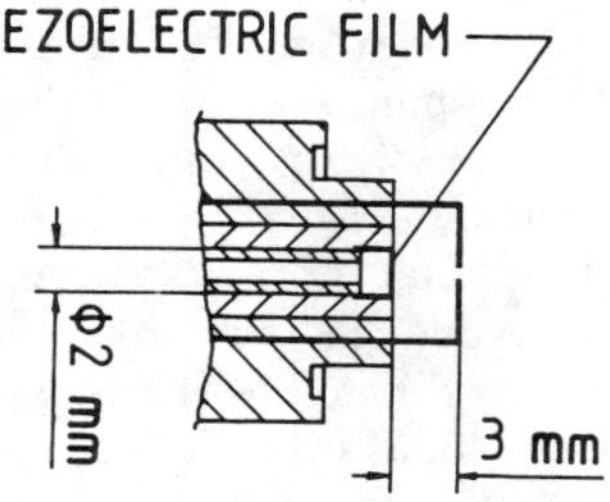

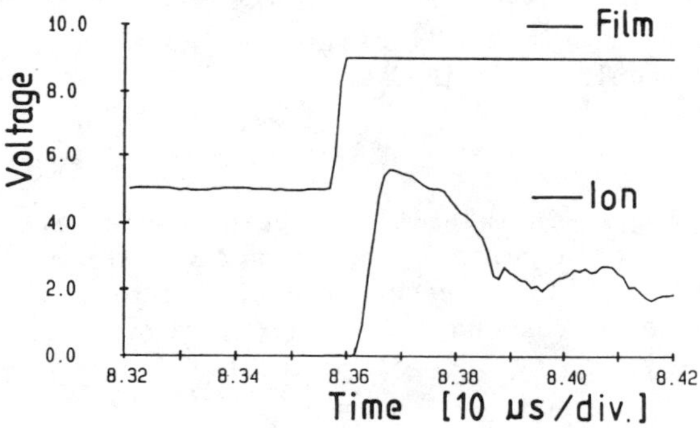

Fig. 3 Double response gauge and its signals: lower - arrival of reaction front; upper - arrival of shock front (signal is overloaded).

mined by the arriving times and the position of the spin head between the two circumferences. D is identical with the experimental detonation velocity. The data from each experiment consist of: D, α, the arrival times and the gauge positions. This set of data permit us to construct the instantaneous wave structure around the tube wall. For equivalence ratio $\varphi = 1$ and initial pressure $p_0 = 0.5$ bar, data from four such experiments have been combined (Fig. 5).

The basic framework of the wave structure in Fig. 5 consists of the primary shock front AEFGA, the transverse shock front BC, the reaction front AEHCB, and the tail wave CI. There are two triple points A and B. The front segment AE at the spin head A is oblique so that the normal velocity of the mass flow into it corresponds to an overdriven detonation with a value of 1.41 D, whereas the segment GA at A has a normal velocity of 0.85 D. Accordingly, the normal flow velocity into the shock front AEFGA varies monotonously from 1.41 D to 0.85 D. The pressure distribution along the shock front AEFGA is shown in Fig. 6 in which the triple point B is treated to be collapsed with A. The transverse wave BC and the segment AE near

A is accompanied by a reaction front that gave the strongest signal of the double response gauge, whereas the reaction behind the section HC near C is very weak. These phenomena can also be distinguished by streak photographs (Fig. 7). These differences of the reaction fronts are caused by the different amount of unburnt mixture influx. The rearward extension of transverse wave BC decays rapidly to a harmonic wave, which can easily be seen from the pressure records in Fig. 4.

The induction zone length behind the shock segment FGA is even longer than the axial distance between the fronts FGA and EHC, because of the angle of incidence of the mixture influx to the reaction front. Because of the two-phase flow effects, the induction zone length $x_i(P)$ behind the point P of the primary shock front should be defined as

$$x_i(P) = \int_0^{\tau_i} u_{px}\, dt \tag{1}$$

Here, the integral is taken along the particle path that begins at P^- and ends at the point where the particle temperature $T_p = T_i$, T_i denotes the ignition temperature of the particle. τ_i is the induction time corresponding to the point $T_p = T_i$. u_{px} is the axial component of the particle velocity.

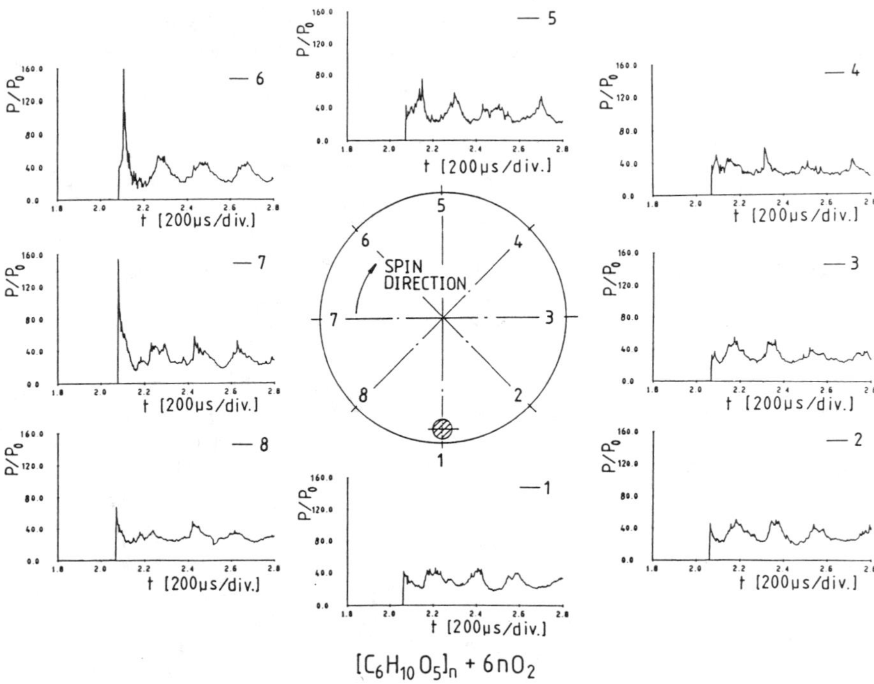

Fig. 4 Pressure profiles around the tube wall. $\varphi = 1$; $p_0 = 0.5$ bar.

For the foremost point F, the shock front is normal to the tube generatrix; thus, the normal flow velocity is equal to D. Accordingly, from Eq. (1) we obtain

$$x_i(F) = V_0 \int_0^{\tau_i} \frac{dt}{\tan\beta}\ , \quad \tan\beta = \frac{u_{py}}{u_{px}} \tag{2}$$

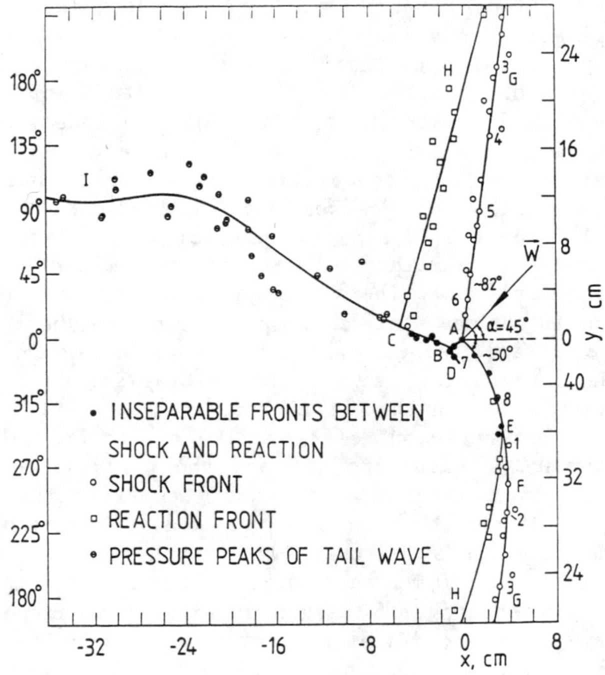

Fig. 5 Instantaneous wave structure around the tube wall. x - axial coordinate; y - circumferential coordinate. Points' label numbers correspond to signal numbers in Fig. 4. φ = 1; p_0 = 0.5 bar.

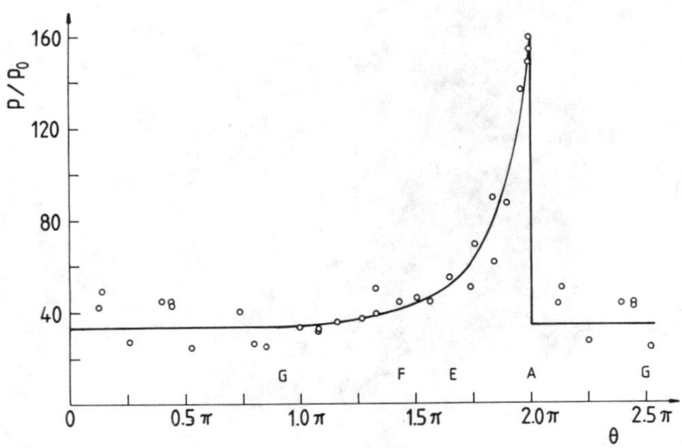

Fig. 6 Pressure values of shock front AEFGA in Fig. 5.

where V_0 is the transverse wave velocity and u_{py} the circumferential component of the particle velocity. In the absence of a detailed calculation only an estimate is made here. From other work[7] one may take $u_{px}(t=\tau_i) = u_{px}(t=0)/2$, so that a value $\tan\beta = 4/3$ is derived; and correspondingly from Fig. 5 one gets $x_i(F) \approx 20$ mm. Thus, a value between 18 mm $[u_{px}(t=\tau_i) = u_{px}(t=0)]$ and 22 mm $[u_{px}(t=\tau_i) = u_{px}(t=0)/7]$ is reasonable.

The single spin structure of the studied two-phase system is strikingly similar to that of marginal gaseous detonations. This supports the point of view that spin detonation appears to be the normal mode of propagation whenever an induction zone of sufficient length exists. In a marginal gaseous detonation, a weaker mixture near the composition limits or a lower initial pressure near the limit causes the induction zone length to increase. In the studied reactive particles-gas mixture, first, the momentum and temperature relaxations are dominant immediately behind the shock front. There the gas velocity and temperature rise at once to the new state corresponding to the shock strength, whereas the particle velocity and temperature increase continuously from the frozen state, i.e., the initial state. This causes the induction zone length to increase.

Second, heterogeneous effects like diffusion and slow reaction between the oxidizing gas and the particles also cause the induction zone length to increase. Comparing the induction zone length between the marginal gaseous detonation and the studied particles-gas mixture is instructive. Schott[3] has estimated $x_i(F) = 5-10$ mm in a mixture containing 5.0% acetylene, 7.5% oxygen and 87.5% argon at initial pressures near 18 Torr in a 4-in.-inner diameter tube. Thus, generally speaking, for a single spin detonation a value of the induction zone

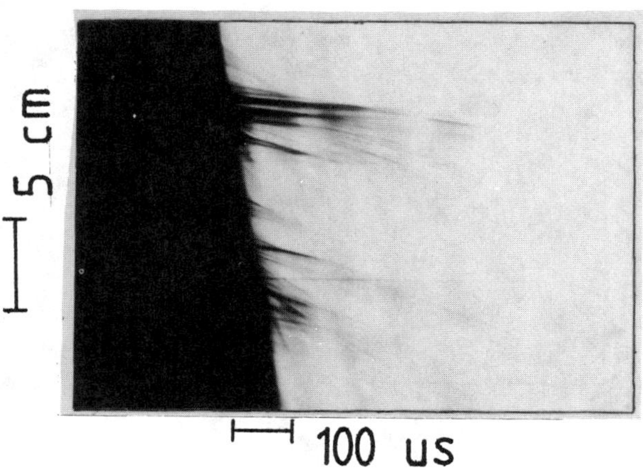

Fig. 7 Streak photograph. $\varphi = 1$; $p_0 = 0.5$ bar.

length between 5-25 mm is acceptable behind the foremost point F.

Because of both hydrodynamic relaxations and heterogeneous effects, the single spin detonation in the present study appears in a wide concentration range with normal initial pressures. Table 1 gives the average experimental states of the spin structure at the tube wall. Although the structure of the studied two-phase system is quite weak, the pressure in the stable combustion region is quite high.

Interior Spin Structure

The structure discussed so far is that near the tube wall. The problem of the interior structure of the spin detonation had been previously unsolved, even for the gaseous mixture. In this section an end-plate is used to try to reveal some details of the interior structure of the single spin detonation. In the end-plate 17 pressure gauges were mounted: eight on a circle of radius $r = 61.5$ mm, eight on another circle of radius $r = 30.5$ mm, and one at the center (see Fig. 8). Experiments were carried out again with equivalence ratio $\varphi = 1$ and initial pressure $p_0 = 0.5$ bar.

The pressure records from one experiment are shown in Fig. 8. Figure 9, derived from two experiments, gives the instantaneous structure of the primary shock front for $r = 0$, 30.5 mm, and 61.5 mm. These results show the following facts:

1) Spin structure also exists on the interior circles.
2) The pressure profiles from both the inner and the outer circle on a radial line differ with respect to their onset but are equal in phase. This means that the spin structure on both circles has the same constant angular velocity Ω.
3) Approaching the center, the pressure difference along the primary shock front and its curvature decrease, the spin head moves forward.
4) Approaching the center, the transverse wave and the tail wave become weaker. At the center the tail wave disappears.

From these facts we may draw the following deductions:

1) Spin structure exists in the whole interior region with the same angular velocity Ω. (Some values of Ω are shown in Table 1.) The spin head stays on one crease. The primary shock front, the transverse wave, and the tail wave get weaker towards the center.
2) At a circle with radius r the transverse wave velocity is

$$V = r\,\Omega \tag{3}$$

and the spin track angle is

$$\alpha = \tan^{-1} r\,\Omega / D \tag{4}$$

Table 1 Properties of the spin structure at the tube wall (T_o = 293± 2 K)

Φ	σ_p g/m^3	p_o bar	n	α	Ω 10^4 rad/s	D m/s	V_o m/s	$\frac{p_{AE}}{p_o}$	$\frac{p_{BD}}{p_o}$	$\frac{p_{AG}}{p_o}$	$\frac{p_{BC}}{p_o}$	$\frac{p_S}{p_o}$
0.64	357	0.5	1	45°	2.492	1757	1757	57	90	37	106	37
1.0	554	0.5	1	45°	2.555	1801	1801	78	143	39	150	47
0.27	300	1.0				1766			97			36
0.45	500	1.0	1	42°	2.376	1860	1675	83	118	37	108	48
1.0	1109	1.0	1			1945		77	145	48	147	55
1.80	2000	1.0				1988					154	65
2.71	3000	1.0	2			1965					214	67
3.61	4000	1.0	2			1961					239	66
4.51	5000	1.0	2			1911					283	65
6.31	7000	1.0	1			1846					238	51
8.11	9000	1.0	1			1795					206	43

Φ - equivalence ratio
σ_p - initial particle concentration
p_o - initial pressure
n - spin head number
α - spin track angle
Ω - spin angular velocity

D - propagation velocity
V_o - transverse wave velocity
p_{AE}, p_{BD}, p_{AG}, p_{BC} - pressures of wave fronts AE, BD, AG, BC near the point A
p_S - pressure peak of the stable combustion region

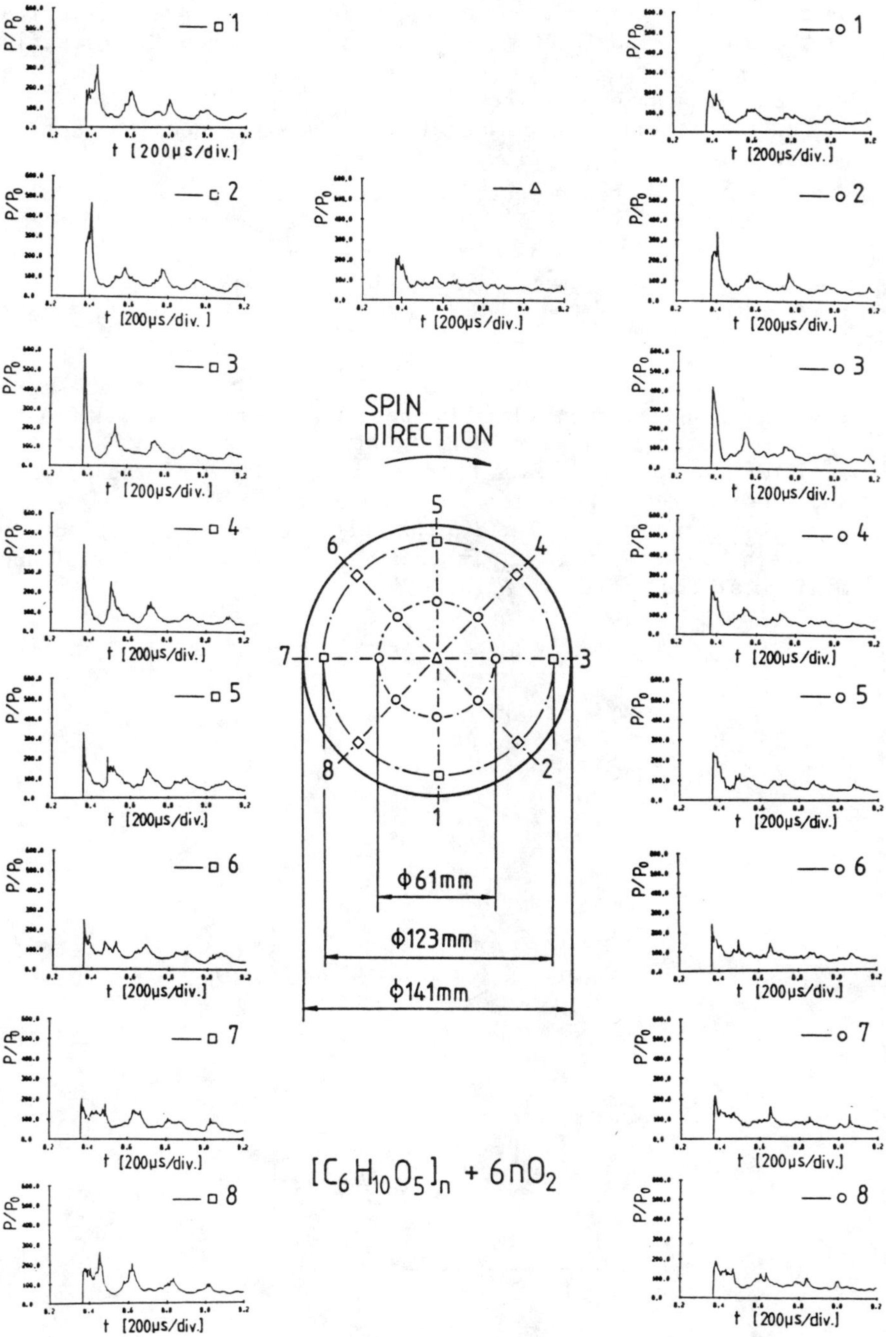

Fig. 8 Pressure profiles at the center and around inner circles. Δ : r = 0; ○ : r = 30.5 mm; □ : r = 61.5 mm. φ = 1; p_0 = 0.5 bar.

where D is the propagation velocity. Thus, the tangent function of the incident angle of mass flux, which equals α, is directly proportional to the radius r. The angle has a maximum at the tube wall and equals zero at the center.

3) For the stable spin detonation, the mixture density distribution of the incident flow in the cross section can be derived with the Navier-Stokes equation as follows:

$$\rho_{m0} = \left[\rho_{m0}^{\Gamma_0 - 1}(0) + \frac{\Gamma_0 - 1}{2A\Gamma_0}\,\Omega^2 r^2\right]^{\frac{1}{\Gamma_0 - 1}} \tag{5}$$

where $\rho_{mo} = \sigma_{p0} + \rho_0$, in which σ_{po} denotes the particle concentration and ρ_0 the gas density ahead of the primary shock front, $\rho_{mo}(0)$ denotes the mixture density at the center, and Γ_0 is the ratio of the specific heats for the mixture ahead of the primary shock front. Here the isentropic state equation $p_0 = A\rho_{mo}^{\Gamma 0}$ for the equilibrium mixture is used and the particle volume is neglected. Equation (5) shows that ρ_{mo} reaches maximum at the tube wall and minimum at the center.

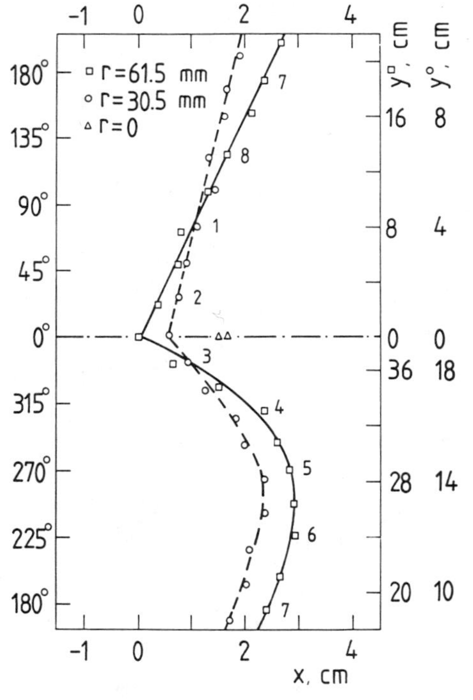

Fig. 9 Instantaneous structure of primary shock fronts for r = 0, 30.5 mm, and 61.5 mm. φ = 1; p_0 = 0.5 bar.

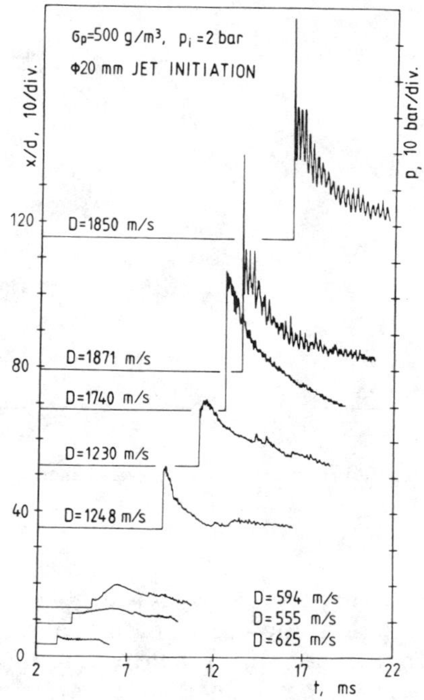

Fig. 10 Transition to detonation. 20 mm-diameter jet initiation with initiator pressure p_i = 2 bar. σ_p = 500 g/m^3; p_0 = 1 bar.

Formation of Stable Detonation

To expand the detonation developing process a jet initiation method is applied. The initiation pulse is still produced by detonating a stoichiometric mixture of hydrogen and oxygen, but by injecting it through a 20-mm hole into the test section. Figure 10 shows the detonation formation process for σ_p = 500 g/m^3 and p_0 = 1 bar. Figure 11 gives the development of the propagation velocity. These results show that a stable self-sustained detonation is able to exist in such a heterogeneous system. The sequence of the process can be distinguished as follows:

1) Initial particle ignition.
2) Self-accelerating reaction-compression process.
3) Nonsteady reaction shock.
4) Stable spin detonation.

Here it is significant that a nonsteady reaction shock exists - for example, at x/d = 67.87 in Fig. 10. Generally, the records of the ion gauge and the photodiode show during the development a squeezing of the induction zone from about 90 μs to about 20 μs behind the shock front. It is apparently self-accelerating.

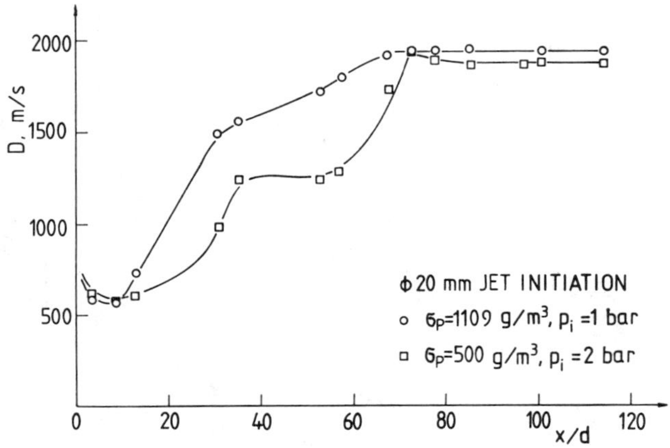

Fig. 11 Wave front velocities vs propagation distance. 20 mm-diameter jet initiation; p_0 = 1 bar.

The pressure records, for example, Fig. 4, show that there is a stable region of pulsating combustion behind the wave front, when the stable spin detonation is achieved. In all experiments it takes about $\tau_0 = 0.8-1.2$ ms. Then apparently the pressure decays. This stable regime length should depend on the balance of the chemical reaction, two-phase relaxations, and wall losses. Although the induction zone length is of the same order of magnitude as in the marginal gaseous detonations, the stable reaction zone length is quite long.

The stable spin detonation possesses two constant parameters, that is, propagation velocity D and angular velocity Ω. Therefore, two conditions are needed for them. One may be some modified Chapman-Jouget condition for the reactive particles-gas mixture.[7,8] The other may be obtained from the conservation law of angular momentum.

Conclusion

The results of the experiments demonstrate that the transverse wave is also the key feature of the structure of the stable detonation in the studied reactive particles-gas mixture. Stable single spin detonation exists also in the interior region and as a whole possesses two constant parameters at the wave front: propagation velocity D and rotation angular velocity Ω.

Acknowledgment

This work was supported by the Stiftung Volkswagenwerk, which is gratefully acknowledged. It is with great pleasure that we thank Professor P. Wolanski for his discussions. The authors are indebted to A. van de Ven and P. Greilich and the technicians in the workshop for their help during the experiments. Special thanks go to Christiane von Hoegen for her competent and timely typing of the manuscript.

References

[1]Zhang, F., and Grönig, H., "Investigation of Detonation in Corn Starch-Oxygen Suspensions," 3rd International Colloquium on Dust Explosion, Szczyrk, Poland, Oct. 1988.

[2]Zhang, F., and Grönig, H., "Detonation Study in Two-Phase Flow (Reactive Particles-Gas)," 17th International Symposium on Shock Waves and Shock Tubes, Bethlehem, PA, July 1989.

[3]Schott, G. L., "Observation of the Structure of Spinning Detonation," Physics of Fluids, Vol. 8, May 1965, pp. 850-865.

[4]Voitsekhovskii, B. V., Mitrofanov, V. V., and Topchian, M. E., "Structure of Detonation Fronts in Gases," Academy of Science, Sibirian Branch, Novosibirsk, USSR, 1963. [English translation: Wright-Patterson Air Force Base Rep. FTD-MT-64-527 (AD-633 821), 1966.]

[5]Soloukhin, R. I., Shock Waves and Detonations in Gases. Mono, Baltimore, MD, 1966.

[6]Fickett, W., and Davis, W. C., Detonation, Univ. of California Press, Berkeley, CA, 1979.

[7]Fan, B. C., and Sichel, M., "A Comprehensive Model for the Structure of Dust Detonations," 22nd International Symposium on Combustion, Seattle, WA, 1988.

[8]Lee, D., "The Structure of Detonations in Dust-Oxidizer Mixtures," doctoral dissertation, Univ. of Michigan, Ann Arbor, MI, 1985.

Influence of Dust Parameters on Detonation Velocity, Structure, and Limits

O. Dahab,
Minia University, Minia, Egypt
and
M. Kapuściński
Marine University, Szczecin, Poland
and
P. Wolański
Warsaw University of Technology, Warsaw, Poland

ABSTRACT

The influence of organic dust parameters on the initiation of detonation, detonation velocity, pressure, temperature, and limits is presented. Egyptian brown coal dust as well as wheat flour and oat dusts were used in the experiments.

Experiments were conducted in a gas oxidizer containing 95% oxygen inside a 4.5-m-long vertical tube with a diameter of 80 mm. Combustion was initiated by a relatively weak ignition source and the flame acceleration was monitored by the streak camera. Pressure variations were measured by piezoelectric transducers and the particle temperature by a three-wavelength pyrometer.

It was found that the transition to detonation was relatively fast for Egyptian brown coal dust, whereas not all experiments with wheat and oat dust mixtures even reached detonation. Pressure measurements show the evidences of transverse waves. Maximum detonation temperatures were higher than 3000 K and velocity exceeded 2100 m/s. When compared with the numerical calculations, the measured detonation parameters show qualitative agreement between the theoretical and experimental results.

INTRODUCTION

Interest in the detonation of dust mixtures has been growing since Cybulski first produced evidence that detonation transition could be obtained even for a coal-dust/air mixtures.[1] Bartkneht proved that for some dust/air mixtures, detonation propagating with the velocity of about 2 km/s could be achieved.[2] Later, Gardner et al. studied in detail the deflagration-to-detonation transition in coal-dust/air mixtures.[3] The possibility of detonative combustion in laboratory equipment for a dust mixture with air and oxygen was demonstrated by Kauffman et al.[4] for different grain dusts.

The modeling research of detonation combustion of dust in an oxygen atmosphere was initially carried out by Nettleton and Stirling[5] and by Wójcicki and Zalesiński.[6] More recently, Wolański et al.[7] calculated the structure of dust detonation. Lee and Sichel[8] and Fan and Sichel[9] later included more factors in those calculations.

Detailed structure in dust detonation was measured first by Kauffman et al.[4,11] and by Fangrat et al.;[12] but Zhang and Groening gave the conclusive evidence that spin structure existed in the detonation of a cornstarch/oxygen mixture.[13]

The present work studies the detonation transitions, limits, and structure in organic dust/oxygen mixtures, as well as the influence on that process of dust parameters and properties.

EXPERIMENTS

Experiments were conducted in a 4.5-m-long, 80-mm dia., vertical detonation tube. Tube construction and operation is described in Reference 11. During the current research, a small auxiliary detonation tube (1-m long with 20-mm dia.) was sometimes used to initiate detonation transition; but for most of the experiments, an electrically activated chemical match was applied. Tube construction allowed for observation of the process of flame acceleration, and continuous direct-streak photographs of that process were taken. At 0.3 m from the tube end, a special observation window for the three-wavelength pyrometer was installed. A piezoelectric transducer was mounted at the same location.

Three different kinds of dust were used in all experiments. Most experiments were carried out for five different fractions of the Egyption brown coal dust, ranging from particles less than 75 μm up to 200 μm. Egyptian brown coal dust is a unique dust. It has a very high lower-heating value of 31 MJ/kg, a high volatile content up to 55%, and a very low amount of ash (less than 4%). For that reason it creates very explosive mixtures. Other dusts used in the experiments were a wheat flour of less than 88 μm dia. and two different fractions of oat dust.

The following procedure was introduced to all experiments. The lower end of the main detonation tube was closed by an aluminum diaphragm and appropriate measures of dust were put into containers. The air was then evacuated from the detonation tube and the dispersion installation. The dispersion containers were filled with oxygen up to the absolute pressure of 0.65 MPa, which enabled a return to a normal atmosphere in the detonation tube just after dust dispersion. Dispersion of the dust was controlled by electromagnetic valves operated automatically by the sequencer. Approximately 1.2 s after dispersion began, the electromagnetic valves were closed and com-

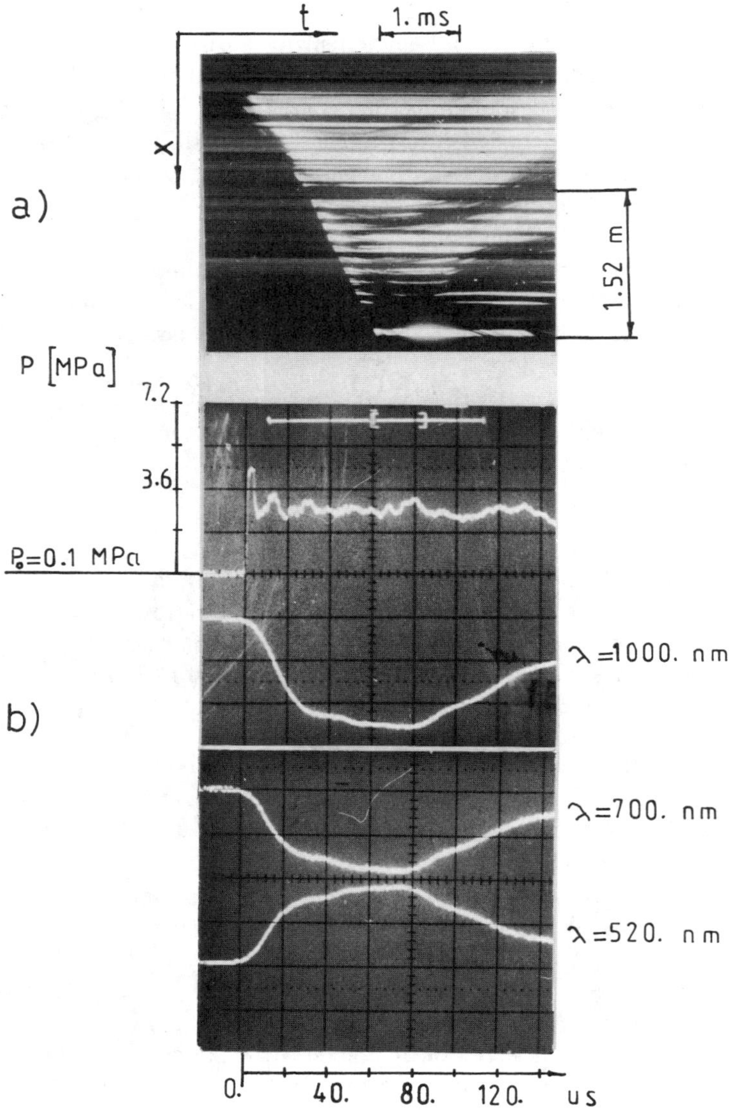

Fig. 1 Detonation of Egyptian brown coal dust, $150 < d < 200\ \mu m$ diameter: $c = 1\ kg/m^3$, $D = 1550\ m/s$, $T_{max} = 2670$ K.

(a) Direct streak photography of transition from deflagration to detonation.

(b) Oscilloscope records of the variations of pressure and light emission in three different wavelengths.

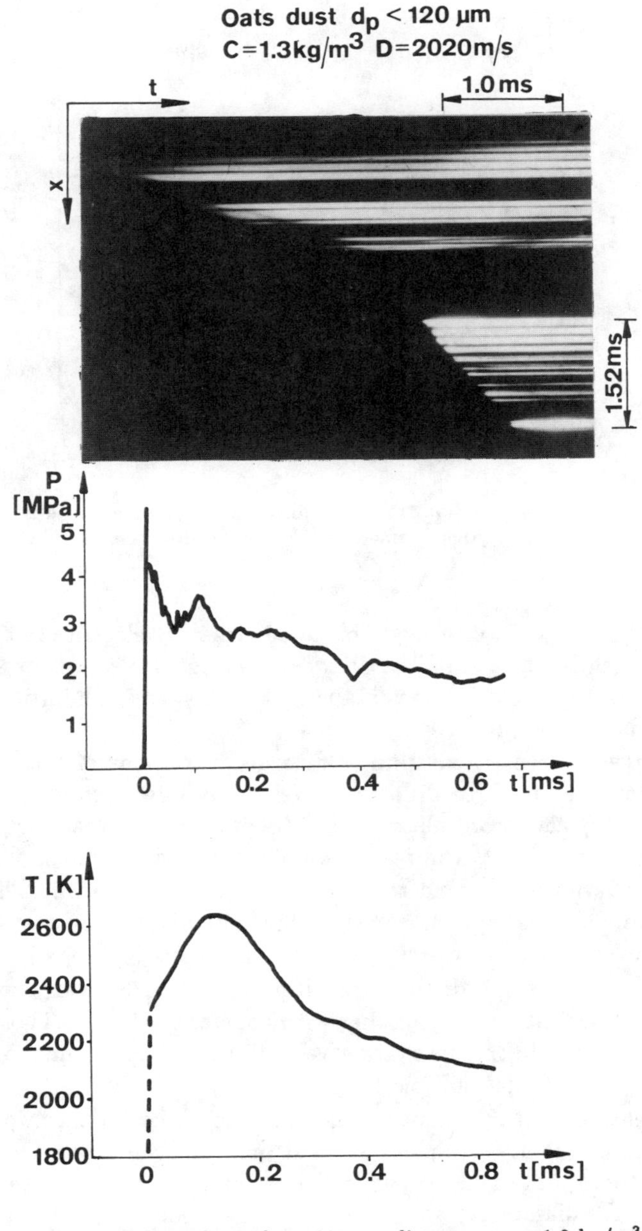

Fig. 2 Detonation of oat dust, $d < 120\ \mu m$, diameter: $c = 1.3$ kg/m^3, $D = 2020$ m/s.

a. Direct streak photography of transition to detonation.

b. Variation of pressure with time.

c. Variation of temperature with time.

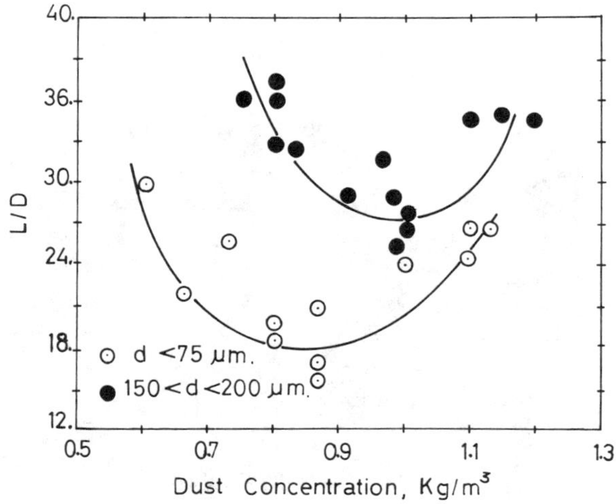

Fig. 3 Dimensionless detonation transition distance as a function of dust concentration for two extreme particle diameters of Egyptian brown coal dust.

bustion of the created mixture was initiated either by electrically controlled chemical match or by a small auxiliary detonation tube. All experiments were carried out in the oxidizer containing 95% oxygen at an initial pressure equal to atmospheric pressure.

For all tests, direct-streak photos recorded variations of pressures, and light emissions in the three different wavelengths were recorded on the oscilloscope. A typical record obtained for Egyptian brown coal dust is shown in Fig. 1. Fig. 2 presents the results obtained for oat dust. In this figure, temperature variations behind the detonation front are shown. These temperatures were calculated from the ratio of emitted light signals at the three different wavelengths.

In these two figures, a distinct transition to detonation can be seen, followed by the detonation propagating with constant velocity. Those photos were used to calculate the detonation velocity as well as the distance necessary for transition to detonation.

The variations of the transition distance, measured for the two extreme particle sizes of the Egyptian brown coal dust, are shown in Fig. 3. The fastest transition, obtained for particles less than 74 μm, occurred at a distance of 15 tube diameters; whereas for the biggest dust particles, transition occurred almost at the end of the tube. All experiments carried out with Egyptian brown coal dust were initiated using weak ignition by chemical match.

The measured detonation velocity, maximum pressures in the detonation front, and the maximum particle temperatures as a function of dust concentration and for different particle sizes are presented in Figs. 4 through 6.

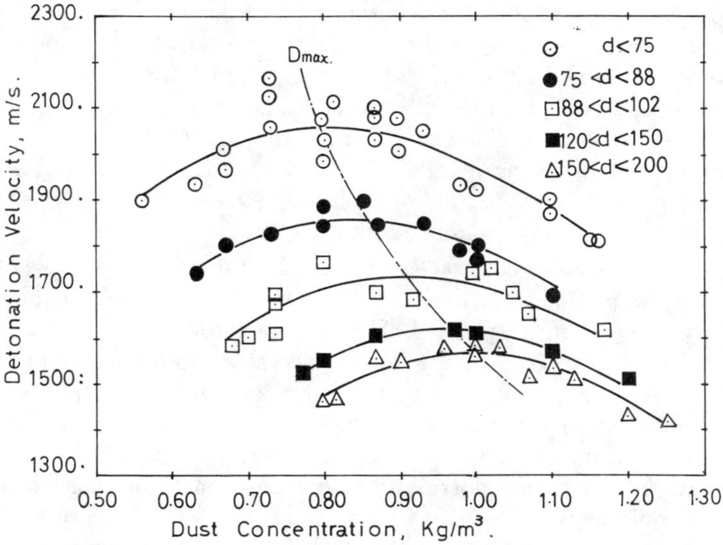

Fig. 4 Detonation velocity as a function of dust concentration for different particle diameters of Egyptian brown coal dust.

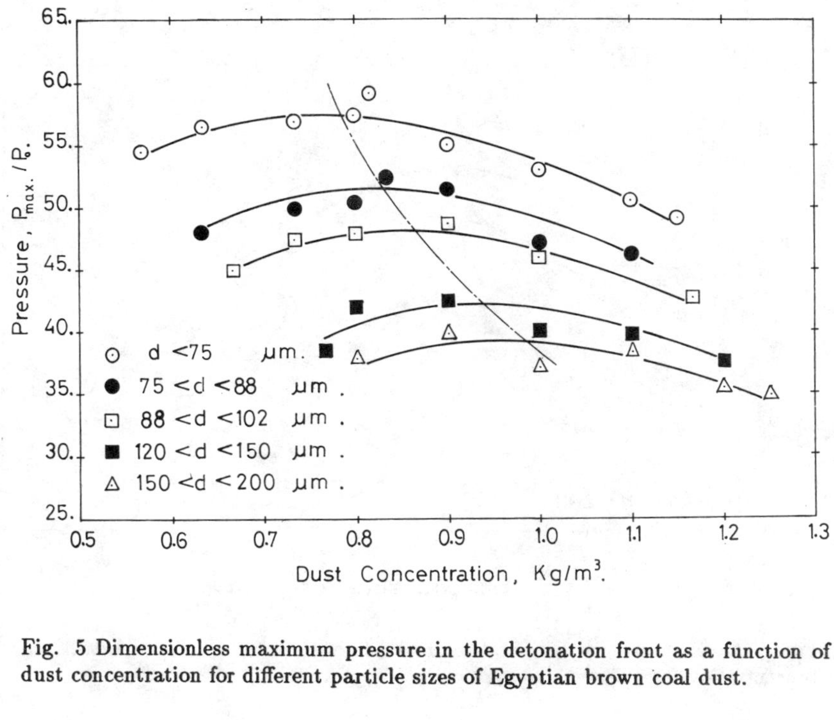

Fig. 5 Dimensionless maximum pressure in the detonation front as a function of dust concentration for different particle sizes of Egyptian brown coal dust.

The highest values of those parameters were obtained for the smallest dust particles. It is also evident that the limiting concentrations supporting detonative combustion in this tube are changing. Similarly, as the limits are changing, the so-called optimum concentration for which maximum detonation parameters are obtained is changing, too, with the change of particle size. Optimum dust concentration for particles less than 74 μm is equal to about 0.8 kg/m^3 and rises to about 1 kg/m^3 for dust particles in the range of 150-μm to 200-μm diameters.

Initiation of detonation was not as easy for the oat dust and wheat flour as for the Egyptian brown coal dust. Detonation of oat dust could be initiated by the electrically ignited chemical matches (Fig. 7), whereas using the small auxiliary detonation tube was necessary to initiate detonation in the wheat flour dust (Fig. 8). The solid line represents transition to detonation in the detonation tube; the dashed line represents observed continuous flame acceleration. For the last cases the velocity at the end of the tube is indicated. It is a strong possibility that transition to detonation in those mixtures could be also obtained with the longer tube. Generally, detonation limits for oat and wheat dust in the tested conditions are much narrower; but the detonation velocity, pressure, and temperature are of the same order as for Egyptian brown coal dust.

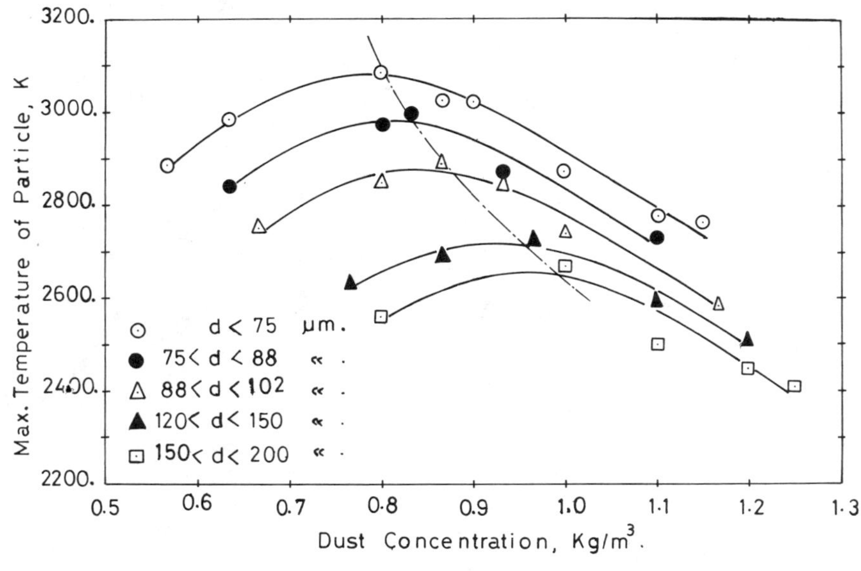

Fig. 6 Maximum particle temperature in the detonation front as a function of dust concentration for different particle sizes of Egyptian brown coal dust.

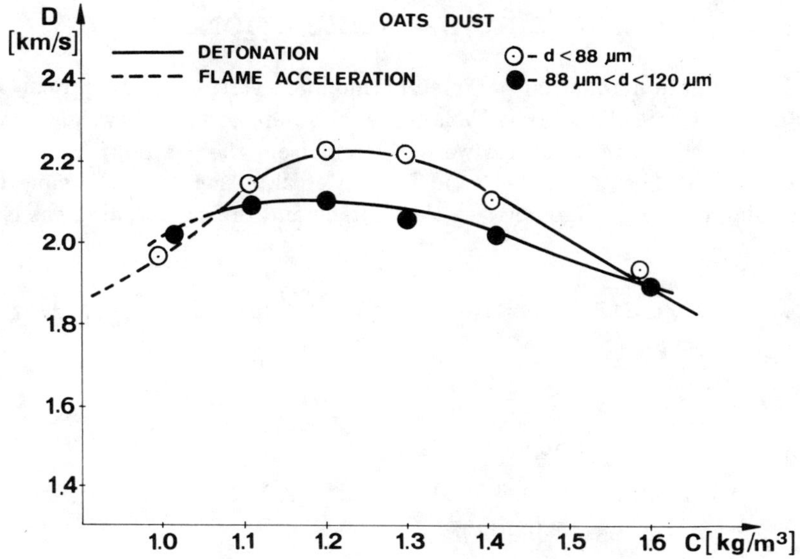

Fig. 7 Detonation velocity for two different particle sizes of oat dust as a function of dust concentration.

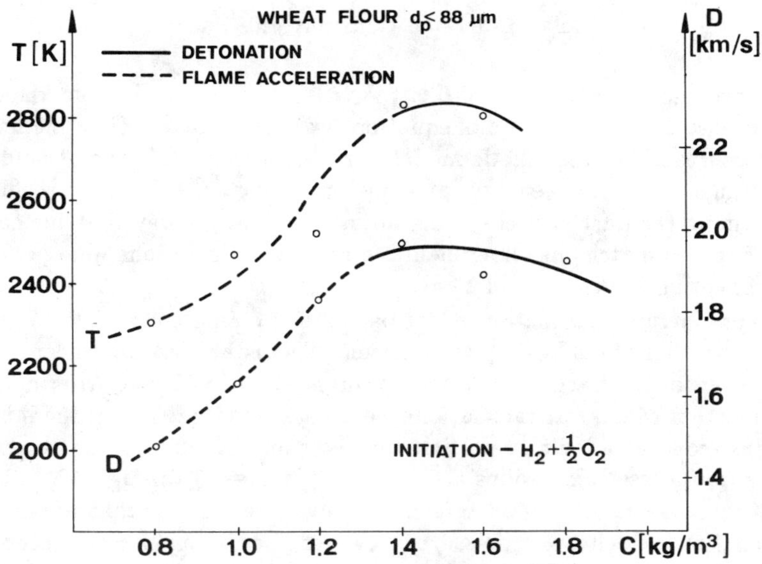

Fig. 8 Detonation velocity and maximum particle temperature in the detonation front as a function of wheat flour concentration.

NUMERICAL CALCULATIONS

To calculate detonation parameters, modified versions of numerical calculations performed earlier by Wolański et al.[7] were applied. It was assumed that the particle temperatures were different from the gas temperature, as introduced by Lee and Sichel[8] and by Fan et al.;[9] therefore, the equation describing heat exchange between the particle and the surrounding gas is as follows:

$$\left(\frac{\pi}{6} \cdot \rho_p \cdot d_p^3\right) \cdot C \cdot \frac{dT_p}{dx} = h' \cdot \pi \cdot d_p^2 (T - T_p) + \frac{\Delta Q}{n} + \sigma' \cdot \varepsilon \cdot \pi \cdot d_p^2 \cdot (T_w^4 - T_p^4)$$

where

C = specific heat of particle
d_p = particle diameter
T = gas temperature
T_p = particle temperature
T_w = Temperature of wall surface
Q = heat release rate per unit volume of mixture
ε = emissivity
ρ_p = particle density
σ' = Stefan-Boltzman Constant

The process of combustion of dust particles was assumed to be kinetically controlled and the whole set of equations was solved by the Gear method.[8,10] The constants for the kinetic model were taken from Reference 7, and some constants—activation energy, preexponential factor, and other coefficients describing the dust particle—were adjusted in such a way that the calculations should match one experimental point. All calculations were performed for Egyptian brown coal dust only.

For matching calculations, dust particles of diameter equal to 45 μm and concentration of 0.8 kg/m^3 were chosen. This is because the mean particle diameter for the fraction less than 75 μm is equal to 45 μm. After matching combustion constants, calculations were performed describing the influence of dust concentration and particle diameter on detonation parameters. Results from these calculations are presented in Figs. 9 through 12. It is seen that there is relatively good agreement between experimental and calculated detonation, velocity, and pressure. However, for the smaller dust concentration where the equivalence ratio equals one, there is a significant deviation of calculations from the experimental results. One possible reason for this is that not all the particles dispersed in the detonation tube will burn, or some will burn only partially. This will affect dust that sticks to the tube walls during dispersion and forms a very rich mixture in the boundary layer region behind the detonation wave.

Figure 9 presents the results of the analytical solution of the C-J detonation obtained for the ideal case without losses. Detonation velocity obtained from such models is much higher than that obtained experimentally and from numerical solutions (which include losses). This is also true for other parameters, such as pressure and temperature.

DISCUSSION AND CONCLUSIONS

In this work very comprehensive research of the detonation of Egyptian brown coal-dust/oxygen mixture was conducted. For the first time, detailed studies of the influence of particle diameter on the detonation of dust/oxygen mixtures are presented. Additional results are given for detonation in wheat flour and oat dusts with oxygen. The experiments show that for very reactive substances, detonation combustion can be obtained even for relatively large particles of 150 to 200 μm. When the particle size increases, detonation limits and maximum detonation parameters shift to higher dust concentrations. This is partly because the specific surface of dust decreases with the increasing diameter of the particles; but it is not possible to reduce all limits to the single value of a particle's reaction surface. One also observes clearly that detonation velocity decreases very rapidly with the increasing particle

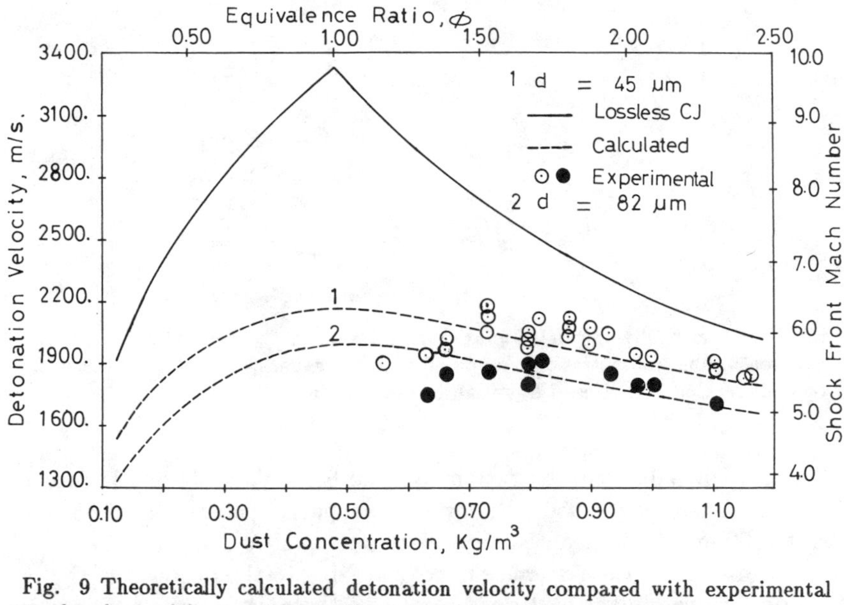

Fig. 9 Theoretically calculated detonation velocity compared with experimental results obtained for two different fractions of Egyptian brown coal dust.

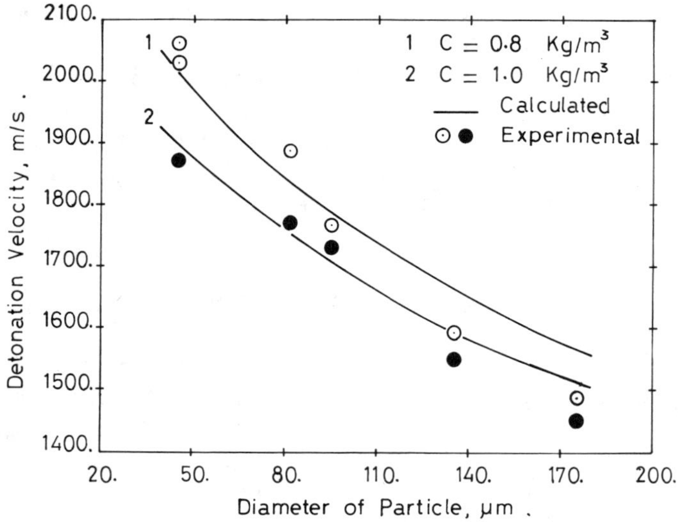

Fig. 10 Numerical calculation of particle-size influence on detonation velocity compared with the experimental data of Egyptian brown coal dust.

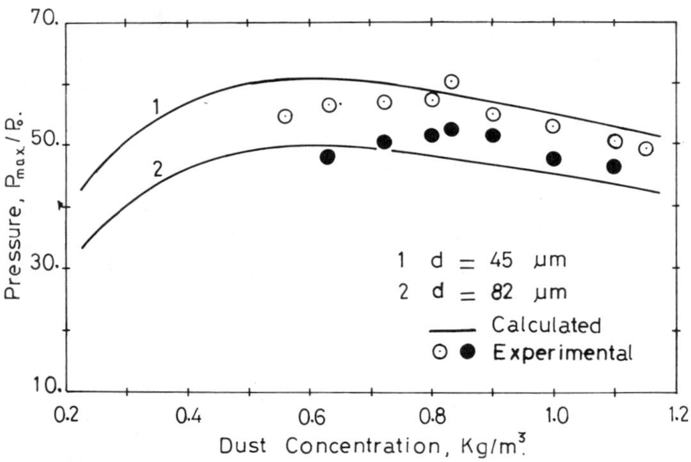

Fig. 11 Numerically calculated maximum pressure in the detonation front, as a function of dust concentration, compared with the experimental data obtained for two different particle sizes of Egyptian brown coal dust.

diameter. For a dust concentration equal to 0.8 kg/m^3, detonation velocity decreases from about 2050 m/s for a mean particle diameter of 45 μm to about 1450 m/s for a mean particle diameter equal to 175 μm. The 30% decrease in the detonation velocity is not only due to the decrease of the total reaction surface of dust particles but also to the bigger momentum losses.

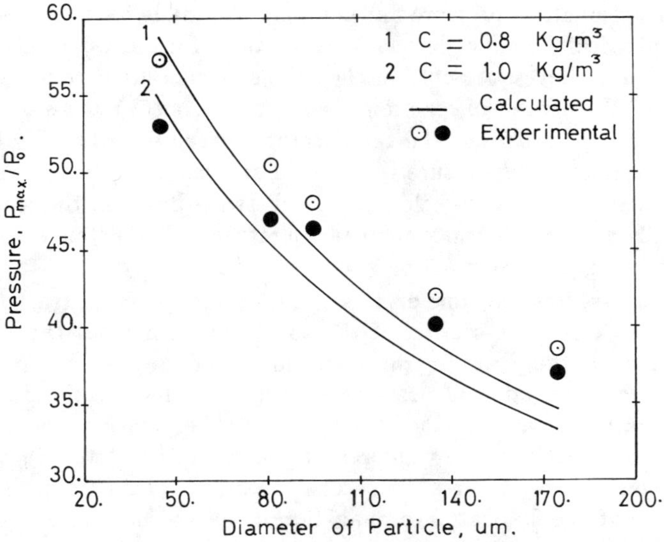

Fig. 12 Numerically calculated influence of particle size on maximum pressure in the detonation front, compared with experimental data obtained for two different dust concentrations of Egyptian brown coal dust.

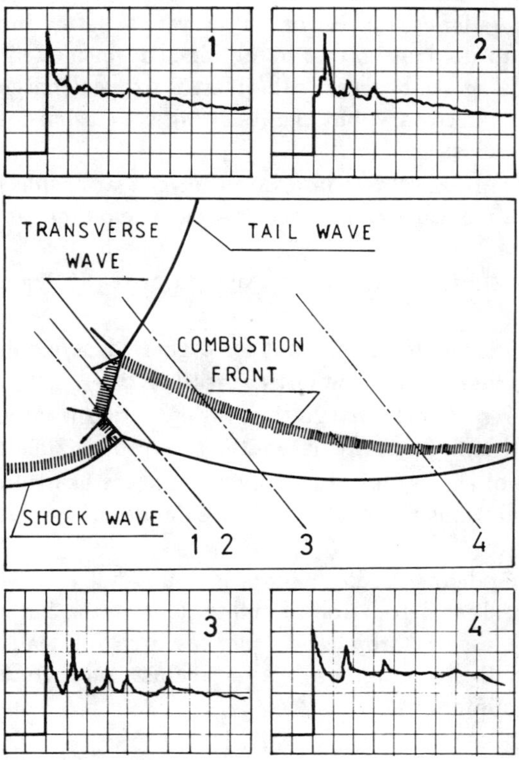

Fig. 13 Schematic diagram of the spin detonation structure constructed from the basic pressure records for the Egyptian brown coal-dust/oxygen mixture, $d < 75\ \mu m$, $C = 0.80\ \mathrm{kg/m^3}$.

Measurements of pressure and temperature made for all tested mixtures indicate basically similar detonation structures for all tested mixtures. The observed differences are only in the values of maximum pressure and temperature. The length of reaction zones varies from 0.1 m to 0.2 m. These values are obtained from the temperature measurements. The time after which maximum temperature is obtained allows one, with the known value of detonation velocity, to calculate the distance between the leading shock wave and the region of maximum temperature. This distance was assumed to be the length of the reaction zone.

In this research only one or, in some runs, two pressure transducers were installed in one cross section of the tube. In such cases the head of the detonation hit the pressure transducer in a random position. Nevertheless, the complete collection of pressure records obtained for the different runs allows one to reconstruct the structure of the detonation wave. Figure 13 presents records from four different runs, with the reconstruction of the typical spin structure. Pressure records obtained for different mixtures also indicate that the detonation propagation in such a mixture has at least one transverse wave.

Experiments conducted so far show that many organic dusts can easily detonate in an oxygen atmosphere. This may indicate that reactive organic dust mixed with air can cause flame acceleration to detonation on a large scale in long corridors, bins, galleries, and so on. However, a better indication of possible transition to detonation can be obtained from modeling studies carried out in an oxidizing atmosphere with different oxygen content. Further research in this area is necessary, but the following conclusions can be drawn from the present studies:

- Many organic dusts can easily detonate in an oxygen atmosphere, and the transition to detonation can readily occur even when there is a relatively weak ignition source.
- Detonation velocities of organic dusts with oxygen are in the range of 1400 to 2100 m/s.
- Maximum pressure in the detonation front for organic dust/oxygen mixtures can rise to 60 times higher than the initial pressure.
- Maximum temperatures of dust particles in the detonation front are in the range of 2400 to 3100 K and, for many detonations, approach 3000 K.
- The detonation front of the organic dust/oxygen mixture has transverse waves, and for some conditions there is very clear evidence of spinning detonation.

It is evident that the modeling studies cannot give a definite answer to the real-world problems of dust detonation at full scale; but the knowledge gathered at significantly less cost from laboratory research will lead to better understanding and finally to better protection against uncontrolled detonations in industry and mining.

ACKNOWLEDGMENTS

Financial support for this work was partially funded by the U.S. Department of Agriculture's Office of International Cooperation and Development, Special Foreign Currency Research Program, under Grant No. FG-Po-370; partially by the Ministry of Agriculture of Poland, under Program No. 10.1; and partially by the Department of Technical Development and Applications of Poland, under Program No. 02.18.

REFERENCES

[1]Cybulski, W., "Detonation of Coal Dust," Buletin de l'Academie des Sciences, Volume 19, 197l, pp. 37–41.

[2]Bartknecht, W., Explosions, Course, Prevention, Protection, Springer-Verlag, Berlin, Heidelberg, New York, 1980.

[3]Gardner, B.R., Winter, R.J., and Moore, M.J., "Explosion Development and Deflagration-to-Detonation Transition in Coal-Dust/Air Suspensions," Proceedings of the Twenty-First International Symposium on Combustion, The Combustion Institute, Pittsburgh, PA, 1986, pp. 335–343.

[4]Kauffman, C.W., Wolański, P., Ural, A., Nicholls, J.A., and Van Dyk, R., "Shock-Wave-initiated Combustion of Grain Dust," Proceedings of the International Symposium on Grain Dust, Manhattan, KS, Oct. 2–4, 1979, pp. 164–190.

[5]Nefleton, M.A. and Stirling, R., "Detonation in Suspensions of Coal Dust in Oxygen," Combustion and Flame, Volume 21, 1973, pp. 307–314.

[6]Wójcicki, S. and Zalesiński, M., "The Mechanism of Transition from Combustion to Detonation in a Mixture of Coal Dust and Gaseous Oxidizer," Proceedings of the Ninth International Shock Tube Symposium, Stanford University Press, Menlo Park, CA, 1973, pp. 821–828.

[7]Wolański, P., Lee, D., and Sichel, M., "The Structure of Dust Detonations," Dynamics of Shock Waves Explosions and Detonations, Progress in Astronautics and Aeronautics, Volume 94, 1984, pp. 241–263.

[8]Lee, D. and Sichel, M., "The Chapman-Jouguet Condition and Structure of Detonation in Dust/Oxygen Mixtures," Dynamics of Explosions, Progress in Astronautics and Aeronautics, Volume 106, 1986, pp. 505–521.

[9]Fan, B.C. and Sichel, M., "A Comprehensive Model for the Structure of Dust Detonations," Twenty-Second International Symposium on Combustion, Combustion Institute, Pittsburgh, PA, 1988, pp. 1741–1750.

[10]Gear, C.W., Numerical Initial Value Problems in O.D.E., Prentice Hall, Englewood Cliffs, NJ, 1971.

[11]Kauffman, C.W., Wolański, P., Arisoy, A., Adams, P.R., Maker, B.N., and Nicholls, J.A., "Dust, Hybrids, and Dusty Detonations," Dynamics of Shock Waves Explosions and Detonations, Progress in Astronautics and Aeronautics, Volume 94, 1984, pp. 221–240.

[12]Fangrat, J., Glinka W., Wolański, P., and Woliński, M., "Detonation Structure in Organic Dust/Oxygen Mixtures," Archivum Combustionis, Volume 7, 1987, pp. 321–332.

[13]Zhang, F. and Groenig, H., "Investigation of Detonation of Cornstarch/ Oxygen Mixtures," Archivum Combustionis, Volume 9, 1989, in press.

Multiple Gauges for In Situ Measurements of Pressure and Particle Velocity in Condensed Materials

P. A. Urtiew and L. M. Erickson
Lawrence Livermore National Laboratory, Livermore, CA

Abstract

In situ measurements of pressure and particle velocity have always been crucial in the development of theoretical models and formulation of equations of state for both inert and reactive materials. Formerly, measurements were made with separate gauge stations placed between layers of investigated material stacked in the direction of shock propagation. This arrangement not only generates a disturbance in the flowfield but also causes interruption in the chemical reaction process being monitored. The introduction of multiple-gauge packages placed on an inclined surface has eliminated most of the problems encountered with separate in-line gauge stations. Such multiple gauges provide the capability of measuring changes in pressure and particle velocity at depth intervals much smaller than those between separate gauge stations. In addition, each consecutive gauge element is not engulfed in the flow wake of the preceding one. The multiple-element gauges described here are ideal for monitoring rapid changes in pressure and particle velocity in processes such as transition to detonation or phase transformation. With this arrangement in a single experiment, one can have as many as twelve signal traces from within a very thin depth in a sample.

Introduction

For many years, in situ gauges have been used very successfully to measure pressure[1-6] and particle velocity[7-12] in dynamically loaded systems. Such measurements were done in both inert and chemically reactive materials. These measurements have provided experimental data for verification and suitable adjustment to reactive flow models.

In our work at the Lawrence Livermore National Laboratory, we have used these gauges extensively to study rates of reaction and the transition to detonation following initiation by shock waves as described by Vantine et al.[13] and Nutt and Erickson.[14]

Such shock initiation is most readily attained in our 4-in. (101-mm) gun facility, where a projectile of known material is accelerated to a desired velocity (up to 2.5 mm/μs) and impacted onto a target. Our experimental setup is illustrated in Fig. 1. The impact generates the desired shock wave,which then propagates through the target and promotes changes in physical parameters (pressure and particle velocity), which we have set out to measure. To account for a possible tilt in the projectile, the target assembly contains six flush-mounted piezoelectric pins equally spaced around the outer edge of the sample and also two other pins extending a known distance ahead of impact to measure projectile velocity and trigger all of the data acquisition systems.

The changes in pressure and particle velocity in the sample are measured with embedded gauges placed at known intervals between layers of the investigated material.

Fabrication of the gauges and their calibration have been described elsewhere by Erickson et al.,[15] Vantine et al.[16,17] The purpose of this communication is to identify some of the problems with the old arrangement of these measuring devices and to describe the new generation of these gauges, which allows more data, cleaner and longer records, and measurements of a less disturbed flowfield.

Old Problems and New Solutions

In the past, our in situ measurements of pressure and particle velocity have been made by laminating appropriate gauge packages between disks of the material to be measured as described by Weingart et al.[8] and Erickson et al.[12] Each gauge station contains a certain thickness of armor that is chemically inert and electrically inactive to protect the gauge from early destruction and to prevent any unwanted response to extraneous electrical

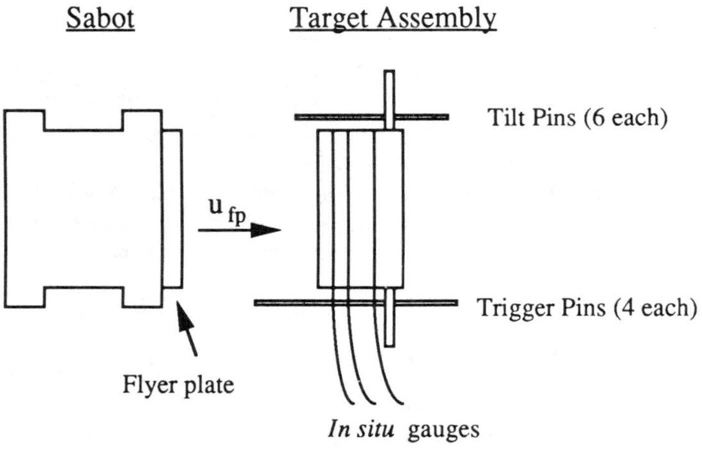

Fig. 1 The 4-in. (101-mm) gun test setup.

signals that may be inherently present in the system under investigation. Placed in such an armor jacket, the gauge package becomes an undesirable disturbance in the flow, which is not only inert when the investigated medium is reacting but is also a source of reflected waves resulting from impedance mismatch of the materials. If one desired several traces of the records on the same shot at different depths of the sample, one was required to place several gauge stations into the shot, introducing several layers of undesired foreign material into an already thin sample of the material of interest.

Thus, the technique of utilizing single-gauge packages has been very successful, but it has its problems, which cannot be overlooked. Among these, the most important are:

1) Separate disks with gauge packages between them cause flow disturbances resulting from impedance mismatch.

2) Several gauge packages composed of inert material interrupt the reaction process.

To overcome these problems, several gauge elements were combined into one single gauge package and placed at an angle to the oncoming flow, as illustrated in Fig. 2. Each gauge element is then at a different depth in the flow, and this depth can easily be controlled by the separation of the elements and by the angle of placement. This then not only eliminates the multiple sources for flow disturbances but also enhances our measuring capability by giving us an unlimited control of depth in positioning the gauge and, at the same time, increases the number of measurements on a single experiment.

Following the same idea, our colleagues at Los Alamos[18] have used a similar technique to make several measurements with a single gauge package. They then combined a multiple-particle velocity gauge with a multiple-impulse

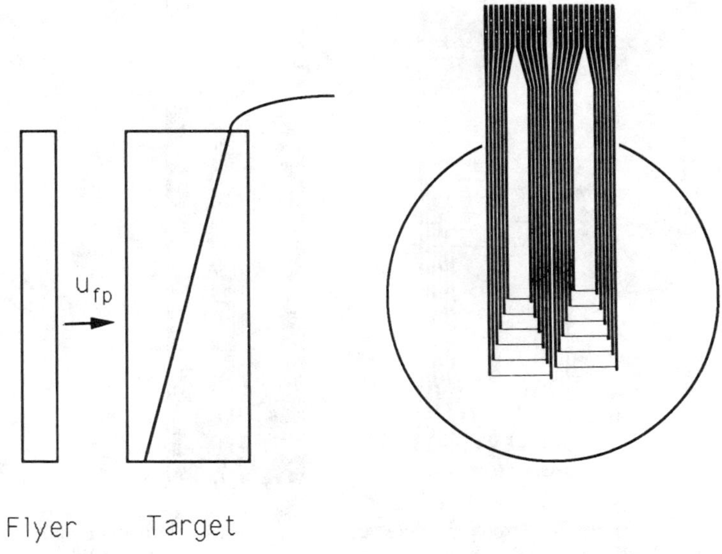

Fig. 2 The target assembly with a multiple gauge placed on an inclined surface.

gauge to provide both particle velocity measurements and pressure. However, we have chosen to make particle velocity and pressure measurements on separate experiments because direct pressure measurements are much more desirable for our needs.

Gauge Construction

As described elsewhere by Erickson et al.[15] all gauges are fabricated in-house by a photolithographic technique etching sensitive elements from a thin foil. The thickness of the gauge elements is about 25 μm. To prevent the explosion products (which are highly ionized and hence conductive) from shorting out the gauges, the entire gauge is insulated with a layer of Teflon, 25-250 μm thick on both sides.

The multiple-particle velocity gauge is available in two designs: one with the common ground lead as shown in Fig. 3a and one with each element having its own ground lead in the gauge package, Fig. 3b. The idea of separating the ground leads within the package is to minimize possible cross-talk and noise during the experiment. There are 12 gauge elements spaced at 2-mm intervals. This interval can be varied depending on the region of interest. Both the spacing between these elements and the wedge angle add to the flexibility of using the gauge at very small depth intervals of the flow.

The multiple manganin gauge is illustrated in Fig. 4a, with details of the gauge element shown in Fig. 4b. This gauge shows six gauge elements

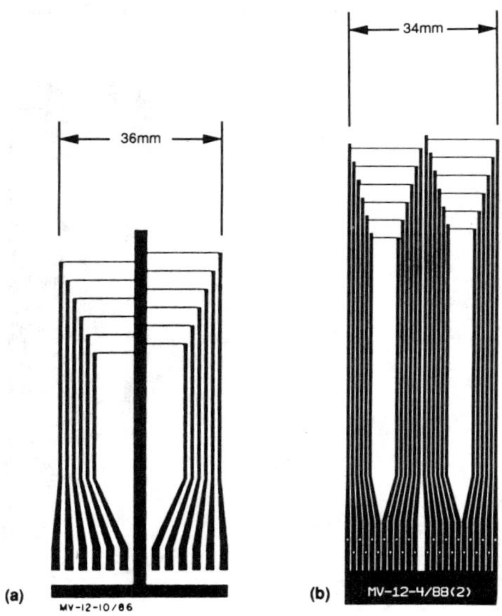

Fig. 3 The multiple-element particle velocity gauge with elements length ranging between 6 and 16-mm: a) common ground lead design, b) design with a separate ground lead for each gauge element.

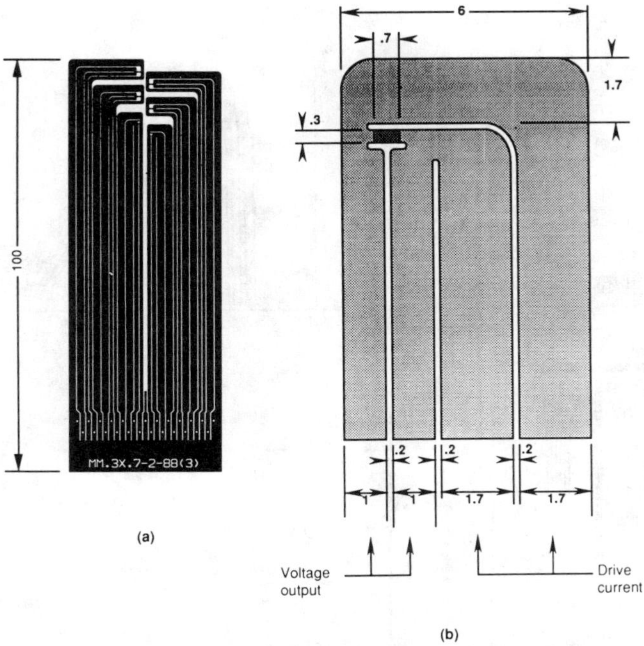

Fig. 4 The multiple-element manganin pressure gauge: a) six element gauge unit, b) details of a single-gauge element with four leads and initial resistance of 60 mΩ.

separated from each other by 3 mm. As in the case of the particle velocity gauge, this spacing can be varied, and we have made gauges with 2-, 5-, and 8-mm intervals as well.

Since the gauges are of very low impedance (0.05-0.06 Ω), a four-terminal measurement is required. Thus, each gauge has two wide leads for the 50-A current drive and two narrow leads for measuring the voltage drop across the active element.

Because the gauge elements are placed on inclined surfaces, the rise time of the signal becomes of interest. In order to prevent this from becoming a detriment, the width of gauge elements is kept very small, i.e., 0.2 and 0.3 mm for particle velocity and manganin pressure gauges, respectively. With these widths on an inclined surface of 24 deg., a shock speed of 5 mm/μs produces a rise time less than 30 ns, which is quite reasonable for these types of measurements.

Experimental Results

As mentioned earlier, the gauge package is placed on a wedge at an angle to the oncoming flow. The results are illustrated in Fig. 5. The flow is generated by the impact of the flyer plate at the front surface of the shot assembly. The flowfield in the time-space diagram is represented with the

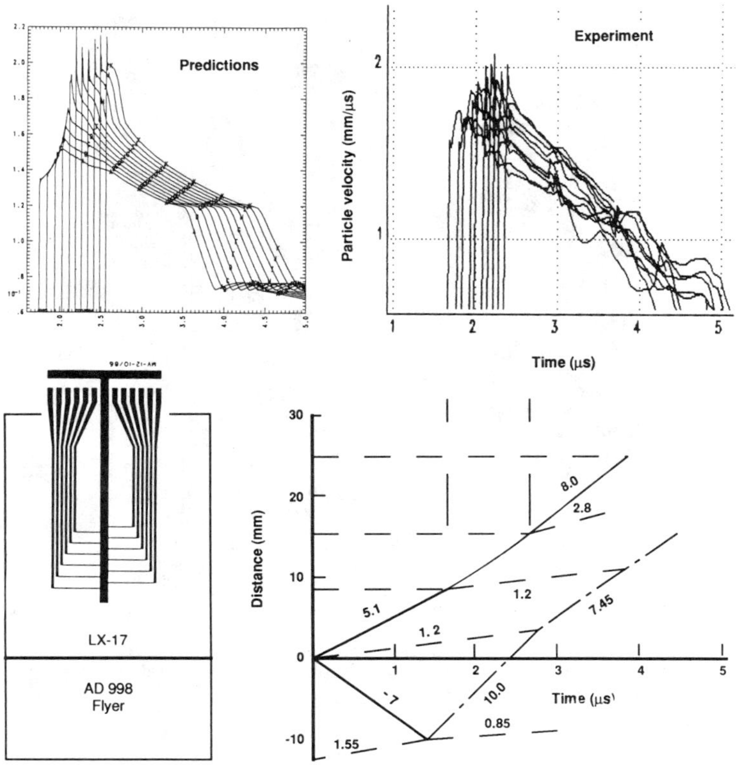

Fig. 5 Experimental results from a 12-element particle velocity gauge in an LX-17 explosive target.

distance ordinate drawn to the same scale as the shot assembly. The moment of impact is taken as the zero time at which two shock waves are generated: one propagating into the sample with embedded gauge package and one traveling back into the flyer plate. The forward running shock wave will travel across the gauge package in a continuous manner but will affect each gauge element at a different time.

In Fig. 5, the gauge package contains 12 particle velocity gauges placed in an LX-17[‡] explosive. As the shock travels across the gauge the gauge elements are set in motion, crossing the magnetic field and thereby generating current, which is then appropriately recorded on an oscilloscope or a transient digitizer. Taking into account the strength of the magnetic field and the length of the sensing element,the signal is then converted into the units of particle velocity. These experimental particle velocity profiles are shown above the space-time diagram. Also shown here is a set of particle velocity profiles that were predicted beforehand by theoretical calculations

‡ 92.5% Triaminotrinitrobenzene, 7.5% Kel-F 800, 1.9 g/cm^3.

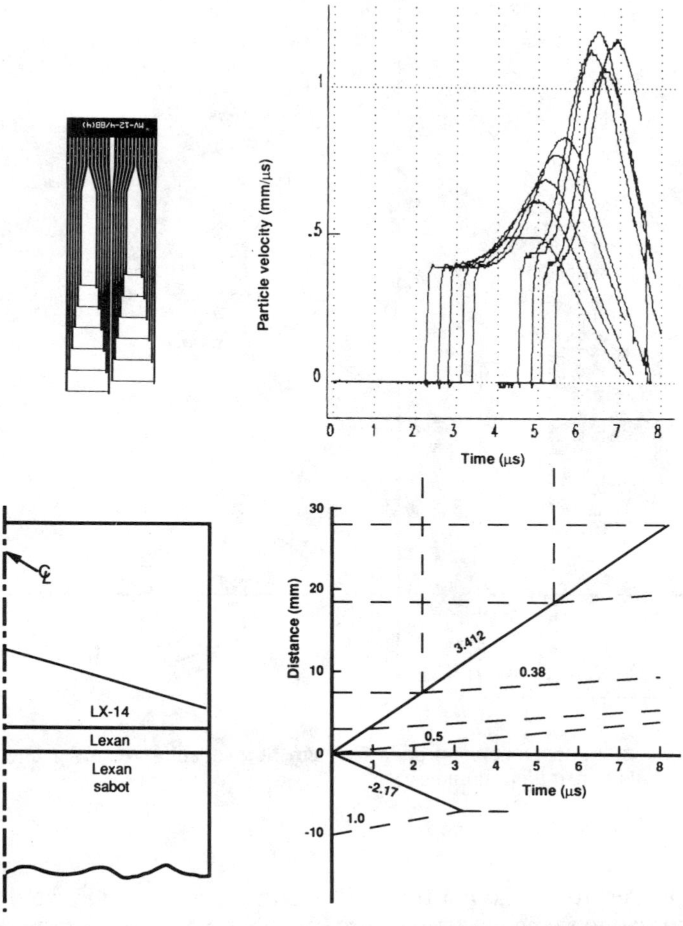

Fig. 6 Experimental results from a 12-element particle velocity gauge in an LX-14 explosive target.

using our LLNL ignition and growth model described by Bahl et al.[19] In this case, the gauge spanned the depth at which the transition to detonation occurred, which is evidenced by both theoretical predictions and experimental results.

Another example of the multiple particle velocity gauge performance is shown in Fig. 6. Here the gauge was placed on a wedge of the LX-14[§] explosive but not deep enough to show the transition to detonation. The records clearly show a continuous rise of particle velocity, indicating that reaction is taking place and that the process is well on its way to turnover to detonation.

§ 95.5% HMX< 4.5% Estane, 1.83 g/cm^3.

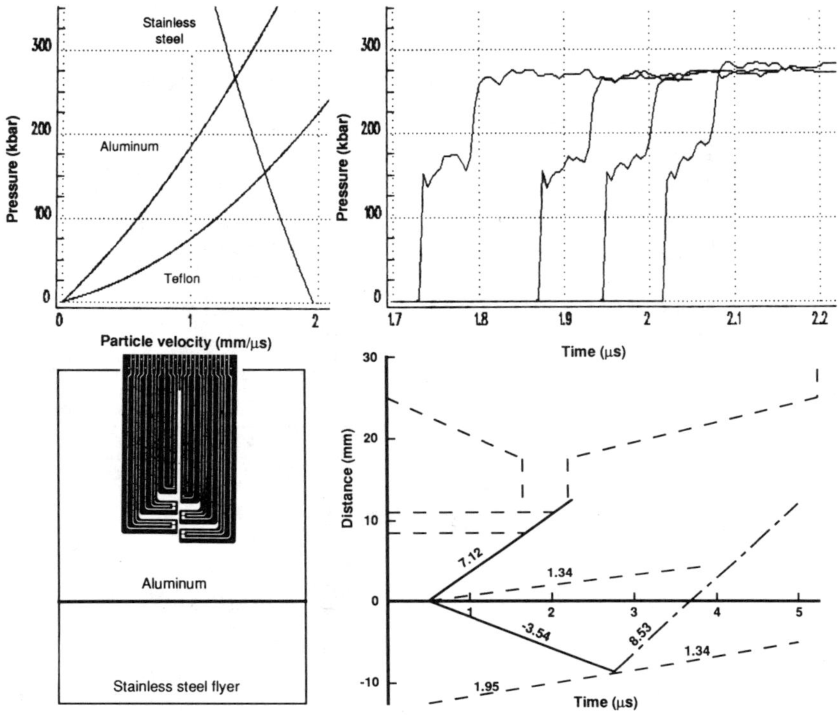

Fig. 7 Experimental results from a six-element manganin pressure gauge placed in an inert aluminum target.

The results of the multiple manganin gauge are illustrated in Fig. 7. Here each gauge responds to pressure by changing its resistance. Since the current flowing through the gauge element is constant, any change in resistance is reflected in a change of the voltage signal. By monitoring both current input and voltage output signals and applying a proper conversion polynomial, one can get an accurate account of the pressure change as seen by each individual gauge element. In this case, only four pressure profiles were obtained from the six gauge elements in one package, and they are shown in Fig. 7, together with the corresponding target assembly and wave diagram in the space-time plane.

In the case at hand, the target was made of aluminum and impacted by a stainless steel flying plate. The armor around the gauge was 250 μm of Teflon on each side. As the shock wave traverses the gauge package, the gauge element initially sees only the pressure reached in Teflon before it rings up to the final pressure in aluminum. This is clearly evidenced by the recorded precursor of about 150 kbar, which clearly agrees with the intersections of shock Hugoniots on the pressure-particle velocity diagram that is also included in the figure.

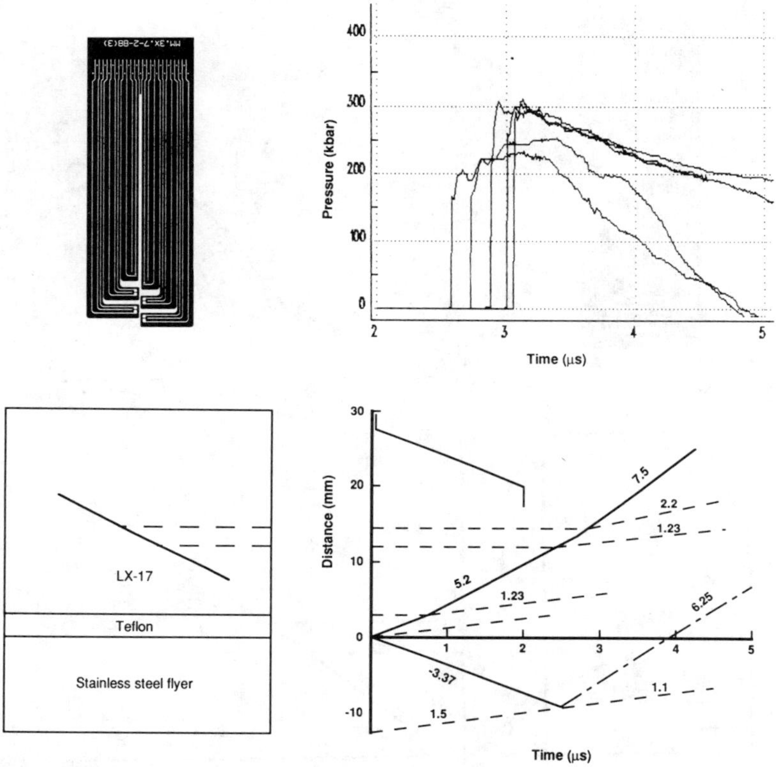

Fig. 8 Experimental results from a six-element manganin pressure gauge placed in an LX-17 explosive target.

When placed in an explosive target such as LX-17, the multiple manganin gauge is very valuable to measure pressure along a Lagrangian space coordinate as the pressure builds up and at some point turns over into the detonation wave. Such a case is illustrated in Fig. 8, where the turnover happened within the 2-mm depth between gauge elements 3 and 4 of the six-element gauge package. In addition to giving an accurate account of the pressure buildup in the predetonation regime, this gauge also provides accurate information on the distance to detonation transition (Pop-plot), which was originally introduced by Ramsey and Popolato.[20] Such pressure buildup is yet better illustrated in Fig. 9, where the six-element gauge was placed in an LX-14 target assembly. All six gauge elements gave us good results and valuable information on the predetonative buildup regime.

Conclusions

Multiple-gauge packages open up a new regime in the resolution of fast-growing processes such as transition to detonation. They eliminate some

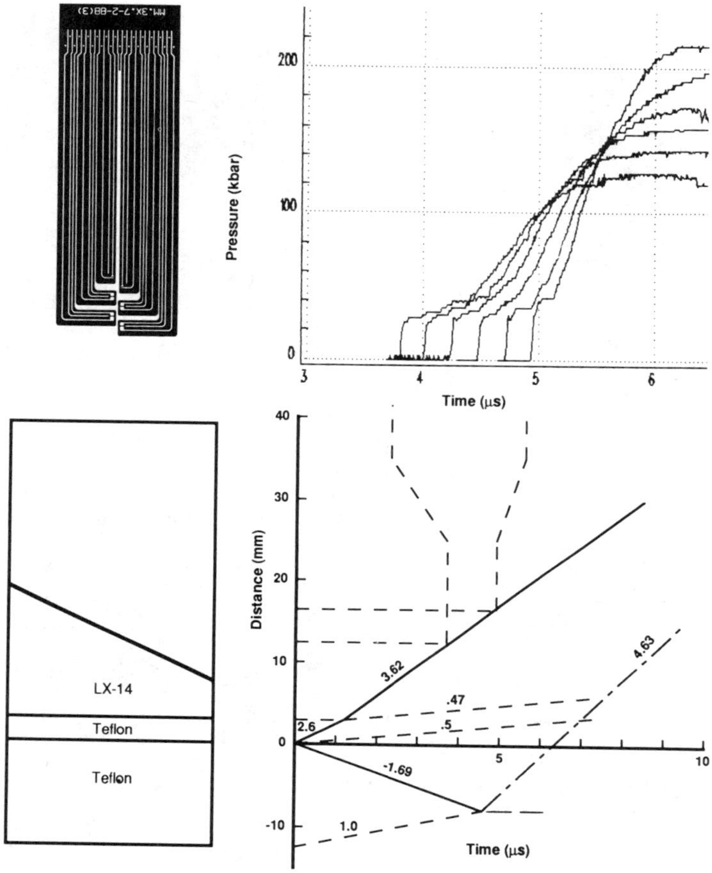

Fig. 9 Experimental results from a six-element manganin pressure gauge placed in an LX-14 explosive target.

of the old problems of multiple interfaces with impedance mismatched materials and at the same time enhance the flexibility of getting several traces within a very small region of interest. By fabricating small elements as described here, the rise time of the gauges can be kept to within a range of less than 30 ns.

Acknowledgments

The authors wish to express their gratitude to all those who, in one way or another, helped them during the process of gauge development: Fred Cutter for etching and fabricating the gauges; George Elliott for encapsulating and preparing the gauge stations; Leona Meegan for assembling the shots; the gun crew, Doug Bakker, Kim Jeter, and Gary Roads for firing; and Tom Cook for helping with the experiment and processing the data. Thanks are also due to our colleagues for numerous discussions on the subject and useful suggestions.

References

[1]Fuller, P.J.A., and Price, J.H., "Electrical Conductivity of Manganin and Iron at High Pressures," Nature, Vol. 103, 1962, p. 262.

[2]Bernstein, D., and Keough, D.D., "Piezoresistivity of Manganin," Journal of Applied Phisics, Vol. 35, 1964, pp. 1471-1474.

[3]Wackerle, J., Johnson, J.O., and Halleck, P.M., "Shock Initiation of High Density PETN," Sixth Symposium (International) on Detonation, Office of Naval Research, White Oak, Silver Spring, MD, ONR ACR 221, 1976, pp. 2-28.

[4]Kanel, G.I., and Dremin, A.N., "Decomposition of Cast TNT in Shock Waves," Fizika Goreniya i Vzryva, Vol. 13, No. 1, 1977, pp. 85-92.

[5]Burrows, K., Chivers, D. K., Gyton, R., Lambourn, B.D., and Wallace, A.A., "Determination of Detonation Pressure Using a Manganin Wire Technique," Sixth Symposium (International) on Detonation, Office of Naval Research, White Oak, Silver Spring, MD,.ONR ACR 221, 1976, pp. 625-636.

[6]Weingart, R.C., Barlett, R., Cochran, S., Erickson, L.M., Chan, J., Janzen, J., Lee, R., Logan, D., and Rosenberg, J.T., "Manganin Stress Gauges in Reacting High Explosive Environment," Proceedings of the Symposium on High Dynamic Pressures, Commissariat à l'Energie Atomique, Paris, France, 1978, pp. 451-461.

[7]Dremin, A.N., Pershin, S.V., and Pogorelov, V.F., "Structure of Shock Waves in KCl and KBr under Dynamic Compression to 200,000 Atmospheres," Combustion, Explosion and Shock Waves, Vol. 1, No. 4, 1965, pp. 1-4 (English Translation)

[8]Edwards, D.J., Erkman, J.O., and Jacobs, S.J., "Electromagnetic Velocity Gauge and Applications to Measure of Particle Velocity in PMMA," Naval Ordnance Laboratory Rept, NULTR-70-79, 1970.

[9]Hayes, B., and Fritz, J.N., "Measurement of Mass Motion in Detonation Products by an Axially-Symmetric Electromagnetic Technique," Fifth Symposium (International) on Detonation, Office of Naval Research, White Oak, Silver Spring, MD, ACR-184, 1970, pp. 447-454.

[10]Erickson, L.M., Johnson, C.B., Parker, N.L., Vantine, H.C., Weingart, R.C., and Lee, R. S., "The Electromagnetic Velocity Gauge: Use of Multiple Gauges, Time Response, and Flow Perturbation," Seventh Symposium (International) on Detonation, Naval Surface Weapons Center, White Oak, Silver Spring, MD, NSWC MP 82-334, 1982, pp. 1062-1071.

[11]Urtiew, P.A., Erickson, L.M., Hayes, B., and Parker, N.L., "Pressure and Particle Velocity Measurements in Solids Subjected to Dynamic Loading," Lawrence Livermore National Laboratory, UCRL-93213, 1985; Fizika Goreniya i Vzryva, Vol. 22, No. 5, 1986, pp. 113-126.

[12]Urtiew, P.A., and Erickson, L.M., "Measurement of Dynamic Pressure in Inert and Reactive Solids by PVF2 and Manganin Gauges," Techniques and Theory of Stress Measurements in Shock Wave Applications, Proceedings of the Summer ASME Meeting, 1987, pp. 29-35.

[13]Vantine, H.C., Rainsberger, R. B., Curtis, W.D., Lee, R. S., Cowperthwaite, M., Rosenberg, J.T., "The Accuracy of Reaction Rates Inferred From Lagrange Analysis and In Situ Gauge Measurements," Seventh Symposium (International) on Detonation, Naval Surface Weapons Center, White Oak, Silver Spring, MD, NSWC MP 82-334, 1981, pp. 466-478, .

[14]Nutt, G.L., and Erickson, L. M., "Reactive Flow Lagrange Analysis in Plastic Bonded Explosives," Journal of Energetic Materials, Vol. 2, 1984, pp. 263-292.

[15]Erickson, L., Weingart, R., Barlett, R., Chan, J., Elliott, G., Janzen, J., Vantine, H., Lee, R., and Rosenberg, J.T., "Fabrication of Manganin Stress Gauges for Use in Detonating High Explosives," Tenth Symposium on Explosives and Pyrotechnics, Franklin Research Center, Philadelphia, PA, 1979.

[16]Vantine, H., Chan, J., Erickson, L.M., Janzen, J., Lee, R. and Weingart, R.C., "Precision Stress Measurements in Severe Shock-Wave Environments with Low Impedance Manganin Gauges," Review of Scientific Instruments, Vol. 51, 1980, pp. 116-122.

[17]Vantine, H.C., Erickson, L.M., and Janzen, J.A. , "Histeresis-Corrected Calibration of Manganin Under Shock Loading," Journal of Applied Physics, Vol. 51, No. 4, 1980, pp. 1957-1962.

[18]Vorthman, J., Andrews, G., and Wackerle, J., "Reaction Rates from Electromagnetic Gauge Data," Eighth Symposium (International) on Detonation, Naval Surface Weapons Center, White Oak, Silver Spring, MD, NSWC MP 86-194, 1986, pp. 99-110.

[19]Bahl, K., Bloom, G., Erickson, L., Lee, R., Tarver, C., Von Holle, W., and Weingart, R., "Initiation Studies on LX-17 Explosive," Eighth Symposium (International) on Detonation, Naval Surface Warfare Center, White Oak, Silver Spring, MD, NSWC MP 86-194, 1986, pp. 1045-1056.

[20]Ramsay, J.B., and Popalato, A., "Analysis of Shock Wave and Initiation Data for Solid Explosives," Fourth Symposium (International) on Detonation, Office of Naval Research, White Oak, Silver Spring, MD, ACR-126, 1965, pp. 233-238.

Author Index

Aiteh, A. 63
Aksenov, V. S. 268
Baker, D. L. 325
Bauer, P. 56
Bauer, P. A. 3
Beeson, H. D. 19
Benedick, W. B. 297
Benz, F. J. 19, 37
Bishop, C. V. 19
Borisov, A. A. 118, 142, 268
Bunker, R. L. 325
Dabora, E. K. 3, 187, 311
Dahab, O. 356
Desbordes, D. 170, 187
Dunand, M. 56
Dupré, G. 63, 156
Edwards, D. H. 256
Erickson, L. M. 371
Frolov, S. M. 133, 241
Gelfand, B. E. 133, 241
Groenig, H. 342
Guerraud, C. 187
Guirguis, R. 202
Herrmann, K.-P. 279
Joannon, J. 156
Jones, D. A. 202
Jones, S. H. M. 256
Kapuściński, M. 356
Khomik, S. V. 118, 142
Knystautas, R. 91, 156, 223, 297
Kosenkov, V. V. 268
Lannoy, A. 170
Lee, J. H. S. 19, 37, 91, 156, 223, 297, 325
Mailkov, A. H. 268
Manson, N. 3
McClenagan, R. D. 19, 37
Medvedev, S. P. 241
Mikhalkin, V. N. 118, 142
Moen, I. O. 91
Murray, S. B. 91
Nzeyimana, E. 77
Oran, E. S. 202
Paillard, C. 63
Péraldi, O. 156
Pitz, W. J. 19
Plaster, M. 37
Plewinsky, B. 279
Presles, H. N. 56
Saneev, E. V. 142
Shepherd, J. E. 37
Sichel, M. 202
Sulmistras, A. 91
Teodorczyk, A. 223
Thibault, P. A. 91
Thomas, G. O. 256
Tieszen, S. R. 297
Tsygranov, S. A. 133
Urtiew, P. A. 371
Van Tiggelen, P. J. 77
Wagner, H. G. 187
Wegener, W. 279
Westbrook, C. K. 19
Wolański, P. 356
Youssefi, S. 63
Zhang, F. 342

PROGRESS IN ASTRONAUTICS AND AERONAUTICS SERIES VOLUMES

*1. **Solid Propellant Rocket Research** (1960)
Martin Summerfield
Princeton University

*2. **Liquid Rockets and Propellants** (1960)
Loren E. Bollinger
Ohio State University
Martin Goldsmith
The Rand Corp.
Alexis W. Lemmon Jr.
Battelle Memorial Institute

*3. **Energy Conversion for Space Power** (1961)
Nathan W. Snyder
Institute for Defense Analyses

*4. **Space Power Systems** (1961)
Nathan W. Snyder
Institute for Defense Analyses

*5. **Electrostatic Propulsion** (1961)
David B. Langmuir
Space Technology Laboratories, Inc.
Ernst Stuhlinger
NASA George C. Marshall Space Flight Center
J.M. Sellen Jr.
Space Technology Laboratories, Inc.

*6. **Detonation and Two-Phase Flow** (1962)
S.S. Penner
California Institute of Technology
F.A. Williams
Harvard University

*7. **Hypersonic Flow Research** (1962)
Frederick R. Riddell
AVCO Corp.

*8. **Guidance and Control** (1962)
Robert E. Roberson, Consultant
James S. Farrior
Lockheed Missiles and Space Co.

*9. **Electric Propulsion Development** (1963)
Ernst Stuhlinger
NASA George C. Marshall Space Flight Center

*10. **Technology of Lunar Exploration** (1963)
Clifford I. Cummings
Harold R. Lawrence
Jet Propulsion Laboratory

*11. **Power Systems for Space Flight** (1963)
Morris A. Zipkin
Russell N. Edwards
General Electric Co.

*12. **Ionization in High-Temperature Gases** (1963)
Kurt E. Shuler, Editor
National Bureau of Standards
John B. Fenn, Associate Editor
Princeton University

*13. **Guidance and Control – II** (1964)
Robert C. Langford
General Precision Inc.
Charles J. Mundo
Institute of Naval Studies

*14. **Celestial Mechanics and Astrodynamics** (1964)
Victor G. Szebehely
Yale University Observatory

*15. **Heterogeneous Combustion** (1964)
Hans G. Wolfhard
Institute for Defense Analyses
Irvin Glassman
Princeton University
Leon Green Jr.
Air Force Systems Command

16. **Space Power Systems Engineering** (1966)
George C. Szego
Institute for Defense Analyses
J. Edward Taylor
TRW Inc.

17. **Methods in Astrodynamics and Celestial Mechanics** (1966)
Raynor L. Duncombe
U.S. Naval Observatory
Victor G. Szebehely
Yale University Observatory

18. **Thermophysics and Temperature Control of Spacecraft and Entry Vehicles** (1966)
Gerhard B. Heller
NASA George C. Marshall Space Flight Center

*Out of print.

***19. Communication Satellite Systems Technology** (1966)
Richard B. Marsten
Radio Corporation of America

***20. Thermophysics of Spacecraft and Planetary Bodies: Radiation Properties of Solids and the Electromagnetic Radiation Environment in Space** (1967)
Gerhard B. Heller
NASA George C. Marshall Space Flight Center

21. Thermal Design Principles of Spacecraft and Entry Bodies (1969)
Jerry T. Bevans
TRW Systems

22. Stratospheric Circulation (1969)
Willis L. Webb
Atmospheric Sciences Laboratory, White Sands, and University of Texas at El Paso

23. Thermophysics: Applications to Thermal Design of Spacecraft (1970)
Jerry T. Bevans
TRW Systems

24. Heat Transfer and Spacecraft Thermal Control (1971)
John W. Lucas
Jet Propulsion Laboratory

25. Communication Satellites for the 70's: Technology (1971)
Nathaniel E. Feldman
The Rand Corp.
Charles M. Kelly
The Aerospace Corp.

26. Communication Satellites for the 70's: Systems (1971)
Nathaniel E. Feldman
The Rand Corp.
Charles M. Kelly
The Aerospace Corp.

27. Thermospheric Circulation (1972)
Willis L. Webb
Atmospheric Sciences Laboratory, White Sands, and University of Texas at El Paso

28. Thermal Characteristics of the Moon (1972)
John W. Lucas
Jet Propulsion Laboratory

29. Fundamentals of Spacecraft Thermal Design (1972)
John W. Lucas
Jet Propulsion Laboratory

30. Solar Activity Observations and Predictions (1972)
Patrick S. McIntosh
Murray Dryer
Environmental Research Laboratories, National Oceanic and Atmospheric Administration

31. Thermal Control and Radiation (1973)
Chang-Lin Tien
University of California at Berkeley

32. Communications Satellite Systems (1974)
P.L. Bargellini
COMSAT Laboratories

33. Communications Satellite Technology (1974)
P.L. Bargellini
COMSAT Laboratories

34. Instrumentation for Airbreathing Propulsion (1974)
Allen E. Fuhs
Naval Postgraduate School
Marshall Kingery
Arnold Engineering Development Center

35. Thermophysics and Spacecraft Thermal Control (1974)
Robert G. Hering
University of Iowa

36. Thermal Pollution Analysis (1975)
Joseph A. Schetz
Virginia Polytechnic Institute
ISBN 0-915928-00-0

37. Aeroacoustics: Jet and Combustion Noise; Duct Acoustics (1975)
Henry T. Nagamatsu, Editor
General Electric Research and Development Center
Jack V. O'Keefe, Associate Editor
The Boeing Co.
Ira R. Schwartz, Associate Editor
NASA Ames Research Center
ISBN 0-915928-01-9

38. Aeroacoustics: Fan, STOL, and Boundary Layer Noise; Sonic Boom; Aeroacoustics Instrumentation (1975)
Henry T. Nagamatsu, Editor
General Electric Research and Development Center
Jack V. O'Keefe, Associate Editor
The Boeing Co.
Ira R. Schwartz, Associate Editor
NASA Ames Research Center
ISBN 0-915928-02-7

39. Heat Transfer with Thermal Control Applications (1975)
M. Michael Yovanovich
University of Waterloo
ISBN 0-915928-03-5

40. Aerodynamics of Base Combustion (1976)
S.N.B. Murthy, Editor
J.R. Osborn,
Associate Editor
Purdue University
A.W. Barrows
J.R. Ward,
Associate Editors
Ballistics Research Laboratories
ISBN 0-915928-04-3

41. Communications Satellite Developments: Systems (1976)
Gilbert E. LaVean
Defense Communications Agency
William G. Schmidt
CML Satellite Corp.
ISBN 0-915928-05-1

42. Communications Satellite Developments: Technology (1976)
William G. Schmidt
CML Satellite Corp.
Gilbert E. LaVean
Defense Communications Agency
ISBN 0-915928-06-X

43. Aeroacoustics: Jet Noise, Combustion and Core Engine Noise (1976)
Ira R. Schwartz, Editor
NASA Ames Research Center
Henry T. Nagamatsu,
Associate Editor
General Electric Research and Development Center
Warren C. Strahle,
Associate Editor
Georgia Institute of Technology
ISBN 0-915928-07-8

44. Aeroacoustics: Fan Noise and Control; Duct Acoustics; Rotor Noise (1976)
Ira R. Schwartz, Editor
NASA Ames Research Center
Henry T. Nagamatsu,
Associate Editor
General Electric Research and Development Center
Warren C. Strahle,
Associate Editor
Georgia Institute of Technology
ISBN 0-915928-08-6

45. Aeroacoustics: STOL Noise; Airframe and Airfoil Noise (1976)
Ira R. Schwartz, Editor
NASA Ames Research Center
Henry T. Nagamatsu,
Associate Editor
General Electric Research and Development Center
Warren C. Strahle,
Associate Editor
Georgia Institute of Technology
ISBN 0-915928-09-4

46. Aeroacoustics: Acoustic Wave Propagation; Aircraft Noise Prediction; Aeroacoustic Instrumentation (1976)
Ira R. Schwartz, Editor
NASA Ames Research Center
Henry T. Nagamatsu,
Associate Editor
General Electric Research and Development Center
Warren C. Strahle,
Associate Editor
Georgia Institute of Technology
ISBN 0-915928-10-8

47. Spacecraft Charging by Magnetospheric Plasmas (1976)
Alan Rosen
TRW Inc.
ISBN 0-915928-11-6

48. Scientific Investigations on the Skylab Satellite (1976)
Marion I. Kent
Ernst Stuhlinger
NASA George C. Marshall Space Flight Center
Shi-Tsan Wu
University of Alabama
ISBN 0-915928-12-4

49. Radiative Transfer and Thermal Control (1976)
Allie M. Smith
ARO Inc.
ISBN 0-915928-13-2

50. Exploration of the Outer Solar System (1976)
Eugene W. Greenstadt
TRW Inc.
Murray Dryer
National Oceanic and Atmospheric Administration
Devrie S. Intriligator
University of Southern California
ISBN 0-915928-14-0

51. Rarefied Gas Dynamics, Parts I and II (two volumes) (1977)
J. Leith Potter
ARO Inc.
ISBN 0-915928-15-9

52. Materials Sciences in Space with Application to Space Processing (1977)
Leo Steg
General Electric Co.
ISBN 0-915928-16-7

53. Experimental Diagnostics in Gas Phase Combustion Systems (1977)
Ben T. Zinn, Editor
Georgia Institute of Technology
Craig T. Bowman, Associate Editor
Stanford University
Daniel L. Hartley, Associate Editor
Sandia Laboratories
Edward W. Price, Associate Editor
Georgia Institute of Technology
James G. Skifstad, Associate Editor
Purdue University
ISBN 0-015928-18-3

54. Satellite Communications: Future Systems (1977)
David Jarett
TRW Inc.
ISBN 0-915928-18-3

55. Satellite Communications: Advanced Technologies (1977)
David Jarett
TRW Inc.
ISBN 0-915928-19-1

56. Thermophysics of Spacecraft and Outer Planet Entry Probes (1977)
Allie M. Smith
ARO Inc.
ISBN 0-915928-20-5

57. Space-Based Manufacturing from Nonterrestrial Materials (1977)
Gerard K. O'Neill, Editor
Brian O'Leary, Assistant Editor
Princeton University
ISBN 0-915928-21-3

58. Turbulent Combustion (1978)
Lawrence A. Kennedy
State University of New York at Buffalo
ISBN 0-915928-22-1

59. Aerodynamic Heating and Thermal Protection Systems (1978)
Leroy S. Fletcher
University of Virginia
ISBN 0-915928-23-X

60. Heat Transfer and Thermal Control Systems (1978)
Leroy S. Fletcher
University of Virginia
ISBN 0-915928-24-8

61. Radiation Energy Conversion in Space (1978)
Kenneth W. Billman
NASA Ames Research Center
ISBN 0-915928-26-4

62. Alternative Hydrocarbon Fuels: Combustion and Chemical Kinetics (1978)
Craig T. Bowman
Stanford University
Jorgen Birkeland
Department of Energy
ISBN 0-915928-25-6

63. Experimental Diagnostics in Combustion of Solids (1978)
Thomas L. Boggs
Naval Weapons Center
Ben T. Zinn
Georgia Institute of Technology
ISBN 0-915928-28-0

64. Outer Planet Entry Heating and Thermal Protection (1979)
Raymond Viskanta
Purdue University
ISBN 0-915928-29-9

65. Thermophysics and Thermal Control (1979)
Raymond Viskanta
Purdue University
ISBN 0-915928-30-2

66. Interior Ballistics of Guns (1979)
Herman Krier
University of Illinois at Urbana-Champaign
Martin Summerfield
New York University
ISBN 0-915928-32-9

***67. Remote Sensing of Earth from Space: Role of "Smart Sensors"**(1979)
Roger A. Breckenridge
NASA Langley Research Center
ISBN 0-915928-33-7

68. Injection and Mixing in Turbulent Flow (1980)
Joseph A. Schetz
Virginia Polytechnic Institute and State University
ISBN 0-915928-35-3

69. Entry Heating and Thermal Protection (1980)
Walter B. Olstad
NASA Headquarters
ISBN 0-915928-38-8

70. Heat Transfer, Thermal Control, and Heat Pipes (1980)
Walter B. Olstad
NASA Headquarters
ISBN 0-915928-39-6

71. Space Systems and Their Interactions with Earth's Space Environment (1980)
Henry B. Garrett
Charles P. Pike
Hanscom Air Force Base
ISBN 0-915928-41-8

72. Viscous Flow Drag Reduction (1980)
Gary R. Hough
Vought Advanced Technology Center
ISBN 0-915928-44-2

73. Combustion Experiments in a Zero-Gravity Laboratory (1981)
Thomas H. Cochran
NASA Lewis Research Center
ISBN 0-915928-48-5

74. Rarefied Gas Dynamics, Parts I and II (two volumes) (1981)
Sam S. Fisher
University of Virginia
ISBN 0-915928-51-5

75. Gasdynamics of Detonations and Explosions (1981)
J.R. Bowen
University of Wisconsin at Madison
N. Manson
Université de Poitiers
A.K. Oppenheim
University of California at Berkeley
R.I. Soloukhin
Institute of Heat and Mass Transfer, BSSR Academy of Sciences
ISBN 0-915928-46-9

76. Combustion in Reactive Systems (1981)
J.R. Bowen
University of Wisconsin at Madison
N. Manson
Université de Poitiers
A.K. Oppenheim
University of California at Berkeley
R.I. Soloukhin
Institute of Heat and Mass Transfer, BSSR Academy of Sciences
ISBN 0-915928-47-7

77. Aerothermodynamics and Planetary Entry (1981)
A.L. Crosbie
University of Missouri-Rolla
ISBN 0-915928-52-3

78. Heat Transfer and Thermal Control (1981)
A.L. Crosbie
University of Missouri-Rolla
ISBN 0-915928-53-1

79. Electric Propulsion and Its Applications to Space Missions (1981)
Robert C. Finke
NASA Lewis Research Center
ISBN 0-915928-55-8

80. Aero-Optical Phenomena (1982)
Keith G. Gilbert
Leonard J. Otten
Air Force Weapons Laboratory
ISBN 0-915928-60-4

81. Transonic Aerodynamics (1982)
David Nixon
Nielsen Engineering & Research, Inc.
ISBN 0-915928-65-5

82. Thermophysics of Atmospheric Entry (1982)
T.E. Horton
University of Mississippi
ISBN 0-915928-66-3

83. Spacecraft Radiative Transfer and Temperature Control (1982)
T.E. Horton
University of Mississippi
ISBN 0-915928-67-1

84. Liquid-Metal Flows and Magnetohydrodynamics (1983)
H. Branover
Ben-Gurion University of the Negev
P.S. Lykoudis
Purdue University
A. Yakhot
Ben-Gurion University of the Negev
ISBN 0-915928-70-1

85. Entry Vehicle Heating and Thermal Protection Systems: Space Shuttle, Solar Starprobe, Jupiter Galileo Probe (1983)
Paul E. Bauer
McDonnell Douglas Astronautics Co.
Howard E. Collicott
The Boeing Co.
ISBN 0-915928-74-4

86. Spacecraft Thermal Control, Design, and Operation (1983)
Howard E. Collicott
The Boeing Co.
Paul E. Bauer
McDonnell Douglas Astronautics Co.
ISBN 0-915928-75-2

87. Shock Waves, Explosions, and Detonations (1983)
J.R. Bowen
University of Washington
N. Manson
Université de Poitiers
A.K. Oppenheim
University of California at Berkeley
R.I. Soloukhin
Institute of Heat and Mass Transfer, BSSR Academy of Sciences
ISBN 0-915928-76-0

88. Flames, Lasers, and Reactive Systems (1983)
J.R. Bowen
University of Washington
N. Manson
Université de Poitiers
A.K. Oppenheim
University of California at Berkeley
R.I. Soloukhin
Institute of Heat and Mass Transfer, BSSR Academy of Sciences
ISBN 0-915928-77-9

89. Orbit-Raising and Maneuvering Propulsion: Research Status and Needs (1984)
Leonard H. Caveny
Air Force Office of Scientific Research
ISBN 0-915928-82-5

90. Fundamentals of Solid-Propellant Combustion (1984)
Kenneth K. Kuo
Pennsylvania State University
Martin Summerfield
Princeton Combustion Research Laboratories, Inc.
ISBN 0-915928-84-1

91. Spacecraft Contamination: Sources and Prevention (1984)
J.A. Roux
University of Mississippi
T.D. McCay
NASA Marshall Space Flight Center
ISBN 0-915928-85-X

92. Combustion Diagnostics by Nonintrusive Methods (1984)
T.D. McCay
NASA Marshall Space Flight Center
J.A. Roux
University of Mississippi
ISBN 0-915928-86-8

93. The INTELSAT Global Satellite System (1984)
Joel Alper
COMSAT Corp.
Joseph Pelton
INTELSAT
ISBN 0-915928-90-6

94. Dynamics of Shock Waves, Explosions, and Detonations (1984)
J.R. Bowen
University of Washington
N. Manson
Université de Poitiers
A.K. Oppenheim
University of California at Berkely
R.I. Soloukhin
Institute of Heat and Mass Transfer, BSSR Academy of Sciences
ISBN 0-915928-91-4

95. Dynamics of Flames and Reactive Systems (1984)
J.R. Bowen
University of Washington
N. Manson
Université de Poitiers
A.K. Oppenheim
University of California at Bereley
R.I. Soloukhin
Institute of Heat and Mass Transfer, BSSR Academy of Sciences
ISBN 0-915928-92-2

96. Thermal Design of Aeroassisted Orbital Transfer Vehicles (1985)
H.F. Nelson
University of Missouri-Rolla
ISBN 0-915928-94-9

97. Monitoring Earth's Ocean, Land, and Atmosphere from Space – Sensors, Systems, and Applications (1985)
Abraham Schnapf
Aerospace Systems Engineering
ISBN 0-915928-98-1

98. Thrust and Drag: Its Prediction and Verification (1985)
Eugene E. Covert
Massachusetts Institute of Technology
C.R. James
Vought Corp.
William F. Kimzey
Sverdrup Technology AEDC Group
George K. Richey
U.S. Air Force
Eugene C. Rooney
U.S. Navy Department of Defense
ISBN 0-930403-00-2

99. Space Stations and Space Platforms – Concepts, Design, Infrastructure, and Uses (1985)
Ivan Bekey
Daniel Herman
NASA Headquarters
ISBN 0-930403-01-0

100. Single- and Multi-Phase Flows in an Electromagnetic Field: Energy, Metallurgical, and Solar Applications (1985)
Herman Branover
Ben-Gurion University of the Negev
Paul S. Lykoudis
Purdue University
Michael Mond
Ben-Gurion University of the Negev
ISBN 0-930403-04-5

101. MHD Energy Conversion: Physiotechnical Problems (1986)
V.A. Kirillin
A.E. Sheyndlin
Soviet Academy of Sciences
ISBN 0-930403-05-3

102. Numerical Methods for Engine-Airframe Integration (1986)
S.N.B. Murthy
Purdue University
Gerald C. Paynter
Boeing Airplane Co.
ISBN 0-930403-09-6

103. Thermophysical Aspects of Re-Entry Flows (1986)
James N. Moss
NASA Langley Research Center
Carl D. Scott
NASA Johnson Space Center
ISBN 0-930403-10-X

104. Tactical Missile Aerodynamics (1986)
M.J. Hemsch
PRC Kentron, Inc.
J.N. Nielsen
NASA Ames Research Center
ISBN 0-930403-13-4

105. Dynamics of Reactive Systems Part I: Flames and Configurations; Part II: Modeling and Heterogeneous Combustion (1986)
J.R. Bowen
University of Washington
J.-C. Leyer
Université de Poitiers
R.I. Soloukhin
Institute of Heat and Mass Transfer, BSSR Academy of Sciences
ISBN 0-930403-14-2

106. Dynamics of Explosions (1986)
J.R. Bowen
University of Washington
J.-C. Leyer
Université de Poitiers
R.I. Soloukhin
Institute of Heat and Mass Transfer, BSSR Academy of Sciences
ISBN 0-930403-15-0

107. Spacecraft Dielectric Material Properties and Spacecraft Charging (1986)
A.R. Frederickson
U.S. Air Force Rome Air Development Center
D.B. Cotts
SRI International
J.A. Wall
U.S. Air Force Rome Air Development Center
F.L. Bouquet
Jet Propulsion Laboratory, California Institute of Technology
ISBN 0-930403-17-7

108. Opportunities for Academic Research in a Low-Gravity Environment (1986)
George A. Hazelrigg
National Science Foundation
Joseph M. Reynolds
Louisiana State University
ISBN 0-930403-18-5

109. Gun Propulsion Technology (1988)
Ludwig Stiefel
U.S. Army Armament Research, Development and Engineering Center
ISBN 0-930403-20-7

110. Commercial Opportunities in Space (1988)
F. Shahrokhi
K.E. Harwell
University of Tennessee Space Institute
C.C. Chao
National Cheng Kung University
ISBN 0-930403-39-8

111. Liquid-Metal Flows: Magnetohydrodynamics and Applications (1988)
Herman Branover, Michael Mond, and Yeshajahu Unger
Ben-Gurion University of the Negev
ISBN 0-930403-43-6

112. Current Trends in Turbulence Research (1988)
Herman Branover, Michael Mond, and Yeshajahu Unger
Ben-Gurion University of the Negev
ISBN 0-930403-44-4

113. Dynamics of Reactive Systems Part I: Flames; Part II: Heterogeneous Combustion and Applications (1988)
A.L. Kuhl
R & D Associates
J.R. Bowen
University of Washington
J.-C. Leyer
Université de Poitiers
A. Borisov
USSR Academy of Sciences
ISBN 0-930403-46-0

114. Dynamics of Explosions (1988)
A.L. Kuhl
R & D Associates
J.R. Bowen
University of Washington
J.-C. Leyer
Université de Poitiers
A. Borisov
USSR Academy of Sciences
ISBN 0-930403-47-9

115. Machine Intelligence and Autonomy for Aerospace (1988)
E. Heer
Heer Associates, Inc.
H. Lum
NASA Ames Research Center
ISBN 0-930403-48-7

116. Rarefied Gas Dynamics: Space-Related Studies (1989)
E.P. Muntz
University of Southern California
D.P. Weaver
U.S. Air Force Astronautics Laboratory (AFSC)
D.H. Campbell
University of Dayton Research Institute
ISBN 0-930403-53-3

117. Rarefied Gas Dynamics: Physical Phenomena (1989)
E.P. Muntz
University of Southern California
D.P. Weaver
U.S. Air Force Astronautics Laboratory (AFSC)
D. Campbell
University of Dayton Research Institute
ISBN 0-930403-54-1

118. Rarefied Gas Dynamics: Theoretical and Computational Techniques (1989)
E.P. Muntz
University of Southern California
D.P. Weaver
U.S. Air Force Astronautics Laboratory (AFSC)
D.H. Campbell
University of Dayton Research Institute
ISBN 0-930403-55-X

119. Test and Evaluation of the Tactical Missile (1989)
Emil J. Eichblatt Jr.
Pacific Missile Test Center
ISBN 0-930403-56-8

120. Unsteady Transonic Aerodynamics (1989)
David Nixon
Nielsen Engineering & Research, Inc.
ISBN 0-930403-52-5

121. Orbital Debris from Upper-Stage Breakup (1989)
Joseph P. Loftus Jr.
NASA Johnson Space Center
ISBN 0-930403-58-4

122. Thermal-Hydraulics for Space Power, Propulsion and Thermal Management System Design (1989)
William J. Krotiuk
General Electric Co.
ISBN 0-930403-64-9

123. Viscous Drag Reduction in Boundary Layers (1990)
Dennis M. Bushnell
Jerry N. Hefner
NASA Langley Research Center
ISBN 0-930403-66-5

124. Tactical and Strategic Missile Guidance (1990)
Paul Zarchan
Charles Stark Draper Laboratory, Inc.
ISBN 0-930403-68-1

125. Applied Computational Aerodynamics (1990)
P.A. Henne
Douglas Aircraft Company
ISBN 0-930403-69-X

126. Space Commercialization: Launch Vehicles and Programs (1990)
F. Shahrokhi
University of Tennessee Space Institute
J.S. Greenberg
Princeton Synergetics Inc.
T. Al-Saud
Ministry of Defense and Aviation Kingdom of Saudi Arabia
ISBN 0-930403-75-4

127. Space Commercialization: Platforms and Processing (1990)
F. Shahrokhi
University of Tennessee Space Institute
G. Hazelrigg
National Science Foundation
R. Bayuzick
Vanderbilt University
ISBN 0-930403-76-2

128. Space Commercialization: Satellite Technology (1990)
F. Shahrokhi
University of Tennessee Space Institute
N. Jasentuliyana
United Nations
N. Tarabzouni
King Abulaziz City for Science and Technology
ISBN 0-930403-77-0

129. Mechanics and Control of Large Flexible Structures (1990)
John L. Junkins
Texas A&M University
ISBN 0-930403-73-8

130. Low-Gravity Fluid Dynamics and Transport Phenomena (1990)
Jean N. Koster
Robert L. Sani
University of Colorado at Boulder
ISBN 0-930403-74-6

131. Dynamics of Deflagrations and Reactive Systems: Flames (1991)
A. L. Kuhl
Lawrence Livermore National Laboratory
J.-C. Leyer
Université de Poitiers
A. A. Borisov
USSR Academy of Sciences
W. A. Sirignano
University of California
ISBN 0-930403-95-9

132. Dynamics of Deflagrations and Reactive Systems: Heterogeneous Combustion (1991)
A. L. Kuhl
Lawrence Livermore National Laboratory
J.-C. Leyer
Université de Poitiers
A. A. Borisov
USSR Academy of Sciences
W. A. Sirignano
University of California
ISBN 0-930403-96-7

133. Dynamics of Detonations and Explosions: Detonations (1991)
A. L. Kuhl
Lawrence Livermore National Laboratory
J.-C. Leyer
Université de Poitiers
A. A. Borisov
USSR Academy of Sciences
W. A. Sirignano
University of California
ISBN 0-930403-97-5

134. Dynamics of Detonations and Explosions: Explosion Phenomena (1991)
A. L. Kuhl
Lawrence Livermore National Laboratory
J.-C. Leyer
Université de Poitiers
A. A. Borisov
USSR Academy of Sciences
W. A. Sirignano
University of California
ISBN 0-930403-98-3

(Other Volumes are planned.)